Measurement Uncertainties

S. V. Gupta

Measurement Uncertainties

Physical Parameters and Calibration of Instruments

Springer

Dr. S. V. Gupta
Sector-8, Rohini 269/B-5
110085 Delhi
India
satyavirji@yahoo.co.in

ISBN 978-3-642-20988-8 e-ISBN 978-3-642-20989-5
DOI 10.1007/978-3-642-20989-5
Springer Heidelberg Dordrecht London New York

Library of Congress Control Number: 2011942600

Printed on acid-free paper

Springer is part of Springer Science+Business Media (www.springer.com)

Dedicated to my wife, Mrs. Prem Gupta,
Children, and Grand Children
as
They all inspire me to live longer

Preface

I have been associated with uncertainty calculations for quite sometime. Asia Metrology Program (APMP) took interest in unification of expressing the measurement results along with uncertainty. In those days the uncertainty components were divided according to source of errors. The errors were named as random and systematic errors. Hence components of uncertainty were named random uncertainty and systematic uncertainty. Around 1980, BIPM took initiative by circulating questioners to the countries who were members of Metre Convention. Every member showed keen interest in expressing the uncertainty in a harmonious way. A small but path-breaking document was produced in 1980, which emphasized that uncertainty is not of different kinds; there are only the ways by which one arrives at the uncertainty value, namely Type A and Type B evaluations of uncertainty. Further the document emphasized that square root of variance or a quantity similar to it is the standard uncertainty.

The book is intended to serve as a guide for expressing the measurement result along with uncertainty. The book conforms to "The Guide to the expression of uncertainty in Measurement" jointly produced by International Organization for Standardization (ISO), International Bureau of Weights and Measures (BIPM), International Federation of Clinical Chemistry (IFCC), International Union of Pure and Applied Chemistry (IUPAC), International Union of Pure and Applied Physics (IUPAP) and International Organization of Legal Metrology (OIML). The book differs from ISO Guide in explaining the basic theory behind relations provided by the Guide. Lots of examples are provided to support the theoretical formulations. All technical and scientific terms used have been explained in the first chapter itself. Various distributions used in uncertainty calculations have been explained in Chaps. 2 and 3. The stress has been given on the properties of Gaussian (Normal) probability distribution. Evaluation of data weather primary or secondary is one special topic discussed in Chap. 4. For each statistical parameter like mean or standard deviation lots of practical examples have been cited. The chapter is highly useful for the nodal laboratories involved in international measurement programmes. Propagation of uncertainty has been discussed by first explaining the Taylor expansion and highlighting the need for the function to be linear. The process of calibration of the measuring instruments at a few points and expressing the whole calibration result in the form of the function of the input quantity has been discussed by citing several examples in Chap. 6. The functions discussed are linear, exponential, power and polynomials types. Detailed steps for arriving at the uncertainty, starting from the modelling of the measurand as a function of input quantities, have been given in Chap. 7. The advantages and limitations of ISO GUM method have been given. Monte Carlo and Bayesian methods of arriving at the uncertainty have been mentioned. The detailed procedure for calibrating the surface plate and theoretical deductions of height at various points along the various designated lines has been given in Chap. 8. The chapter includes the uncertainty calculations between the points on the same line and also on different lines. The uncertainty calculation as per NABL (National Accreditation Board for Testing and Calibration Laboratories) requirement has also been given in calibration of surface Chap. 8. The uncertainty calculations in mass measurement have been dealt with in Chap. 9. While discussing various sources of errors such as that of buoyancy correction, uncertainty requirements in measurement of various environmental parameters have also been cited. Uncertainty in volume measurements by various methods has been detailed. Uncertainty in the calibration of volumetric glassware by gravimetric method, larger capacity measures by volumetric method and storage tanks by dimensional measurements have been discussed in detail in Chap. 10. The uncertainty calculation in the calibration hydrometers by comparison method has also been given. Chapter 11 deals with the uncertainty calculation in the measurement of and calibration of measuring instruments for length, pressure, temperature and luminous flux. Chapter 12 deals with electrical parameters; uncertainty in measurement of and calibration of measuring instruments has been detailed. Uncertainty calculations of vector measurands which are the function of dependent input quantities have been discussed. Some important Tables for Normal (Gaussian) probability distributions, Student's t distribution, χ^2 and Fisher's F values for different percentage points have been given. The limits for mean and standard deviations for various degrees of freedom have also been tabulated. A bibliography of recent papers, books and documents on uncertainty in measurement has been appended.

I would like to thank Mr. Vivek Bagga for discussing the calibration of surface plates. My earnest thanks are due to Mrs. Reeta Gupta, Scientist National Physical Laboratory, for logistic support. I wish to record my profound appreciation for the keen interest and strenuous efforts put in by Dr. Habil. Claus Ascheron, the Executive Editor Physics in making it possible to bring the manuscript to the desired level and to Springer Verlag GmbH, Germany, in bringing this book to light.

New Delhi
July 2011

S.V. Gupta

Contents

Chapter 1
Some Important Definitions

1.1 Introduction

The very nature of all physical measurements suggests that it is impossible to carry out a measurement of any physical quantity with no error. Hence, whenever the value of a physical quantity is determined through a measurement process, it is only the best estimate of the value of the physical quantity obtained from the given experimental data. The estimated value may be slightly less or more than the true value of the physical quantity. In an experimental work, basically four major elements are involved, namely (1) instruments, (2) observer, (3) measurement process, and (4) statistics. The instruments include environment conditions and influence quantities. Even when appropriate corrections for known or suspected sources of errors have been applied, there still remains an uncertainty, that is, a doubt about how well the result of a measurement represents the true value of the quantity being measured. During a measurement, the errors may creep in due to inherent error in instruments, effect of environment on an instrument reading, the error in reading the instrument by the observer and errors incurred due to a particular process of measurement. So when giving a measured value of any quantity, one will never be sure enough to give a specific value but would like to say that measured value of the same quantity may lie in a certain range.

While entering into the further details of statistics, measuring instruments and other details, we are likely to come across a few terms not well known to everybody so we would like to define them. Efforts have been made to classify them according to their association and nature.

In general a measurement of a quantity and estimating the uncertainty involve:

- Quantity
- Measurement process
- Statistics involved
- Instruments and standards used
- Influence quantities
- Special mathematical functions

S.V. Gupta, *Measurement Uncertainties*, DOI 10.1007/978-3-642-20989-5_1,

It is proposed to discuss various terms likely to appear in the forthcoming chapters under the aforesaid categories.

1.2 Terms Pertaining to Quantity

1.2.1 Quantity

Quantity is the property of a phenomenon, body, or substance, to which a number may be assigned with respect to reference.

The reference can be a unit of measurement, a measurement procedure, or a reference material.

For example, mass of a body is a quantity which is the property of that body and can be assigned a number with respect to the measurement unit, namely kilogram. Hardness is the property of a material, which can be assigned a number with reference to a particular procedure, for example, Rockwell scale. The values of the hardness of same material (same quantity) will be different on Rockwell and Vickers scales, and hence are dependent upon the procedure. The value of hardness may also be stated in terms of material, if stated in mho scale. Reference material, in this case, is diamond having the value 10 on mho scale.

There are quite a large number of quantities such as length, mass, volume, acceleration, momentum, electric charge, current, and potential inductance. However, most of them are interrelated. Hence, a much smaller number of quantities are required to represent all other quantities.

1.2.2 System of Base Quantities

It is a set of base quantities such that every other quantity can be expressed in terms of base quantity or their combination of. For example, all mechanical quantities can be expressed in terms of mass, length, and time. The International system of units is based on seven base quantities, namely mass, length, time, electric current, temperature, intensity of illumination, and mole.

Base quantities with their respective symbols and symbols of their dimensions, as adopted in International system of units, are given in Table 1.1.

1.2.3 Derived Quantity

A quantity in a system of quantities, which is defined in terms of base quantities, is known as a derived quantity. For example, velocity is the ratio of length and time, while kinetic energy is the product of mass of the moving body and square of its velocity.

Table 1.1 Base quantities symbols and their units with respective symbols

Quantity	Symbol	Name of unit	Symbol of unit	Symbol of dimension
Length	L, x, r, etc.	Metre	m	L
Mass	m	Kilogram	kg	M
Time	T	Second	s	T
Electric current	I, i	ampere	A	I
Intensity of illumination	I_v	candela	cd	J
Temperature	T	kelvin	K	Θ
Mole	n	Mole	mol	N

1.2.4 Quantity Equation

Quantity equation is a mathematical relationship between the quantities (base as well as derived). If a quantity Q_1 is a product of two quantities Q_2, Q_3 and number n, then quantity equation of Q_1 is

$$Q_1 = n \times Q_2 \times Q_3. \tag{1.1}$$

1.2.5 Dimension of a Quantity

If a quantity is expressed as a product of several base quantities with symbols as given in Table 1.1 as

$$Q = L^{\alpha} M^{\beta} T^{\gamma} N^{\delta} J^{\zeta}. \tag{1.2}$$

Here, α, β, γ, δ, and ζ are exponents of the base quantities which may be positive, negative, or zero. These are called the dimensional exponents.

There are two versions of the definition of dimension, which are equally prevalent. According to one version dimensions of a quantity are the powers to which base quantities must be raised to represent that quantity. That is powers (exponents) of base quantities α, β, γ, δ, etc. are the dimensions of the quantity Q. So a quantity having each exponent as zero is called dimensionless quantity.

According to second version as given in [1] dimension of a quantity is the expression representing the quantity in terms of base quantities raised to the integral powers. That is dimension of Q given in (1.2) is expressed as

$$\text{Dim}\, Q = L^{\alpha} M^{\beta} T^{\gamma} N^{\delta} J^{\zeta}. \tag{1.3}$$

If each of the exponents α, β, γ, δ, etc. is zero than the dimension of the quantity will be as

$$\text{Dim}\, Q = L^0 M^0 T^0 N^0 J^0 = 1. \tag{1.4}$$

Thus, a quantity having each of the exponents of base quantity as zero has the dimension 1.

We see here that a quantity having each of the exponents as zero will be called dimensionless or of dimension 1 respectively if the dimensions of the quantity are defined as exponents or as an expression consisting of the products of base units raised to an integral power. In the literature, therefore, we may come across both words namely "dimensionless" and "of dimension 1" for one and the same quantity.

1.2.6 Measurand

Quite often the quantity under measurement is called measurand [2] (Specific quantity subject to measurement).

Specific Quantity: Quantity is taken in a general sense, for example length, mass, temperature, amount of substance. Quantity in general is generic in nature like energy; it may be mechanical, electrical, or light. Among the mechanical energy it may be kinetic or potential. However, the quantity is also taken in a specific sense, in that the quantity is associated with a specific object, for example length of a given rod, mass of a specific object, or concentration of alcohol in a given mixture. Quantity in general is independent of system of measuring units. Specific quantity is the quantitative value of energy of a specific body or system with reference to a measurement system.

1.2.7 True Value of a Quantity

The value that characterizes the quantity perfectly defined at the instant at which it was measured is known as a true value. This is something which is not achievable, except in cases of defining the base units. For example: True value of quantity of mass of the International Prototype of Kilogram is one kilogram. Please remember, it is not a measured value but only an internationally agreed value of that specific platinum iridium cylinder. Similar is the case for the quantity value of speed of light in vacuum. When we come to experiments next best to true value of the quantity is the mean value of the quantity, which has been measured a large number of times. It has been assumed that quantity value being measured remains constant in the duration of experimentation. This value is often called as conventional true value.

1.2.8 Conventional True Value of a Quantity

[2]: The value of a quantity, which for a given purpose, may be substituted for the true value of the quantity.

For example: At a given location the value assigned to the specific quantity realized by a reference standard may be taken as a conventional true value.

The CODATA (2002) recommended value for Avogadro Constant $(6.0221415 \pm 0.0000010) \times 10^{23}$ may also be taken as conventional true value.

1.2.9 Measured Value

The value of the quantity obtained after proper measurements and applying all necessary corrections due to the instruments including the standards and due to environmental conditions is the measured value. Quite often the whole measurement process is repeated several times.

1.2.10 Relation in Between Measured Value and True or Conventional True Value

True or conventional true value tends to the mean of measured values, when the whole process of measurement is repeated infinitely large number of times.

1.3 Terms Pertaining to Measurement

1.3.1 Measurement

[2]: The set of operations having the object of determining the value of a specific quantity.

1.3.2 Method of Measurement

The logical sequence of operations, in generic terms, used in the performance of measurements according to a given principle.

Methods of measurement may be further qualified in various ways for example as given below.

1.3.3 Substitution Method

A method of measurement by direct comparison in which the value of the quantity to be measured is replaced by a known value of the quantity chosen in such a way that the effect on the indicating device of these two values is same.

1.3.4 Differential Method

A method of measurement by comparison, based on comparing the quantity under measurement with the quantity of same kind having known value only slightly different from that of the quantity to be measured, and measuring the difference between the values of these quantities. A weight under test is compared against a standard weight of known mass of same denomination, so that effect on balance indication is small and measurable.

1.3.5 Null Method

A method of measurement by comparison, based on balancing the quantity under measurement against the quantity of known value such that indication is zero. However, by changing the known quantity even slightly there should be some indication. In null method Type A uncertainty due to repeated observations will be zero, but due to resolution of the instrument under observation, the uncertainty is to be considered and evaluated by Type B evaluation method.

1.3.6 Measurement Procedure

The set of operations, in specific terms, used in the performance of a particular measurement according to a given method.

Note: A measurement procedure should usually be recorded in the certificate of calibration/testing or examination.

1.3.7 Result of Measurement

The value attributed to the measurand by measurement and after applying due corrections. Example mass of weight obtained against a reference weight after applying buoyancy corrections.

1.3.8 Error

The error is the measured value of the quantity minus the conventional value of the same quantity. The errors arise due to inaccuracy, non-repeatability, and resolution (threshold value) of instruments involved. The errors may also be due to the observer, environmental conditions, and the measurement process. (Accuracy is a qualitative concept and it should not be confused with the term precision.)

1.3.9 Spurious Error

Spurious errors are due to mistakes by the observer, malfunctioning of an instrument and these invalidate the observation. Reversing the digits in recording the observation, having loose connection in an electrical measurement, and the presence of air pockets in fluid flow measurements are some examples of spurious errors. Observations with such errors are not to be incorporated in any statistical analysis. If there is a reason to believe the existence of such errors with an observation, that observation should be discarded.

1.3.10 Relative Error

The error of measurement divided by a true value of the measurand. Since a true value of the measurand cannot be determined; in practice a conventional true value is used.

1.3.11 Random Error

An error is the quantity, which varies in an unpredictable manner in both magnitude and sign. When a large number of measurements of the same value of a quantity are made under essentially the same conditions, random error approaches to zero. The random error follows the Gaussian (normal) distribution with zero mean. For small sample (smaller no. of observations), the results used to be corrected by means of Student's t factor. However, the assumption remains that results belong to normal distribution. These errors may be due to uncontrollable environmental conditions, personal judgement of the observer, and inherent instability of the measuring instrument or any other cause of random nature.

1.3.12 Systematic Error

An error which in replicate measurements remains constant and cannot be reduced by taking larger number of observations if the equipment and conditions of measurement remain unchanged. These errors may be due to the inability in detection of the measuring system, constant bias, error in the value of the standard, a physical constant, and property of the medium or conversion factor used. The value and the sign of this error do not change with the given measuring system. Systematic errors can be broadly classified as (1) constant and (2) variable. Constant systematic errors are those which do not change with time but sometimes may vary with the magnitude of the measured quantity. Zero setting error in an instrument is a constant systematic error, while inaccuracy in the calibration scale may depend upon the magnitude of the quantity measured. Variable systematic errors do depend upon the time, say value of a resistor, which may vary with time because of ageing effect. The systematic errors may also occur due to insufficient control of environmental conditions.

1.3.13 Accuracy of Measurement

Closeness of the agreement between the result of a measurement and the true value of the measurand.

1.3.14 Precision of Measurement Result

The precision of an instrument reflects the number of significant digits in the stated result. The result reported to larger places to the right of decimal is supposed to be more precise. A result of acceleration due to gravity given as 9.5671 ms^{-2} is more precise than the result 9.80 ms^{-2}, though the latter is more accurate than the former. An instrument may have a better repeatability but less precision and vice versa. For example an ammeter graduated in ampere and always showing the same result for a given constant input is more repeatable and less precise as it reads only in terms of amperes. An ammeter reading in mA but giving not repeatable values is more precise but less repeatable. Any unbiased ammeter will give better accurate results. A good instrument should be more precise, more repeatable, and least away from the true value of the input quantity.

1.3.15 Repeatability

The repeatability is the closeness between the results of the successive measurements of the same measurand carried out in

- Same measurement procedure
- The same observer
- Same conditions (environmental)
- The same location
- Repetitions are carried out for a short period of time

1.3.16 Reproducibility (of Measurement Results)

Reproducibility is closeness of agreement between the results of measurements of the same measurand, where the measurements are carried out under

- Changed conditions
- Different principle or method of measurement
- Different observer
- Different locations
- Different conditions of use
- Different time

A valid statement of reproducibility requires specifications of the condition changed. The reproducibility may be expressed quantitatively in terms of dispersions between the results.

1.3.17 Correction

Correction is a small quantity which is to be added algebraically to the observed value. It may be pertaining to

- An instrument or the standard used (Certificate Correction).
- To bring the measured value to the reference environmental conditions like temperature, pressure humidity, etc. All length measurements are normally reduced to 20°C.
- Different physical properties of standard used and the under-test. For example buoyancy correction, when a weight having a density different from that of the standard and is compared in air.

1.4 Terms Pertaining to Statistics

1.4.1 Observation

Observation is a value of the quantity, under measurement, as read out from a measuring instrument. Any observation for the purpose of mathematical manipulation is often called a variable.

1.4.2 Independent Observations

Two observations are independent if the occurrence of one observation provides no information about the occurrence of the other observation. A simple example is measuring the height of everyone in your sample at a single point of time. These are unrelated observations. However, if you were to measure one child's height over a certain period of time, these observations would be dependent because the height at each point of time would depend upon the height at the previous occasion. It will be slightly more than the previous value.

1.4.3 Population

The total set of all observations that one wants to analyse to assign the numerical value to the quantity under measurement (measurand).

1.4.4 Sample

A subset of the population usually selected randomly. In practice only a few observations are taken to quantify the given measurand. Such ensemble of observations is also known as sample.

1.4.5 Measurement

The process of experimentally obtaining one or more values that can reasonably attributed to the quantity under measurement. Measurement is also defined as the observation after application of all corrections. Sometimes a numerical value of a quantity is calculated from the observations taken from a set of instruments and then necessary corrections are applied to the observation of each instrument. Resistance of a resistor is calculated by taking observations of electric current passing through it and potential difference across it. The due corrections are applied to the ammeter and voltmeter observations if necessary.

Quite often more than one observation or set of observations is taken to quantify a quantity. Ideally infinite number (a very large) number of observations should be taken to finally assign a numerical value to a quantity under measurement.

1.4.6 Population of Measurement

An infinite number of independent measurements, carried out for determination of the value of a certain quantity, constitute a population.

1.4.7 Sample of Measurements

In practice, only a finite number of measurements are carried out for the determination of a certain quantity which constitutes a sample.

1.4.8 Frequency/Relative Frequency

In a sample some observations may occur more than once, number of times an observation repeats itself is known as its frequency. Sometimes observations are divided into groups; each group has a certain range. This range in called the interval. Normally intervals or ranges of sub-groups in a sample are equal. Number of observations lying in a given interval is called as the group frequency. The relative frequency is the ratio of frequency of a certain observation to the total number of observations (Sum of all frequencies). If n is the frequency of certain group of observations and total number of observations is N, then relative frequency is n/N.

1.4.9 Mean

Sum of all observations divide by the number of observations.

1.4.10 Sample Mean

If $x_1, x_2, x_3, \ldots, x_n$ be n measurements then Sample Mean $\bar{X}$ is defined as

$$\bar{X} = \frac{\sum_{p=1}^{p=n} X_p}{n}. \tag{1.5}$$

1.4.11 Population Mean

The limiting value of sample mean as number of measurements tends to infinity is the population mean.

$$\mu = \mathrm{Lim}_{n \Rightarrow \infty} \frac{\sum_{p=1}^{p=n} X_p}{n}. \tag{1.6}$$

1.4.12 *Merits and Demerits of Arithmetic Mean [3]*

1.4.12.1 Merits

1. It is well defined
2. Based on all observations
3. All observations are equally important
4. It is amenable to algebraic manipulations for example mean of the set of observations is derivable from the means and sizes of its subsets as given below:

$$\bar{X} = \frac{\sum_{q=1}^{q=m} n_q \bar{x}_q}{\sum_{q=1}^{q=m} n_q}. \qquad (1.7)$$

5. Out of all the averages, arithmetic mean is least affected by fluctuations of sampling.

1.4.12.2 Demerits

1. It cannot be determined by inspection or cannot be located by graphical means.
2. The arithmetic mean is not applicable for qualitative data, like intelligence or colour. Data should be in quantitative terms.
3. Even if single observation is missing or not legible, the arithmetic mean cannot be determined unless the observation is left out of the set.
4. The arithmetic mean is affected most by extreme values of observations. Mistakes giving extreme observations affect it most.
5. The arithmetic mean sometimes gives a value which is not meaningful in practical life.

1.4.13 *Median*

Median of a distribution is the value of the measurement which divides it into two equal parts. Median, in a set of observations, is the value of observation such that number of observations below it is the same as the number of observations above. Thus, median is a positional average.

1.4.14 *Quartiles*

The quartiles divide the distribution into four equal parts. First quartile divides the distribution in the ratio of 1:3; second quartile 2:2 obviously is its median.

Third quartile divides the distribution into 3:1. First and third quartiles are normally indicated by Q_1 and Q_3, respectively.

1.4.15 *Dispersion*

Numerical designations of how closely data cluster about the mean or other measure of central tendency is the dispersion.

Dispersion may be for the observations, for example biggest observation minus the smallest observation; semi-quartile deviations, i.e. $(Q_3–Q_1)/2$ or semi-inter quartile range; deviation may also be from any measure of central tendency; for example the deviation may be from mean, mode, or median.

Deviation is zero from the arithmetic mean. The absolute values of deviations from the arithmetic mean are minimum.

1.4.16 *Standard Deviation*

It is the square root of the average of squares of deviations from the arithmetic mean.

1.4.17 *Variance*

It is the square of the standard deviation, i.e. mean of the square of deviations from the arithmetic mean.

Like arithmetic mean, mean deviation, standard deviation, and variance use all observations. Mean deviation has a step of considering every deviation as positive which appears to be bizarre. However, taking the squares of deviations removes this step. Moreover like the arithmetic mean, variance is also amenable to arithmetic calculations. Combined variance of two sets of data of sizes n_1, n_2 means $\bar{x}_1, \bar{x}_2$, and SD s_1 and s_2 is given by

$$S^2 = [n_1(s_1^2 + d_1^2) + n_2(s_2^2 + d_2^2)]/(n_1 + n_2), \quad (1.8)$$

where

$$d_1 = (\bar{x}_1 - \bar{x}),$$
$$d_2 = (\bar{x}_2 - \bar{x}),$$

and

$$\bar{x} = \frac{n_1\bar{x}_1 + n_2\bar{x}_2}{n_1 + n_2}. \quad (1.9)$$

Equation (1.8) is true for any number of samples. If all samples have a common mean, i.e. d_1, d_2, etc. are zero, then mean variance is the mean of variances. The observed value of a given quantity from different samples must be same; that is mean is same; hence variance of the large number of samples from the same population is the weighted mean of all the variances. Size of each sample is taken as its weight factor.

1.4.18 Sample Standard Deviation

Sample standard deviation s is defined as

$$s = \left[\sum_{p=1}^{p=n} (x_p - \bar{x})^2/n\right]^{1/2}. \tag{1.10}$$

Or variance s^2 is

$$s^2 = \sum_{p=1}^{p=n} (x_p - \bar{x})^2/n. \tag{1.11}$$

1.4.19 Population Standard Deviation

The limiting value of sample standard deviation as number of measurements tends to infinity is the population standard deviation and is normally denoted as σ and is given by

$$\sigma^2 = \mathrm{Lim}_{n=\infty} \sum_{p=1}^{p=n} (x_p - \bar{x})^2/n. \tag{1.12}$$

As n cannot be made infinite, that is one cannot take infinite number of measurements, one can have only an estimate of population standard deviation.

1.4.20 Estimate of Population Standard Deviation

The best estimate of population standard deviation S is given by

$$S = \left[\frac{\sum_{p=1}^{p=n} (x_p - \bar{x})}{n-1}\right]^{1/2}. \tag{1.13}$$

1.4.21 Estimate of Population and Sample Standard Deviations-Relation

Since population standard deviation cannot be determined by practical measurements an estimate S of the population standard deviation is obtained from sample standard deviation s as

$$S = s\sqrt{\frac{n}{(n-1)}}. \tag{1.14}$$

1.4.22 Independent Variable

The variable that causes or predicts the dependent variable is an independent variable. Any observed measurement is an independent variable, for example observed difference of mass between the two weights obtained by comparison on the balance. It is also called as input variable or quantity.

1.4.23 Dependent Variable or Response Variable

The variable that is caused or predicted by the independent variable is the dependent variable. It is a function of n independent variables, where n is a natural number. For example, resistance is a dependent variable, which is a function of two independent variables, namely current through it and the potential difference across it.

Two variables may be independent of each other or have some sort of dependence on each other. For example height of man and financial status are two independent variables. But height of growing boy and his age are correlated variables.

1.4.24 Correlation

It is the relationship between two or several variables within a distribution.

1.4.25 Correlation Coefficient

Correlation coefficient is the ratio of the covariance of two random variables to the product of their standard deviations. That is correlation coefficient $r(x_1, x_2)$ is given as

$$r\,(x_1, x_2) = \frac{\mathrm{cov}\,(x_1, x_2)}{s\,(x_1) \times s(x_2)}, \tag{1.15}$$

where $s(x_1)$ and $s(x_2)$ are variances, and $\text{cov}(x_1, x_2)$ is a covariance of two variables x_1 and x_2.

1.4.26 Covariance

The sum of products of the deviations x_{1p} and x_{2p} from their respective averages divided by one less than the number of observed pairs.

$$\text{Cov}(x_1, x_2) = \frac{\sum_{p=1}^{p=N} (x_{1p} - \bar{x}_1)(x_{2p} - \bar{x}_2)}{N - 1}. \tag{1.16}$$

1.4.27 Random Variable

Any real number which is outcome of a random experiment is the random variable. In other words, a variable which takes any of the values of a specified set of values and which is associated with a probability distribution (ISO 3534-1,13) is a random variable.

1.4.28 Discrete Random Variable

A random variable which takes only isolated values is said to be a discrete values, for example outcome of tossing a coin several times.

1.4.29 Continuous Random Variable

A random variable that takes any value within a given interval (finite or infinite) is said to be continuous variable.

1.4.30 Probability

A real number in the scale of 0–1 attached to the occurrence of a random event. It is also equal to the relative frequency of occurrence of a particular value of the random variable.

From the definition of frequency, we see that relative frequency is same as the probability of happening of any of the equal observations. Here n is the frequency of a certain observation and N is the total number of observations.

1.4.31 Probability Distribution

A function giving the probability that a random variable takes within a given interval. The interval may be finite or infinite. The probability sum of the entire set of random variables (finite or infinite) is one. Or the sum of probabilities of occurrence of each and every random variable in the given interval is 1. For example if $P(x)$ is the probability of happening of any random variable x within the given interval (a, b) then

$$\int_a^b P(x)\,\mathrm{d}x = 1. \tag{1.17}$$

1.4.32 Normal Distribution

A bell-shaped curve or distribution indicating that the variable x at the mean occurs with highest probability and that the probability of occurrence progressively decreases as observations deviate from the mean. If x is variable μ and σ are respectively the mean and standard deviation of the population then

$$P(x) = \left\{1/\sigma\sqrt{2\pi}\right\} \mathrm{Exp}\left[-(x-\mu)^2/(2\sigma^2)\right]. \tag{1.18}$$

1.4.32.1 Alternative Definition

An ensemble of random continuous variables predicting the value of a given quantity is a normal distribution. It is a bell-shaped curve as indicated in Fig. 1.1 such that observations at or close to the mean occur with highest probability, and

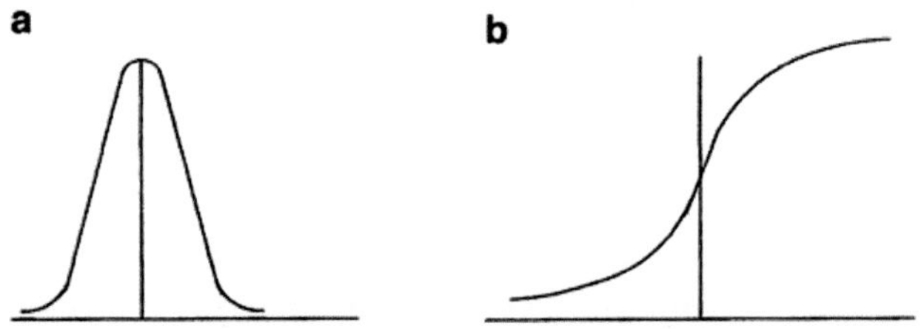

Fig. 1.1 (**a**) Probability distribution, (**b**) Cumulative distribution

that the probability of occurrence progressively decreases as observations deviate from the mean.

A normal probability density function $P(x)$ of variable x with mean μ and standard deviation σ, the probability function is given as

$$P(x) = \left\{1/\sigma\sqrt{2\pi}\right\} \operatorname{Exp}\left[-(x-\mu)^2/(2\sigma^2)\right].$$

1.4.33 Properties of Normal Distribution

The semi-range (from mean to extreme value on either side of mean) is almost 3 times the standard deviation.

The range of $\pm 3\sigma$ covers 99.73% of all observations.
The range of $\pm 2\sigma$ covers 95.45% of all observations.
The range of $\pm \sigma$ covers 68.27% of all observations.
The range of $\pm 0.6745\sigma$ covers 50% of all observations.
0.6745 σ is called as probable error.

1.4.34 Probable Error

The amount by which the arithmetic mean of a sample is expected to vary because of chance alone (50% probability) is the probable error. The value of probable error is 0.6745 times the standard deviation of a normal population.

1.4.35 Range

A measure of dispersion and is equal to absolute difference between the largest and the smallest values of the variable in a given distribution.

1.4.36 Confidence Level

Confidence level is a measure of the degree of reliability with which a result is expressed. If a result is reported with say 95% level of confidence, it means that the true or conventional true value will lie within the specified range with a probability of 0.95.

Confidence level is also referred as *Probability Coverage*.

1.4.37 Confidence Interval

Confidence interval is the range of the measurand between which the measured value is likely to lie with the specified level of confidence. If X is the measured value of the measurand with a semi-range U at the confidence level of 95%, it implies that the probability for the true value lying in between $X - U$ and $X + U$ is 0.95. In other words, if the same quantity is measured by any other observer with an unbiased instrument a large number of times, then 95% of values of the measurand will lie in between $X - U$ and $X + U$. This is often called coverage probability.

Or

The upper and lower boundaries that the estimator is $P\%$ sure that the estimated value will fall within the stated range of the measurand. This is also called as coverage interval.

1.4.38 Outlier

An extreme value in a frequency distribution, which has a disproportionate influence on the mean.

1.4.39 Parameter

A measure used to summarize characteristic of a population based on all items in the population. Mean is one of the parameter of the population. Another very frequently met parameter is variance.

1.4.40 Random Selection

It is the method of selecting an item from its population such that chance of selection of any item is equal. An ensemble of such items is called as a random sample.

1.4.41 Sample Statistic

Measures that summarize a sample are called sample statistics. Mean, mode, or median each is an example of a statistic. These are also known as measures of central tendency. Range is also a statistic.

1.4.42 Error

An error pertains to a measurement and not to an instrument. An error is the difference between the value obtained on the basis of a set of measurements and the conventional true value of the quantity measured.

1.4.43 Standard Error, or Standard Deviation of the Mean

Standard error is an estimate of the standard deviation of the sampling distribution of means, based on the data from one or more random samples.

Numerically, it is equal to the estimated standard deviation divided by the square root of n, where n is the size of the sample for which standard deviation is taken. If S is the best estimate of the population derived from the sample size n, then Standard Error S_E is given as

$$S_{\mathrm{E}} = \frac{S}{\sqrt{n}} = \frac{\sum_{p=1}^{p=n} (x_p - \bar{x})^2}{\sqrt{n\,(n-1)}}. \tag{1.19}$$

Standard error is also the standard uncertainty evaluated by Type A method.

1.4.44 Uncertainty

The uncertainty of a measurement is the range about the measured value within which the true value or the conventional true value of the measured quantity is likely to lie at the stated level of confidence. The value of semi-range U, apart from instruments, observer and process used for measurement, depends upon the confidence level with which the measured result is stated. For example for the same experimental result the value of U with 66.6% confidence level is σ, while the value of U with 95.45% confidence level is 2 σ. Strictly speaking, uncertainty can be calculated only when true (population) standard deviation is known or it can be estimated from the standard deviation calculated from finite number of observations having Gaussian (Normal) distribution.

1.4.45 Evaluations of Uncertainty

The uncertainty in the result of measurement generally consists of several components, which may be grouped into two categories according to the way in which their numerical values are estimated.

- Type A evaluation of uncertainty:

Type A evaluation of standard uncertainty applies to the observed data. Arithmetic mean and variance of the data are calculated by usual statistical methods. The estimate of the standard deviation of the mean is the standard uncertainty. That is Type A evaluation deals with the primary measurement data obtained by the experimenter.

- Type B Evaluation of uncertainty:

Type B evaluation of uncertainty applies to those input quantities for which mean and variance have not been obtained by repeated observations, but the variance $u^2(x_p)$ is obtained by judgement using all relevant information on the possible variability. The pool of information may include

- Previous measurement data
- Experience and general knowledge of the behaviour of the material
- Instruments specifications
- Manufactures specifications
- Data provided by the calibration laboratory

1.4.46 Random Uncertainty(e_r)

The value of S – the estimate of population standard deviation from the mean – is used to express random component of uncertainty (e_r) as

$$e_r = t \times S/\sqrt{n} = t \times s/\sqrt{(n-1)}. \tag{1.20}$$

Here t is Student's "t" factor and $S/\sqrt{n}$ is the standard error of the mean.

The above calculations are based upon the assumption that all measurements follow the Gaussian (Normal) distribution $f(x)$ and $f(x)$ is represented as

$$f(x) = (l/\sigma\sqrt{\pi})\exp[-(x-\mu)^2/2\sigma^2]. \tag{1.21}$$

Random uncertainty is that part of uncertainty in assigning the value of a measured quantity which is due to random errors. The value of the random uncertainty is obtained on multiplication of the standard deviation of the mean by the student factor $\sim t'$. The value of factor $\sim t'$ depends upon the sample size from which the standard deviation has been determined, and the confidence level at which the results of measurement are to be expressed.

Note: This term was in very much use before the guidance prepared jointly by BIPM, ISO and OIML was issued.

1.4.47 Systematic Uncertainty(U_S)

Systematic uncertainty is that part of uncertainty which is due to systematic errors and cannot be experimentally determined unless the equipment and environmental conditions are changed. It is obtained by suitable combination of all systematic errors arising due to different components of the measuring system.

Note: This term was in very much use before the guidance prepared jointly by BIPM, ISO and OIML was issued.

1.4.48 Standard Uncertainty

Standard Uncertainty is numerically equal to the square root of weighted sum of variances due to all sources. In other words, standard uncertainty is the uncertainty of the measurement expressed as standard deviation [4].

1.4.49 Expanded Uncertainty

Expanded uncertainty is the product of standard uncertainty and the coverage factor. The factor is normally greater than 1 so that experimenter has more confidence in stating that the true or conventional true value of the measurand lies within the stated range, For example, coverage factor is 1.96 for confidence level of 95.45% and 3 for confidence level of 99.7%.

1.4.50 Expressing Uncertainty of Measurement

Uncertainty of measurement can be expressed in two ways, namely in terms of absolute uncertainty or relative uncertainty. For example absolute measurement uncertainty in the measured value of the resistance of one ohm resistor is $\pm 1\,\mu\Omega$.

For relative uncertainty is the ratio of the absolute uncertainty to the nominal value of the measurand. In the above example it is 1×10^{-6}. However, quite often the qualifying word "relative" to the word "uncertainty" is not written.

1.4.51 Coverage Interval

The interval containing the set of true quantity values of a measurand with stated probability, based on the information available.

A coverage interval does not need to be centered on the measured quantity value.

A coverage interval should not be termed "confidence interval" to avoid confusion with the stoical concept.

A coverage interval can be derived from an expanded measurement uncertainty.

1.4.52 Coverage Probability

The probability that the set of true quantity values of a measurand is contained within specified coverage interval.

1.4.53 Central Limit Theorem

If the output variable Y is a linear function of n input quantities X_i, such that all X_i are characterized by normal distributions, and is expressed as

$$Y = c_1X_1 + c_2X_2 + c_3X_3 + \cdots + c_NX_N = \sum_{p=1}^{p-N} c_pX_p, \tag{1.22}$$

then Central Limit theorem states that convolved distribution of Y is also a normal distribution. Even if the distributions of the X_p are not normal, the distribution of Y may be approximated to a normal distribution with

$$E(Y) = \sum_{p=1}^{p=N} c_pE(X_p). \tag{1.23}$$

Here $E(X_p)$ is the expectation (mean) of X_p. The variance $V(Y)$ or $\sigma^2(Y)$ is given as

$$\sigma^2(Y) = \sum_{p=1}^{p=N} c_p^2\sigma_p^2. \tag{1.24}$$

1.5 Influence Quantity

Quantity that is not included in the specification of the measurand but nonetheless affects the result of the measurement like

- Temperature in linear measurements
- Temperature, pressure humidity and composition of air in mass measurement

- Frequency in the measurement of an alternating current
- Air density in interferometric measurements
- Bilirubin concentration in the measurement of haemoglobin concentration in human blood plasma

1.6 Instruments and Standards

1.6.1 *Repeatability of an Instrument*

It is the ability of the measuring instrument to give identical indications or responses for repeated applications of the same value of the input quantity, under stated conditions of use. Quantitative measurement of repeatability of an instrument is carried out by finding the standard deviation from the mean of large number of measured values of the same quantity under essentially the same conditions of use.

1.6.2 *Precision of the Instrument*

It is the ability of the instrument to indicate the smallest value of the stimulus. A balance able to read directly in terms of 1 mg at 1 kg load is more precise than the balance which reads up to 1 g at 1 kg level. It represents essentially how fine the scale is graduated. The result reported to larger places to the right of decimal is supposed to be more precise. A result of acceleration due to gravity given as 9.805671 ms^{-2} is more precise than the result 9.80 ms^{-2}. An instrument may have a better repeatability but less precision and vice versa. For example, an ammeter graduated in ampere and always showing the same result for a given constant input is more repeatable and less precise as it reads only in terms of amperes. An ammeter reading in mA, but not giving repeatable values, is more precise but less repeatable. Any unbiased ammeter will give better accurate results. A good instrument should be more precise, more repeatable and least away from the true value of the input quantity.

1.6.3 *Accuracy of an Instrument*

The accuracy of an instrument is its ability to give correct results. The accuracy and repeatability are two different properties of an instrument. Accuracy is a measure of an instrument's ability to tell the truth, while repeatability is a measure of its ability to indicate the same value of the measured quantity. Instruments, like some people, are capable of telling the same lie over and over again. Consequently, good repeatability is not a guarantee of good accuracy. Although poor repeatability is a sure sign of poor accuracy, but good repeatability is no sign of good accuracy. In mathematical sense, one may say that good repeatability of the instrument is a necessary but not a sufficient condition of good accuracy. The accuracy of an instrument may be found out by combining the measures of its repeatability and systematic errors by using quadrature (root mean square) method, namely

$$\text{Accuracy} = \left\{(\text{repeatability})^2 + (\text{inaccuracy})^2 + (\text{systematicerror})^2\right\}^{1/2}.$$

1.6.4 Accuracy of a Standard

Normally this term means the tolerance within which the true value of quantity of an artefact lies. The accuracy of a kilogram standard is ±1 mg, which means that mass value of that kilogram will be anywhere within 1 kg ± 1 mg. Quite often the term accuracy is clubbed with uncertainty. Apparently these appear to be similar, but these are opposite in sense; more accuracy means lesser value of uncertainty.

1.6.5 Difference Between Uncertainty and Accuracy

So we have seen that there are two terms namely accuracy and uncertainty; one cannot be replaced by the other. To achieve better experimental results, one would like to have more and more accuracy but lesser uncertainty. Accuracy pertains to an instrument or a standard of measurement. Moreover, accuracy means how close the instrument indication is to the true or its conventional true value of the quantity. Or how close is the quantity value of the standard to its nominal value. For example the mass of a one kilogram standard is 999.99998 g ± 04 mg; the standard is accurate within 0.02 mg. Quite often people state the ability of a measurement laboratory confusingly in terms of accuracy and uncertainty simultaneously, which is not correct. The confusion starts from the fact that accuracy is often expressed as by the statement that standard is accurate within 2 parts per million. The instrument is accurate within 0.01% of measured value or balance is accurate within 0.0001% of the range.

1.6.6 Difference Between the Correction, Error and Uncertainty

The calibration certificate of an instrument gives a correspondence between its indication and the quantity it is most likely to measure. The difference between them is the correction. This correction is to be invariably applied. However, there will be an element of doubt in the value of the correction so stated. This doubt is quantitatively expressed as an overall uncertainty in assigning the value to the correction stated and will be one component of the uncertainty of that instrument. For example, in case of a metre bar, the distance between the zero and 1,000 mm graduation marks may be given as 1,000.045 ± 0.005 mm. Then −0.045 is correction and 0.005 mm is the uncertainty in the value of the metre bar. In addition to this component, other components of uncertainty (Type B) may be there, for example, due to the finite width of the graduation lines of this metre bar.

1.6.7 Correction Factor

The correction factor is a number by which the uncorrected result of a measurement is multiplied. Sometimes a correction factor is given by the calibrator of the instrument.

1.6.8 Discrimination Threshold

The smallest change in the stimulus which produces a perceptible change in the response of a measuring instrument is the discrimination threshold of the instrument. The discrimination threshold may depend upon electrical noise, mechanical friction, air damping, inertia, or quantization. Discrimination threshold should be taken into account while estimating uncertainty by Type B evaluation.

1.7 Some Special Integrals and Functions

1.7.1 Gamma Function

The Gamma function is a definite integral given as

$$\int_0^\infty \exp(-x) \times x^{n-1} \mathrm{d}x = \Gamma n. \tag{1.25}$$

Γn is a Gamma function

$$\Gamma 1 = 1, \ \Gamma 1/2 = \sqrt{\pi}.$$

Recurrence formula for Gamma Function

$$\Gamma m/2 = (m/2 - 1)\ \Gamma\left(m/2 - 1\right).$$

It is similar to factorial n which is true for all natural numbers

$$\Gamma m/2 = (m/2-1)\,(m/2-2)\cdots 3\times 2\times 1 \quad \text{if } m \text{ is even}$$

$$\Gamma m/2 = (m/2-1)\,(m/2-2)\cdots 3/2\times 1/2\times \sqrt{\pi}.$$

1.7.1.1 Gamma Probability Density Function

A probability density function defined as

$$f(x) = \frac{\exp(-x)\times x^{n-1}}{\Gamma n} \tag{1.26}$$

is a Gamma probability density function for $0 < x < \infty$.

1.7.2 Beta Function of First Kind B(m,n)

The function given as

$$B(m,n) = \int_0^1 x^{m-1}(1-x)^{n-1}\mathrm{d}x \quad 0 < x < 1, \tag{1.27}$$

$$= \frac{\Gamma m \Gamma n}{\Gamma(m+n)}. \tag{1.28}$$

1.7.2.1 Beta Probability Functions of First Kind

A function given as

$$f(x) = \frac{x^{m-1}(1-x)^{n-1}}{B(m,n)} \text{ for all positive } m,n \text{ and } x: 0 < x < 1$$

$$= 0 \text{ otherwise} \tag{1.29}$$

is a Beta probability function of first kind.

1.7.3 Alternative Form of Beta Function

In (1.27) put $x = \sin^2\theta$, giving $\mathrm{d}x = 2\sin\theta\cos\theta\,\mathrm{d}\theta$, hence (1.27)

$$\int_0^{\pi/2} \sin^{2m-2}(1-\sin^2\theta)^{n-1} 2\sin\theta\,\cos\theta\,\mathrm{d}\theta,$$

$$2\int_0^{\pi/2}(\sin\theta)^{2m-1}(\cos\theta)^{2n-1}\mathrm{d}\theta = \frac{\Gamma m\Gamma n}{\Gamma(m+n)}. \tag{1.30}$$

By writing $2m = M + 1$
and $2n = N + 1$
we get

$$\int_0^{\pi/2}\sin^M\theta\cos^N\theta\,\mathrm{d}\theta = \frac{\Gamma(M+1)/2\Gamma(N+1)/2}{2.\Gamma(M+N+2)/2}. \tag{1.31}$$

The above function is another form of Beta Function.

1.7.4 Beta Function of Second Kind B (m,n)

$$B(m,n) = \int_0^{\infty}\frac{x^{m-1}}{(1+x)^{m+n}}\mathrm{d}x. \tag{1.32}$$

1.7.4.1 Beta Probability Functions of Second Kind

A function of continuous random variable given as

$$\begin{aligned} f(x) &= \frac{x^{m-1}}{B(m,n)(1+x)^{m+n}} \text{ for } m, n \text{ and } x \text{ positive and } 0 < x < \infty \\ &= 0 \text{ otherwise} \end{aligned} \tag{1.33}$$

is a Beta function of second kind.

1.7.5 Cauchy Distribution

A probability density function of continuous random variable expressed as

$$f(x) = \frac{1}{\pi}\times\frac{b}{(x-m)^2+b^2} \tag{1.34}$$

is the Cauchy's Function with parameters m and b. It is also a bell-shaped curve with mean m, but it is more peaked at the centre and has flatter tails than a normal probability curve. The probability function is shown in Fig. 1.2a. Its cumulative distribution is shown in Fig. 1.2b and mathematical expression is given as follows:

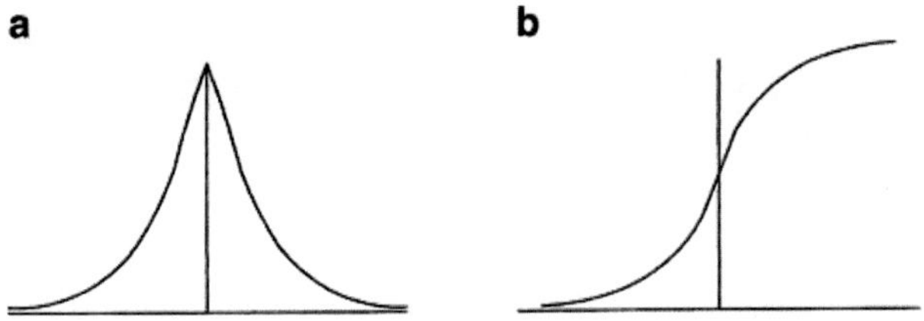

Fig. 1.2 (**a**) Cauchy Function (PDF), (**b**) Cumulative Cauchy distribution

$$F(x) = \frac{1}{2} + \frac{1}{\pi}\tan^{-1}\{(x - m)/b\}. \tag{1.35}$$

1.7.6 Arc Sine(U-Shaped) Distribution

If a quantity X is known to be cyclic sinusoidal, with unknown phase ϕ, between the limits $-a$ to a, then according to the principle of maximum entropy, a rectangular distribution $R(0, 2\pi)$ would be assigned to ϕ. The distribution assigned to X is the sine inverse (Arc sign) distribution given as

$$X = b\sin(\phi). \tag{1.36}$$

The PDF for X

$$\begin{aligned} g_x(\xi) &= \frac{1}{\pi\sqrt{a^2 - \xi^2}}, \quad \text{for all } \xi \text{ between } -\text{a to a} \\ g_x(\xi) &= 0, \quad \text{for all other values of } \xi \end{aligned} \tag{1.37}$$

PDF gives a U-shaped curve.

References

1. BPM, *International System of Units*, 8th edn. (BIPM, Sevres, France, 2006)
2. BIPM, *International Vocabulary of Metrology (VIM)* (BIPM, Sevres, France, 2006)
3. S.C. Gupta, V.K. Kapoor, *Fundamentals of Statistics* (Sultan Chand, New Delhi, 1994)
4. NABL, *Guidelines for Estimation and Expression of Uncertainty in Measurement* (NABL 141-National Accreditation Board for testing and calibration laboratories], New Delhi, 2000)

Chapter 2
Distribution Functions

2.1 Introduction

In the process of calculating uncertainty in measurements, we come across quite a few statistical terms such as random variable, independent event, distribution function and probability density function. In this chapter, we deal with random variable, distribution functions, and probability density functions, discrete and continuous functions. The normal (Gaussian) probability density function and its properties are discussed.

The word "density" used in probability density function is seldom used. In loose terms every cumulative distribution function or probability density function is called as distribution. It is only in context to other things that a distinction between cumulative and density function is made.

2.2 Random Variable

Random variable is a real number representing an outcome of any random experiment. For example, tossing of n unbiased coins and counting number of heads appearing therein, then any one of the possible outcomes of this experiment will represent a random variable. Any outcome ω will be a natural number like 0, 1, 2, $\ldots, n$.

If S is the sample space then

$$\omega \in S. \tag{2.1}$$

That is for each ω there exist $X(\omega)$ – the probability of happening belonging to a set of real numbers R.

For example, ω in the above example may be r number of heads, then $X(\omega)$ will be ${}^nC_r/2^n$ for n unbiased coins. The value of probability is obtained by using the binomial distribution.

S.V. Gupta, *Measurement Uncertainties*, DOI 10.1007/978-3-642-20989-5_2,

Similarly any observation indicated by an unbiased instrument is an outcome ω of a random experiment. Here, ω belongs to the set of observations S indicated by the instrument. Then $X(\omega)$ is also a random variable belonging to the set of real numbers. In case of measurement with unbiased instruments, $X(\omega)$ will belong to a normal distribution.

Formal definition of a random variable may be as follows:

For the sample space S associated with a random experiment, there exists a real-valued variable $X(\omega)$ belonging to a real space $(-\infty, +\infty)$.

This is an example of one-dimensional random variable. If the functional values are ordered pair then it is called as a two-dimensional random variable. In general for an n-dimensional random variable whose domain is S has $X(\omega)$ is the collection of n-tuples of real numbers (vectors in n-space).

2.3 Discrete and Continuous Variables

Discrete variable is that which takes only finite number of values in a given interval. Similarly continuous variable is that which takes infinite number of values in a given interval. The interval may be large or small. That is discrete variable will take only certain values in small steps.

2.4 Discrete Functions

2.4.1 Probability Distribution of a Random Variable

Probability distribution of a random variable is a function giving probability that a random variable takes any given value, which belongs to a given set of values.

2.4.2 Discrete Probability Function

In case of statistical data of discrete variables, each variable x_i will have a specific frequency f_i. It means that a particular variable x_i will occur f_i times.

If N is the total frequency then

$$\sum_{i=1}^{i=n} f_i = N. \qquad (2.2)$$

Then f_i/N is the probability for the existence of the variable x_i.

2.5 Distribution Function

Distributions function is the sum of all probabilities of a random variable such that the random variable is less than the given value. In fact, it is the cumulative sum of all frequencies such that X is less than or equal to the given value.

2.5.1 *Continuous Distribution Function*

A function $F(x)$ giving, for every value of a random variable x, the probability that the random value of X be less than or equal to x is a continuous distribution function. It is expressed as

$$F(x) = \Pr(X \leq x). \tag{2.3}$$

The distribution function defined in this way is also called a cumulative distribution function. The word cumulative is seldom used before the cumulative continuous distribution function.

2.5.2 *Discrete Distribution*

In statistical data of discrete variable, the set of relative cumulative frequencies (cumulative frequency divided by total frequency) is a distribution function.

For example, if there are n independent variables $x_1, x_2, x_3, \ldots, x_n$ with $f_1, f_2, f_3, \ldots, f_n$ with

$$\sum_{i=1}^{i=n} f_i = N,$$

then f_i/N is the relative frequency and $\sum_{i=1}^{i=r} f_i/N$ is the relative cumulative frequency of x_i for all values of i. Such a set of relative cumulative frequencies is a discrete distribution function of $x_1, x_2, x_3, \ldots, x_n$. In this case also, the word cumulative is seldom used before the cumulative discrete distribution function.

2.6 Probability Density Function

For a continuous random variable X, the probability density function is the derivative (if it exists) of its distribution function $F(X)$ i.e.

$$f(x) = \mathrm{d}F(x)/\mathrm{d}x. \tag{2.4}$$

$f(x)\mathrm{d}x$ is the probability element such that the random variable X lies in between x and $x + \mathrm{d}x$.

Mathematically

$$f(x) = \Pr(x < X < x + \mathrm{d}x). \tag{2.5}$$

The integral or the sum of all the probabilities of a continuous variable taking every value in between $-\infty$ and $+\infty$ is a certainty; hence

$$\int_{-\infty}^{\infty} f(x)\mathrm{d}x = 1. \tag{2.6}$$

2.6.1 *Discrete Probability Function*

If a discrete random variable X can take values $x_1, x_2, x_3, \ldots, x_n$, with probabilities $p_1, p_2, p_3, \ldots, p_n$, such that

$$p_1 + p_2 + p_3 + \cdots + p_n = 1 \tag{2.7}$$

and

$$p_i \geq 0 \quad \text{for all } i, \tag{2.8}$$

then these two sets constitute a discrete probability distribution.

A function P_r for each value of x_i of discrete random variable X, giving the probability p_i when the random variable takes the value x_i such that

$$p_i = \Pr(X = x_i) \tag{2.9}$$

is the probability function.

2.7 Discrete Probability Functions

2.7.1 *Binomial Probability Distribution*

Binomial distribution is one of the most important probability function used in practical applications. The applications range from sampling inspection to the failure of rocket engines.

Suppose that a series of n independent trials have been made, each of which can be a success with probability p or a failure with probability $(1 - p)$. The number of success, which is observed, will be any natural number between 0 and n.

An event with r successes necessarily means an event with r successes and $(n - r)$ failures. Such an event is denoted as $p^r(1-p)^{n-r}$, but r successes and $n-r$

failures may be arranged in ^nC_r ways, so the probability of the event $p^r(1-p)^{n-r}$ is $^nC_r p^r(1-p)^{n-r}$. If this probability is denoted as p_r then $p_1, p_2, p_3, \ldots, p_n$ are the respective probabilities of $1, 2, 3, \ldots, n$ successes, giving

$$\begin{aligned} {}^nC_0(1-p)^n + {}^nC_1p(1-p)^{n-1} &+ {}^nC_2p^2(1-p)^{n-2} \\ &+ \cdots {}^nC_{n-1}p^{n-1}(1-p) \\ &+ {}^nC_np^n \end{aligned} \tag{2.10}$$

Binomial probability distribution is applicable whenever a series of trials is made satisfying the following conditions:

Each trial has only two outcomes, which are mutually exclusive. One of the two outcomes is denoted as success then other is failure, for example head and tail in a coin, go and not go, and defective and non-defective in industrial production

1. Probability p of a success is constant in each trial. This also means that probability of failure $(1-p)$ is also constant.
2. The outcomes of successive trials are independent.

Larger is the sample size the outcomes will fit better to the binomial function.

2.7.1.1 Probability of the Binomial Distribution

It may be noticed that independent variable, in case of binomial distribution, is r with relative frequency f_r, which is same as the probability p_r for r success and $n-r$ failures

$$P_r = {}^nC_r p^r q^{n-r} = f_r \quad \text{for all from 1 to } n. \tag{2.11}$$

Here q, for the sake of brevity, is written for $(1-p)$.

2.7.1.2 Moments

In general, $\sum_{r=1}^{r=n} r^k f_r$ is called the kth moment of the random variable r. The arithmetic mean is the first moment. Second moment in conjunction of first moment will give variance. If arithmetic mean is zero then second moment is the variance.

2.7.1.3 Arithmetic Mean

Hence the mean μ of the binomial distribution is given as

$$\sum_{r=0}^{r=n} r \cdot {}^nC_r p^r q^{n-r}, \tag{2.12}$$

but

$$^{n}C_r = \frac{n!}{r!(n-r)!} = \frac{n(n-1)!}{r \times (r-1)! \times \{(n-1)-(r-1)\}!} = ^{n-1}C_{r-1}(n)/r.$$

Substituting this value of $^{n}C_r$ in (2.12) gives us

$$\begin{aligned}\mu &= np \sum {}^{n-1}C_{r-1} p^{r-1} q^{n-1-(r-1)} \\ &= np\,(p+q)^{n-1} \\ &= np\,(p+1-p)^{n-1} \\ &= np.\end{aligned} \quad (2.13)$$

2.7.1.4 Standard Deviation

Similarly one can find the standard deviation of the binomial distribution.

$$\text{Second moment} = \sum r^2 \cdot {}^{n}C_r p^r q^{n-r}.$$

Following steps twice as we have done for arithmetic mean above, we get

$$\text{Second moment} = np(n-1)p + np,$$

$$\text{Standard deviation} = \left[\text{second moment} - (\text{first moment})^2\right]^{1/2},$$

giving us

$$\left[np(n-1)p + np - (np)^2\right]^{1/2} = [np\{np - p + 1 - np\}]^{1/2}.$$

Standard deviation of a binomial distribution is

$$\{np(1-p)\}^{1/2}. \quad (2.14)$$

2.7.2 *Poisson's Distribution*

Another important discrete distribution is the Poisson's distribution. When a probability of happening of an event is very small, i.e. p is small and n is quite large such that np the mean in binomial distribution is finite, then binomial distribution reduces to Poisson's distribution with np as the parameter. Examples are found in industrial production, for example defective blades in a blade-manufacturing factory. Overfilling of packages with an automatic filling machine in a packaging industry is another example.

Poisson's distribution with r as discrete random variable is given as

$$P_r = \frac{(np)^r}{r!} \exp(-np). \tag{2.15}$$

2.7.2.1 Mean of the Poisson's Distribution

$$\begin{aligned}
\text{Mean} &= \sum_{r=0}^{r=\infty} r\{(np)^r/r!\} \exp(-np) \\
&= np \exp(-np) \sum_{r=1}^{r=\infty} (np)^{r-1}/(r-1)! \qquad (2.16) \\
&= np \exp(-np) \exp(np) \\
&= np.
\end{aligned}$$

Arithmetic mean μ of Poisson's distribution is np.

2.7.2.2 Standard Deviation of the Poisson's Distribution

Variance of Poisson's distribution V is written as

$$\begin{aligned}
V &= \text{Second moment} - (\text{first moment})^2 \\
&= \{\exp(-np) \sum_{r=0}^{r=\infty} r^2 (np)^r/r!\} - (np)^2 \\
&= \exp(-np) \sum_{r=1}^{r=\infty} \{r(r-1) + r\}(np)^r/r!\} - (np)^2 \\
&= \exp(-np) \left[(np)^2 \sum_{r=2}^{r=\infty} (np)^{r-2}/(r-2)! \right. \\
&\quad \left. + np \sum_{r=1}^{r=\infty} (np)/(r-1) \right] - (np)^2 \qquad (2.17) \\
&= (np)^2 + (np) - (np)^2 \\
&= np.
\end{aligned}$$

Hence standard deviation of the Poisson's Distribution is $\sqrt{np}$ and mean is np.

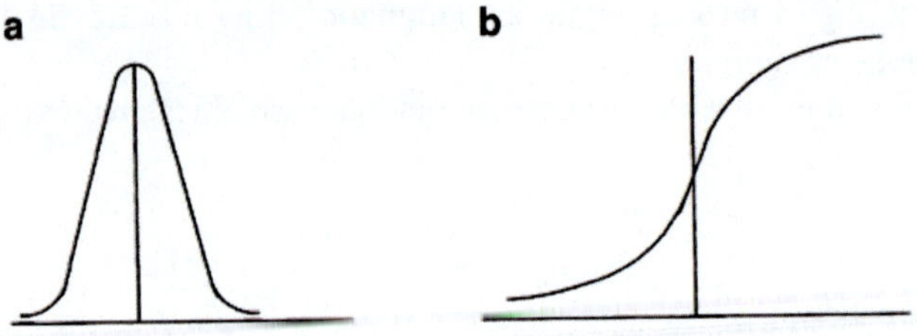

Fig. 2.1 (**a**) Normal Probability Function, (**b**) Normal Cumulative Distribution

2.8 Continuous Probability Distributions

2.8.1 *Normal Probability Function*

A binomial distribution, in which non of p or $(1 - p)$ is small and n approaches to ∞, reduces to normal or Gaussian distribution. This bell-shaped distribution is most well known and is most widely used.

Figure 2.1a represents the normal probability function. The curve is symmetrical about the mean μ. Taking σ as standard deviation, the Gaussian probability density function can mathematically be expressed as

$$P(x) = \frac{1}{\sigma\sqrt{2\pi}} \exp\left\{\frac{-(x-\mu)^2}{2\sigma^2}\right\}. \tag{2.18}$$

The cumulative distribution curve for the Gaussian probability function is shown in Fig. 2.1b and is mathematically expressed as

$$P(X \leq x_1) = \int_{-\infty}^{x_1} P(x)\mathrm{d}x, \tag{2.19}$$

$P(x)$ being the probability function of a random variable x; hence by definition,

$$\int_{-\infty}^{\infty} P(x)\mathrm{d}x = 1.$$

The Gaussian function has the following properties:

The curve for the normal (Gaussian) probability distribution is bell shaped and is symmetrical about the line $x = \mu$.

- Mean, mode and median of a normal distribution are the same.
- $P(x)$ decreases rapidly as the numerical values of x increases.
- $P(x)$ is maximum at $x = \mu$ and is equal to $1/\sigma\sqrt{2\pi}$.
- The x-axis is an asymptote to the curve.
- The points of inflexion are at $x = \mu \pm \sigma$ and the ordinates of the points are $\frac{1}{\sigma\sqrt{2\pi}} \exp(-1/2)$.

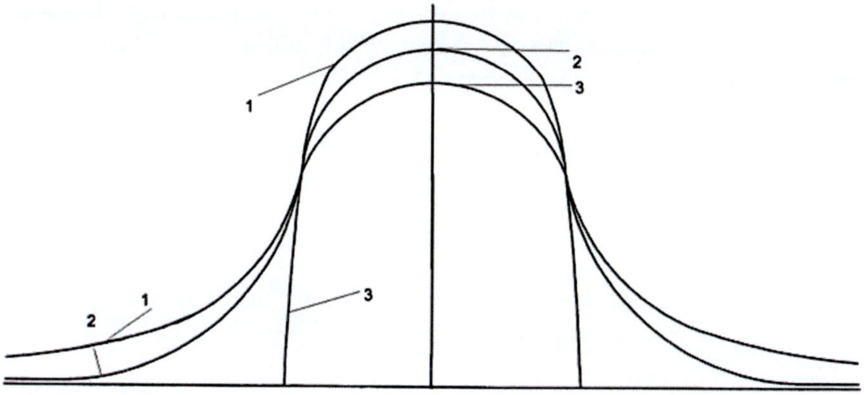

Fig. 2.2 Normal and similar curves

- As the curve represents a probability, which cannot be negative so no portion of the curve will lie below the x-axis.
- The semi-range (from mean to extreme value on either side of mean) is almost three times the standard deviation σ.
- The range of $\mu \pm 3\sigma$ covers 99.73% of area covered by the curve and x-axis; i.e. the probability of the random variable lying between $\mu \pm 3\sigma$ is 0.9973. Hence 99.73% of all normal variates will lie in this interval.
- The range of $\mu \pm 2\sigma$ covers 95.45% of all normal variates; i.e. the area covered in between $\mu \pm 2\sigma$ is 95.45% of the total area.
- The range of $\mu \pm \sigma$ covers 68.27% of normal variates.
- The range of $\mu \pm 0.6749\sigma$ covers 50% of the normal variates; 0.6749σ is called as probable error.

From the property of the area covered for different values of x helps us in deciding as to which curve is normal and which is not. In Fig. 2.2, though all the three curves have same mean and standard deviation, but only one of them represents the normal distribution. Making use of the aforesaid properties about the area covered between various ordinates, we can say that only curve 2 represents the normal curve because it has very small area (about 0.03%) covered beyond $x = 3\sigma$.

The curve 1 is not a normal curve as the area covered by it beyond $x = 3\sigma$ is much more than 0.03%. Similarly the curve 3 is also not a normal curve as all the area is covered between $x = -2\sigma$ and $x = 2\sigma$.

2.8.2 *Cumulative Distribution of the Normal Probability Function*

Distribution or cumulative function $F(X \leq x)$ means relative cumulative frequency or the total area of the normal curve covered by it with the x-axis from x equal to

$-\infty$ to the ordinate at x_1. In any experiment with an unbiased instrument this also represents the percentage of observations likely to fall within the limit when x varies from $-\infty$ to x.

$$F(X \leq x_1) = (1/\sigma\sqrt{2\pi}) \int_{-\infty}^{x_1} \exp(-(x-\mu)^2/2\sigma^2\, dx. \qquad (2.20)$$

And

$$F(-\infty < x < \infty) = \frac{1}{\sigma\sqrt{2\pi}} \int_{-\infty}^{\infty} \exp(-(x-\mu)^2/2\sigma^2 dx = 1. \qquad (2.21)$$

Putting $z = (x-\mu)/\sigma$ in (2.21) gives us

$$F(-\infty < z < \infty) = \frac{1}{\sqrt{2\pi}} \int_{-\infty}^{\infty} \exp(-z^2/2) dz. \qquad (2.22)$$

As the integrand is an even function, (2.22) can be written as

$$\begin{aligned} F(0 < z < \infty) &= 2(1/\sqrt{2\pi})S \int_{0}^{\infty} \exp(-z^2/2)\, dz = 1 \\ \text{or } &= (1/\sqrt{2\pi}) \int_{0}^{\infty} \exp(-z^2/2)\, dz = 1/2 \\ &= (1/\sqrt{2\pi}) \int_{-\infty}^{0} \exp(-z^2/2)\, dz. \end{aligned} \qquad (2.23)$$

Limits in above integrals are z equal to zero to z equal to ∞ and $z = -\infty$ to $z = 0$, but $z = (x-\mu)/\sigma$; hence corresponding limits of x in the second integral will be $x = -\infty$ to $x = \mu$. So the last integral in (2.23) can be written as

$$F(-\infty < X < \mu) = \int_{-\infty}^{x=\mu} \exp\left\{\frac{(x-\mu)^2}{2\sigma^2}\right\} = 1/2. \qquad (2.24)$$

2.8.3 *Normal Distribution and Probability Tables*

Table A.1 gives the probability of happening for the given value of variable z. The values of z in steps of 0.01 have been taken from 0 to 3.49.

$z = (x-\mu)/\sigma$, $z = 1$ corresponds one standard deviation.

Table A.2 gives the cumulative frequency (area covered) from $-\infty$ to different value of $z \left\{\frac{1}{\sqrt{2\pi}} \int_0^z \exp(-z^2/2)dz\right\}$.

In fact, the table gives the cumulative normal distribution against deviation of the variate from the mean expressed in terms of standard deviation.

Table A.3 gives the area covered by the variable from 0 to z. In fact, Table A.3 can be derived from Table A.2 by subtracting 0.5 from each entry.

Table A.4 gives the probability interval for the given value of z. It is the area covered by the variables from $-z$ to $+z$. For given value of z, every entry in Table A.4 is twice the entry in Table A.3.

Table A.5 gives the values of z for the given probability interval.

2.8.4 Mean and Variance of a Linear Combination of Normal Variates

Let z be a linear combinations of two normal variates and is given by

$$z = ax + by. \tag{2.25}$$

Then the probability distribution of z will also be a normal distribution giving

$$\text{Mean of } z = \mu_z = a\mu_x + b\mu_y \tag{2.26}$$

and

$$\text{Variance of } z = \sigma_z^2 = a^2\sigma_x^2 + b^2\sigma_y^2. \tag{2.27}$$

Generalizing the above statements, if z is a linear combination of n normal variates given as

$$z = \sum_{p=1}^{p=n} a_p x_p. \tag{2.28}$$

Then probability distribution of z will also be a normal distribution with mean and standard deviation given by

$$\text{Mean} = \mu_z = \sum_{p=1}^{p=n} a_p \mu_p, \tag{2.29}$$

$$\sigma_z = \left(\sum_{p=1}^{p=n} a_p^2 \sigma_p^2 \right)^{1/2}. \tag{2.30}$$

2.8.5 Standard Deviation of Mean

Let there be n normal variates $x_1, x_2, x_3, \ldots, x_n$, then mean $\bar{x}$ of these variates is given by

$$\bar{x} = (1/n) \sum_{p=1}^{p=n} x_p. \tag{2.31}$$

Following (2.30) and taking $1/n = a_p$ for all values of p, then standard deviation from (2.30) is given by

$$\sigma_{\bar{x}} = \left(\sum_{p=1}^{p=n} \sigma_p^2 / n^2 \right)^{1/2}. \tag{2.32}$$

If these n normal variates belong to the same population for example observations of an unbiased measuring instrument, then each variate will have the same σ, giving us

$$\sigma_{\bar{x}} = (n\sigma^2/n^2)^{1/2} = \sigma/\sqrt{n}. \tag{2.33}$$

2.8.6 *Deviation from the Mean*

From (2.16), the mean deviation is given by

$$|x_r - \mu| = \frac{1}{\sigma\sqrt{2\pi}} \int_{-\infty}^{\infty} |(x-\mu)| \exp\{-(x-\mu)^2/2\sigma^2\}\, \mathrm{d}x.$$

Putting

$$y = (x-\mu)/\sigma\sqrt{2},$$
$$\text{we get } \mathrm{d}y = \mathrm{d}x/\sigma\sqrt{2} \text{ or}$$
$$\sigma\sqrt{2}\,\mathrm{d}y = \mathrm{d}x,$$

giving

$$\frac{1}{\sigma\sqrt{2\pi}} \int_{-\infty}^{\infty} \sigma\sqrt{2}\,|y| \exp(-y^2) \left\{(\sigma\sqrt{2})\mathrm{d}y\right\} \tag{2.34}$$

$$= 2\sigma\sqrt{\frac{1}{2\pi}} \int_{-\infty}^{\infty} |y| \exp(-y^2)\, \mathrm{d}y$$

$$= 2\sigma\sqrt{\frac{1}{2\pi}} \int_{-\infty}^{0} -y \exp(-y^2)\mathrm{d}y + 2\sigma\sqrt{\frac{1}{2\pi}} \int_{0}^{\infty} y \exp(-y^2)\mathrm{d}y. \tag{2.35}$$

In the first integral, putting

$$y = -z,$$
$$\mathrm{d}y = -\mathrm{d}z$$

lower limit of $z = -\infty$ and upper limit of $z = 0$, the first integral becomes

$$= 2\sigma\sqrt{\frac{1}{2\pi}} \int_{0}^{\infty} z \exp(-z^2)\mathrm{d}z$$

$$= 2\sigma\sqrt{\frac{1}{2\pi}}\int_0^\infty z\exp(-z^2)\mathrm{d}z.$$

Hence (2.35) becomes

$$= 4\sigma\sqrt{\frac{1}{2\pi}}\int_0^\infty z\exp(-z^2)\mathrm{d}z = 4\sigma\frac{1}{\sqrt{2\pi}}\left[\frac{\exp(-z^2)}{-2}\right]_0^\infty$$

$$= -\sigma\sqrt{\frac{2}{\pi}}[0-1] = \sigma\sqrt{\frac{2}{\pi}}. \tag{2.36}$$

2.8.7 *Standard Deviation of Standard Deviation*

We will first find out the probability density function for the standard deviation from the first principle. In deriving the expression, we will apply the fact that any linear combination of normal random variables has a normal distribution. We will then compare it with normalized normal probability function and get the corresponding value of the standard deviation of the standard deviation.

We know σ the standard deviation is calculated by the formula

$$\sigma^2 = \sum_{p=1}^{p=\infty}\frac{(x_p - \bar{x})^2}{n}.$$

This formula is valid if n is very large say more than 200. For smaller values of n the best estimate of σ is s and is given by

$$s^2 = \sigma_v^2 = \sum_{p=1}^{p=n}\frac{(x_p - \bar{x})^2}{n-1}.$$

We know that $\bar{x}$ is sum of n number of normal variable; hence $\varepsilon_p = (x_p - \bar{x})$ is also a normal variable and therefore will follow a normal probability distribution. If there are n such deviations each will follow the normal distribution; hence, if σ_1 is the standard deviation, the probability P of occurrence of all the n deviation is given by

$$P = \frac{\exp\left(-\sum_{p=1}^{p=n}\varepsilon_p^2/2\sigma_1^2\right)}{\sigma_1^n(2\pi)^{n/2}}. \tag{2.37}$$

Similarly if standard deviation is $\sigma_1 + \delta$, where δ is a small quantity, then probability P_1 is given by

$$P_1 = \frac{\exp\left(-\sum\limits_{p=1}^{p=n} \varepsilon_p^2/2(\sigma_1+\delta)^2\right)}{(\sigma_1+\delta)^n(2\pi)^{n/2}}.$$

Thus, the ratio of $P/P_1 = Q$ is given by

$$\begin{aligned} Q &= \left(1+\frac{\delta}{\sigma_1}\right)^{-n} \exp\left[\frac{1}{2}\sum_{p=1}^{p=n}\varepsilon_p^2\left\{\frac{1}{\sigma_1^2}-\frac{1}{(\sigma_1+\delta)^2}\right\}\right] \\ &= \left(1+\frac{\delta}{\sigma_1}\right)^{-n} \exp\left\{\frac{1}{2}\sum_{p=1}^{p=n}\varepsilon_p^2\frac{(2\delta\sigma_1+\delta^2)}{\sigma_1^2(\sigma_1+\delta)^2}\right\} \\ &= \exp\left\{\frac{1}{2}\sum_{p=1}^{p=n}\varepsilon_p^2\frac{(2\delta\sigma_1+\delta^2)}{\sigma_1^2(\sigma_1+\delta)^2} - n\log\left(1+\frac{\delta}{\sigma_1}\right)\right\} \end{aligned} \tag{2.38}$$

Next if σ_1 is the value, which makes P to be maximum, then partial derivative of P with respect σ_1 must be zero.

Taking log of P in (2.37), we get

$$\log P = -n\log\sigma_1 - \sum_{p=1}^{p=n}(\varepsilon_p^2/2\sigma_1^2) - \frac{n}{2}\log(2\pi). \tag{2.39}$$

Differentiating and putting it to zero, we get

$$\frac{1}{P}\frac{\mathrm{d}p}{\mathrm{d}\sigma_1} = -\frac{n}{\sigma_1} - \frac{\sum\limits_{p=1}^{p=n} -2\varepsilon_p^2}{\sigma_1^3} = 0,$$

giving us

$$\sigma_1^2 = \frac{\sum\limits_{p=1}^{p=n}\varepsilon_p^2}{n}. \tag{2.40}$$

Substituting for $\sum\limits_{p=1}^{p=m}\varepsilon_p^2$ from (2.40) in (2.38), we get

$$Q = \exp\left\{\frac{1}{2}n\frac{(2\delta\sigma_1+\delta^2)}{(\sigma_1+\delta)^2} - n\log\left(1+\frac{\delta}{\sigma_1}\right)\right\}. \tag{2.41}$$

Expanding the exponent in terms of δ/σ_1 and neglecting terms containing δ^3 and higher powers, we get Q in simplified form as

$$Q = \exp(-n\delta^2/\sigma^2).$$

Thus, the probability that the value of σ_1 lies between $\sigma_1 + \delta$ and $\sigma_1 + \delta + d\delta$ is given by

$$Q_1 = K \exp(-n\delta^2/\sigma_1^2)\, d\delta. \tag{2.42}$$

K is to be such that total probability of Q_1 is unity when δ varies from $-\infty$ to $+\infty$ is 1, giving us

$$\int_{-\infty}^{\infty} K \exp(-n\delta^2/\sigma_1)\, d\delta = 1. \tag{2.43}$$

Put

$$\delta\sqrt{n}/\sigma_1 = y,$$

giving

$$d\delta = \frac{\sigma_1}{\sqrt{n}} dy.$$

Hence (2.41) becomes

$$K\frac{\sigma_1}{\sqrt{n}} \int_{-\infty}^{\infty} \exp(-y^2)\, dy = 1,$$

but $\int_{-\infty}^{\infty} \exp(-y^2)\, dy = \sqrt{\pi}$.

Hence giving us

$$K = \frac{\sqrt{n}}{\sigma_1\sqrt{\pi}}.$$

Substituting the value of K in (2.40), we get

$$\begin{aligned} Q_1 &= \frac{\sqrt{n}}{\sigma_1\sqrt{\pi}} \exp(-n\delta^2/\sigma_1^2)\, d\delta \\ &= \frac{1}{\left(\sigma_1/\sqrt{2n}\right)\sqrt{2\pi}} \exp\left[-\left\{\frac{\delta^2}{2\left(\sigma_1/\sqrt{2n}\right)^2}\right\}\right]. \end{aligned} \tag{2.44}$$

Comparing (2.42) with the standard form of normal distribution namely

$$\frac{\exp(-x^2/2\sigma^2)}{\sigma\sqrt{2\pi}}.$$

we note that standard deviation of the density function in (2.40) is

$$\sigma_1/\sqrt{2n}.$$

Hence standard deviation of standard deviation σ is

$$\frac{\sigma 1}{\sqrt{2n}}. \tag{2.45}$$

2.8.8 Nomenclature for Normal Distribution

Normal distribution is characterized by its mean μ and variance σ^2; hence quite often it is denominated as $N(\mu, \sigma^2)$. A normal distribution designated as $N(12, 5)$ is mathematically equivalent to

$$f(x) = \frac{1}{\sqrt{10\pi}} \exp\left\{(x-12)^2/10\right\}.$$

We know that the sum of two normal variates is also a normal variate. Hence normal distribution of the sum of two normal variates having designations $N(\mu_1, \sigma_1^2)$ and $N(\mu_2, \sigma_2^2)$ will be $N(\mu_1+\mu_2, \sigma_1^2+\sigma_2^2)$ and will be mathematically expressed as

$$f(x_1+x_2) = \frac{1}{(\sigma_1+\sigma_2)\sqrt{2\pi}} \exp -\{(x_1+x_2)-(\mu_1+\mu_2)\}/2(\sigma_1^2+\sigma_2^2). \tag{2.46}$$

2.8.9 Probability Function of the Ratio of Two Normal Variates [1]

Let x and y be two normal variables with means μ_1 and μ_2 and standard deviations σ_1 and σ_2, respectively. We wish to derive a probability distribution of z where z is given as

$$z = \frac{(x-\mu_1)}{(y-\mu_2)}. \tag{2.47}$$

Equation (2.47) may be written as

$$\frac{\sigma_2^2}{\sigma_1^2} z^2 = \frac{(x-\mu_1)^2/\sigma_1^2}{(y-\mu_2)^2/\sigma_2^2}, \tag{2.48}$$

but $(x-\mu_1)^2/\sigma_1^2$ and $(y-\mu_2)^2/\sigma_2^2$ are the squares of independent standardized normal variables and σ_2^2/σ_1^2 is an independent χ^2 variable with 1 degree of freedom. But ratio of the squares of two independent variables is also a χ^2 variable of degree 1. Thus, $z^2/(\sigma_1^2/\sigma_2^2)$ is the quotient of two independent χ^2 variables each with 1 degree of freedom.

We know that if χ_1^2 and χ_2^2 are two independent χ^2 variables with n_1 and n_2 degrees of freedom, respectively, then

χ_1^2/χ_2^2 is a $\beta_2(\mu/2, \nu/2)$ variate, whose probability density function $f(x)$ is given by definition

$$f(x)\mathrm{d}x = \frac{1}{B(\mu,\nu)}\frac{x^{\mu-1}}{(1+x)^{\mu+\nu}}\mathrm{d}x \quad \text{for positive values of } \mu,\ \nu \text{ and } x$$
$$= 0 \text{ otherwise.} \tag{2.49}$$

Here $B(\mu,\nu) = \frac{\Gamma\nu\Gamma\mu}{T(\nu+\mu)}$, and Γ stands for Gamma function.

Hence, the probability function of $z^2(\sigma_1^2/\sigma_2^2)$, which is the ratio of two χ^2 variables each having 1 degree of freedom; hence its probability density function is given by

$$f\left(\frac{z^2}{\sigma_1^2/\sigma_2^2}\right) = \frac{\Gamma(1/2+1/2)}{\Gamma(1/2)\Gamma(1/2)} \cdot \frac{\left(\frac{\sigma_2^2 z^2}{\sigma_1^2}\right)^{1/2-1}}{\left(1+\sigma_2^2 z^2/\sigma_1^2\right)} \cdot \sigma_2^2 dz^2/\sigma_1^2. \tag{2.50}$$

Now

$\Gamma 1 = 1$ and $\Gamma(1/2) = \sqrt{\pi}$; substituting these values in (2.50), we get

$$f(z) = \frac{\sigma_1\sigma_2}{\pi(\sigma_1^2+\sigma_2^2 z^2)z}\mathrm{d}z^2 \text{ for } 0 \le z^2 \le \infty,$$

giving

$$f(z) = \frac{2\sigma_1\sigma_2}{\pi(\sigma_1^2+\sigma_2^2 z^2)}\mathrm{d}z \quad \text{for } 0 \le z \le \infty$$

or

$$f(z) = \frac{\sigma_1\sigma_2}{\pi(\sigma_1^2+\sigma_2^2 z^2)}\mathrm{d}z \quad \text{for} -\infty \le z \le \infty. \tag{2.51}$$

This probability function is important for the experiment in which the output quantity is the ratio of two independent normal variables. For example, measurement of resistance of a resistor by measuring current A passing through it and potential difference V across it. V and A are normal variables and resistance R is related to A and V as

$$R = \frac{V}{A}.$$

Hence

$$f(R)\,\mathrm{d}R = \frac{\sigma_1\sigma_2}{\pi(\sigma_1^2+\sigma_2^2 R^2)}\mathrm{d}R. \tag{2.52}$$

Here σ_1^2 and σ_2^2 are the variances of voltage and current measurements, respectively.

2.8.10 *Importance of Normal Distribution*

Most of the discrete distributions such as binomial, Poisson, and hypergeometric approach to the normal distribution.

Many of the sampling distributions such as Student's t, Snedecor's F, Chi square etc. tend to be a normal distribution for larger samples (of size greater than 10).

Quite often even if the variable is not normally distributed, it can be made to follow normal distribution by simple transformation. For example if the distribution of a variable X is skewed, the distribution of $\sqrt{X}$ might become a normal distribution.

Distributions of sample mean and sample variance tend to follow normal distribution.

The entire theory of small sample tests, for example, Student's t, Snedecor's F and Chi square, is based on the assumption that parent population from which samples are drawn follows normal distribution.

All readings indicated by an unbiased measuring instrument belong to the normal distribution. Random errors of every unbiased instrument follow the normal distribution with zero mean. All industrial products manufactured by automatic devices tend to follow normal distribution. Hence, normal distribution finds largest applications in statistical quality control in industry.

2.8.11 *Collation of Data from Various Laboratories [2]*

2.8.11.1 Most Probable Mean of the Data

All measuring unbiased instruments indicate the value of the measurand, which follows the normal distribution. The problem of collating the data given by different laboratories is quite common. Each laboratory gives the value of the measurand along with the uncertainty. The problem is to find the best estimate of the mean value and the uncertainty associated with it. For example, if there are n independent normal variates $x_1, x_2, x_3, \ldots, x_n$, then $\sigma_1, \sigma_2, \sigma_3, \ldots, \sigma_n$ respectively are their standard deviations. Assuming that X is the most probable value of x, then deviations from the probable value are $X - x_1, X - x_2, X - x_3, \ldots, X - x_n$. As each variate follows normal distribution, the probability of precisely this set of deviations, therefore, is the product of n normal probability functions for the aforesaid values of deviations, giving us

$$\frac{\exp(-(x_1 - X)/2\sigma_1^2}{\sigma_1\sqrt{2\pi}} \frac{\exp-(x_2 - X)/2\sigma_2^2}{\sigma_2\sqrt{2\pi}} \frac{\exp-(x_3 - X)/2\sigma_3^2}{\sigma_3\sqrt{2\pi}} \cdots \frac{\exp-(x_n - X)/2\sigma_n^2}{\sigma_n\sqrt{2\pi}}.$$

This expression may be written as

$$\frac{\exp\left\{-\sum\limits_{p=1}^{p=n}(x_p - X)^2/2\sigma_p^2\right\}}{(2\pi)^{n/2}\prod\limits_{1}^{n}\sigma_p}, \tag{2.53}$$

where $\prod\limits_{1}^{n}\sigma_p = \sigma_1 \times \sigma_2 \times \sigma_3 \times \cdots \times \sigma_n$.

Most probable value of X will be such that the above expression becomes maximum, which means that the exponent expression becomes a minimum. That is

$$\sum_{p=1}^{p=n}(x_p - X)^2/2\sigma_p^2 \quad \text{is a minimum.} \tag{2.54}$$

The expression is a minimum if its first differential with respect to X is zero. Differentiating it with respect to X and putting it equal to zero, we get

$$\sum -\frac{2(x_p - X)}{2\sigma_p^2} = 0, \tag{2.55}$$

giving us

$$X\sum_{p=1}^{p=n}\frac{1}{\sigma_p^2} = \sum_{r=1}^{r=n} x_p/\sigma_p^2. \tag{2.56}$$

Differentiating (2.56) again with respect to X, we get

$$\sum_{p=1}^{p=n}\frac{1}{\sigma_p^2}. \tag{2.57}$$

This is always positive. Hence the expression in (2.54) is a minimum. Hence giving the most probable value of X from (2.56) as

$$X = \frac{\sum\limits_{p=1}^{p=n} x_p/\sigma_p^2}{\sum\limits_{r=1}^{r=n}\frac{1}{\sigma_p^2}}. \tag{2.58}$$

If x_p is replaced by $\bar{x}_p$ the mean of the pth sample of n_p observations, then standard deviation of single observation should be replaced by standard deviation of the mean, which is equal to

$$\sigma_p/\sqrt{n_p}. \tag{2.59}$$

Hence most probable mean of results of several laboratories is given by

$$\bar{X} = \frac{\sum_{p=1}^{p=n} \left\{ \bar{x}_p n_p / \sigma_p^2 \right\}}{\sum_{p=1}^{p=n} (n_p / \sigma_p^2)}. \tag{2.60}$$

Weighted mean of $x_1, x_2, x_3, \ldots, x_n$ with respective weights of $w_1, w_2, w_3, \ldots, w_n$ is given as

$$\bar{X} = \sum_{p-1}^{p=n} w_p x_p \Bigg/ \sum_{p=1}^{p=n} w_p. \tag{2.61}$$

Comparing (2.60) and (2.61), we get

$$\text{The weight factor } w_p \text{ of } \bar{x}_p = \left\{ n_p / \sigma_p^2 \right\} \Bigg/ \sum_{p=1}^{p=n} (n_p / \sigma_p^2). \tag{2.62}$$

Hence in the above equation n_p/σ_p^2 is the weight factor of $\bar{x}_p$. Hence the collating laboratory must know about the number of observations taken for calculating the mean value by each laboratory.

We have seen that most probable value of the mean is not simple arithmetic means of the estimated values but a weight mean. The weight factor given in (2.62) is proportional to the number of observations and inversely proportional to the variance of each laboratory. It may be noticed that variance here is to be calculated by normal statistical means (Type A evaluation of standard uncertainty).

2.8.11.2 Standard Deviation of the Most Probable Mean

Here we have seen that weight factor of the mean is $\frac{n_p/\sigma_p^2}{\sum_{p=1}^{p=n} n_p/\sigma_p^2}$; we further know that if

$$\bar{X} = \sum_{p=1}^{p=n} a_p \bar{x}_p, \tag{2.63}$$

then variance of $\bar{X} = \sum_{p=1}^{p=n} a_p^2 \sigma_p^2 / n_p$.

We know that σ_p^2/n_p is the variance of the mean $\bar{x}_p$ and $a_p = \frac{n_p/\sigma_p^2}{\sum_{p=1}^{p=n} n_p'\sigma_p^2}$, giving

$$\text{Variance of } \bar{X} = \frac{\sum_{p=1}^{p=n} \left(n_p/\sigma_p^2\right)^2 \sigma_p^2/n_p}{\left(\sum_{p=1}^{p=n} n_p/\sigma_p^2\right)^2} \tag{2.64}$$

and

$$\begin{aligned} \text{Variance of } \bar{X} &= \sum_{p=1}^{p=n} \left[\frac{\{(n_p/\sigma_p^2)^2\}}{n_p/\sigma_p^2}\right] \Big/ \left(\sum_{p-1}^{p=n} n_p/\sigma_p^2\right)^2 \\ &= \frac{\sum_{p=1}^{p=n} n_p/\sigma_p^2}{\left[\sum_{p=1}^{p=n} n_p/\sigma_p^2\right]^2} \\ &= \frac{1}{\sum_{p=1}^{p=n} n_p/\sigma_p^2}. \end{aligned} \tag{2.65}$$

$$\text{Standard deviation of the mean } \bar{X} = \left[\frac{1}{\sum_{p=1}^{p=n} n_p/\sigma_p^2}\right]^{1/2}. \tag{2.66}$$

The data sent for such collation not only contain the estimated value of the parameter and standard deviation but should also contain the number of observations taken by each laboratory.

References

1. S.C. Gupta, V.K. Kapoor, *Fundamentals of Mathematical Statistics*, 9th edn. (Sultan Chand, New Delhi, 1994), p. 13.27
2. C.F. Dietrich, *Uncertainty, Calibration and Probability* (Adam Hilger, New York, 1991), pp. 39–41
3. R.B. Frenkel, Fiducial inference applied to uncertainty estimation when identical readings are obtained under low instrument resolution. Metrologia **46**, 661–667 (2009)

Chapter 3
Other Probability Functions

3.1 Introduction

In addition to normal probability function, quite often other probability functions such as rectangular, triangular and trapezium functions are quite often used in a measurement laboratory. Small sample functions play a vital role in the calculation of uncertainty in a measurement laboratory. In this chapter, we discuss rectangular, triangular and trapezium functions with special reference to obtain standard uncertainty from the given semi-range of these functions. Small sample probability functions such as Student's t, χ^2 and Fisher distributions along with their application are also discussed.

3.2 Important Distributions

3.2.1 Rectangular Distribution

Let us consider a probability function $f(X)$ defined as follows:

$$f(x) = \frac{1}{2a} \quad \text{for all values of } X - a \leq x \leq a$$
$$f(x) = 0 \quad \text{for all other values of } x. \tag{3.1}$$

This means that the probability of finding the variable x in between $-a \leq x \leq a$ is same and is equal to $1/2a$. This function is also called as uniform probability function. If a value of certain parameter of an object lies within the tolerance of $\pm a$, then it is obvious that the actual value of the parameter will lie anywhere within the range $-a$ to $+\,a$ with equal probability; i.e. the parameter follows rectangular probability function. In the literature, especially the one obtained from Handbooks,

S.V. Gupta, *Measurement Uncertainties*, DOI 10.1007/978-3-642-20989-5_3,

the value of the parameter is stated together with a specific range. No other information about the nature of the stated value, procedure of measurement or obtaining the semi-range is given. In that case, we can safely assume that the probability of the stated value lying anywhere within the stated range is equal. We see a little later how one can calculate the standard uncertainty from the given semi-range.

A measurement laboratory calibrates a parameter of certain object and issues a certificate stating the value of the parameter and the uncertainty of measurements. The user laboratory of the object assumes that the stated value may lie between the limits of uncertainty with equal probability. For example, mass of a kilogram standard of mass (weight) is stated as 1,000.0025 g with uncertainty ±0.0001 g. In the absence of any information as to which probability distribution the stated value follows or the method of obtaining the semi-range, the user laboratory may assume that the true mass of the mass standard may be anywhere in between 1,000.0024 and 1,000.0026 g. The chance of the stated value of mass lying anywhere between the stated ranges is the same.

Taking mean as origin the rectangular distribution looks like as shown in Fig. 3.1.

3.2.1.1 Mean of the Rectangular Function

Mean $\bar{x}$ is given as

$$\bar{x} = \int_{-\infty}^{\infty} x f(x)\mathrm{d}x = \int_{-a}^{a} (x/2a)\mathrm{d}x = \frac{1}{2a}\left[x^2/2\right]_{-a}^{a} = 0.$$

3.2.1.2 Variance of Rectangular Function

We know that variance σ^2 of a probability function in cases of zero mean is given by

$$\int_{-\infty}^{\infty} x^2 f(x)\mathrm{d}x.$$

In this case, as mean is zero x^2 represents the square of the deviations from the mean. Hence, the variance σ^2 is given by

$$\sigma^2 = \int_{-\infty}^{\infty} x^2 f(x)\mathrm{d}x$$

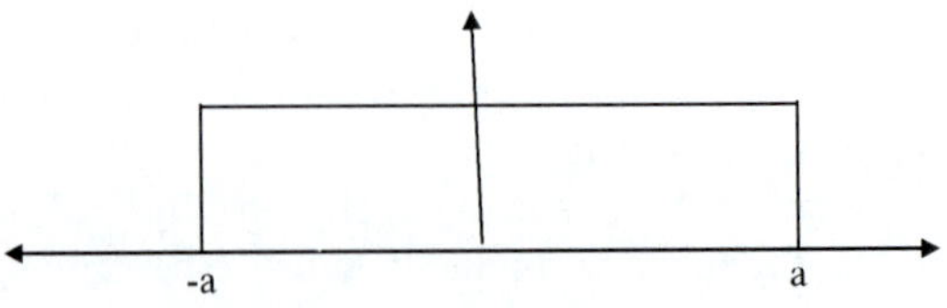

Fig. 3.1 Rectangular probability distribution

$$
\begin{aligned}
&= \int_{-a}^{a} (x^2/2a)\mathrm{d}x \\
&= \frac{1}{2a}[x^3/3]_{-a}^{a} \\
&= \frac{1}{2a}[2a^3/3] \\
&= a^2/3.
\end{aligned} \tag{3.2}
$$

Hence standard uncertainty, which is equal to standard deviation, is $a/\sqrt{3}$.

3.2.2 *Triangular Probability Function*

In this case, probability of true value at the stated value is a maximum and the probability uniformly decreases away from the stated value and becomes zero at the range point. Taking true value as origin and semi-range of $\pm a$, the probability of lying the true value increases uniformly from $-a$ to zero, becomes maximum at the stated value and uniformly decreases to zero at $x = a$. The probability is zero at $x = \pm a$.

Mathematically the triangular probability function $f(x)$ is defined as follows:

$$
\begin{aligned}
&f(x) = (x + a)/a^2 \text{ for all } x \text{such that} - a \le x \le 0 \\
&f(x) = (a - x)/a^2 \text{ for all } x \text{such that } 0 \le x \le a \\
&f(x) = 0 \text{ for all other values of } x.
\end{aligned} \tag{3.3}
$$

Such a function is shown in Fig. 3.2.

3.2.2.1 Mean of the Triangular Probability Function Is Given as

$$
\bar{x} = \int_{-\infty}^{\infty} xf(x)\mathrm{d}x = \int_{-a}^{0} \frac{x^2 + ax}{a^2}\mathrm{d}x + \int_{0}^{a} \frac{ax - x^2}{a^2}\mathrm{d}x
$$

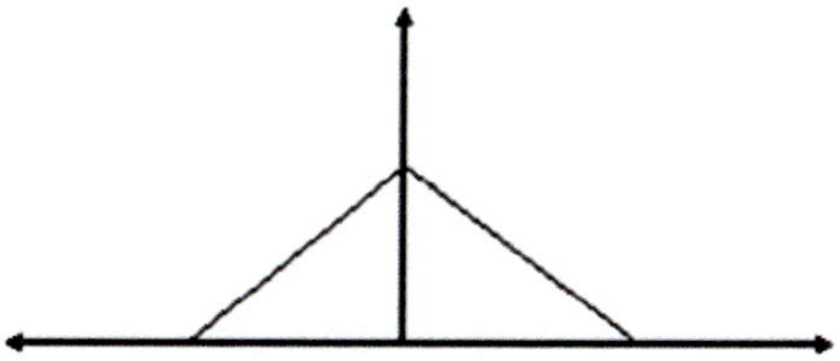

Fig. 3.2 Triangular function

$$= \left[\frac{x^3/3 + ax^2/2}{a^2}\right]_{-a}^{0} + \left[\frac{ax^2/2 - x^3/3}{a^2}\right]_{0}^{a}$$

$$= \left[\frac{a^3/3 - a^3/2}{a^2}\right] + \left[\frac{a^3/2 - a^3/3}{a^2}\right] = 0.$$

3.2.2.2 Variance σ^2 of Triangular Distribution

$$\sigma^2 = \int x^2 f(x)\mathrm{d}x.$$

Here again, as mean is zero, x^2 is the square of the deviation from mean. The integral is taken over the range for which $f(x)$ is nonzero. Giving us

$$\sigma^2 = \int_{-a}^{0} x^2 \frac{x+a}{a^2}\mathrm{d}x + \int_{0}^{a} \frac{x^2(a-x)\mathrm{d}x}{a^2}$$

$$\sigma^2 = (1/a^2)\left[x^4/4 + ax^3/3\right]_{-a}^{0} + (1/a^2)\left[ax^3/3 - x^4/4\right]_{0}^{a} = a^2/6. \quad (3.4)$$

Hence standard uncertainty, which is equal to standard deviation, is $a/\sqrt{6}$.

3.2.3 *Trapezoidal Probability Function*

Trapezium distribution is some sort of compromise between triangular and rectangular distributions. Rectangular distribution assumes that the true value may lie anywhere in the range $-a$ to $+a$ with equal probability. The triangular distribution assumes that the probability of true value lying at the stated value is a maximum and decreases uniformly on either side of the stated value. However, in many realistic cases, it is more reasonable to assume that true value can lie anywhere within a narrower interval around the midpoint with the same probability, while the probability of true value lying outside this small interval on either side uniformly decreases to zero at the extremities of the interval $(-a, a)$. For such cases, probability distribution curve is a symmetrical trapezium with range $2a$ as one side and other side parallel to it is taken as $2a\beta$. Here β is a fraction lying between 1 and 0. The β equal to zero reduces the distribution to triangular one, while β equal to one makes it a rectangular distribution.

Referring to Fig. 3.3, the height of the trapezium k will be given by the fact that area of the trapezium is equal to one.

$$\text{Area} = (a - a\beta)k/2 + 2a\beta k + (a - a\beta)k/2 = a(1+\beta)k.$$

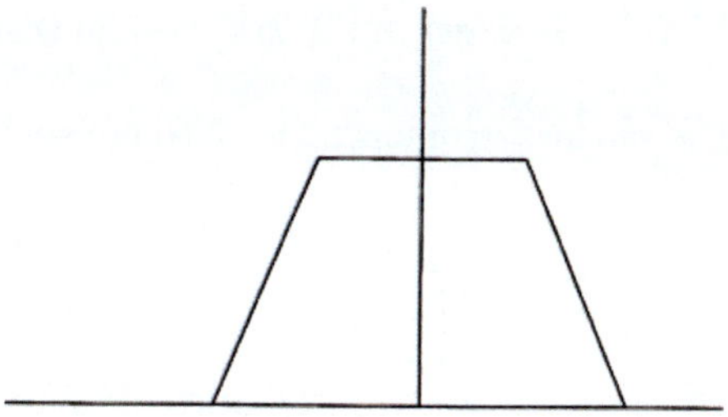

Fig. 3.3 Trapezoidal distribution

Equating the area equal to 1, we get

$$k = 1/a(1 + \beta). \tag{3.5}$$

Equation of probability distribution function represented by a straight line passing through the points $(-a, 0)$ and $(-a\beta, k)$ is

$$y = k(a + x)/a(1 - \beta), \text{ giving}$$
$$y = (a + x)/a^2(1 - \beta^2), \tag{3.6}$$

the horizontal line through the points $(-a\beta, k)$ and $(a\beta, k)$ is

$$y = k = 1/a(1 + \beta) \tag{3.7}$$

and other slant line through the points $(a, 0)$ and $(a\beta, k)$ is given as

$$y = (a - x)/a^2(1 - \beta^2). \tag{3.8}$$

Combining (3.6), (3.7) and (3.8), the probability distribution $f(x)$ having symmetrical trapezium as its curve is given by

$$\begin{aligned}
f(x) &= \frac{a + x}{a^2(1 - \beta^2)} \quad \text{for all values of } x \text{such that} - a \le x \le -a\beta \\
f(x) &= \frac{1}{a(1 + \beta)} \quad \text{for all values of } x \text{ such that} - a\beta \le x \le a\beta. \\
f(x) &= \frac{a - x}{a^2(1 - \beta^2)} \quad \text{for all values of x such that } a\beta \le x \le a \\
f(x) &= 0 \quad \text{for any other values of } x
\end{aligned} \tag{3.9}$$

3.2.3.1 Mean of the Trapezoidal Distribution

In this case, also mean of the function can be shown to be zero. Hence x^2 is the square of the deviation from the mean.

3.2.3.2 Variance of the Trapezoidal Distribution

The variance $\sigma^2$of this trapezoidal probability function is given by

$$\sigma^2 = \int_{-a}^{-a\beta} \frac{a+x}{a^2(1-\beta^2)} x^2 \mathrm{d}x + \int_{-a\beta}^{a\beta} \frac{a(1-\beta)}{a^2(1-\beta^2)} x^2 \mathrm{d}x + \int_{a\beta}^{a} \frac{a-x}{a^2(1-\beta^2)} x^2 \mathrm{d}x$$

$$\sigma^2 = \frac{1}{a^2(1-\beta^2)} \left[\left[ax^3/3 + x^4/4\right]_{-a}^{-a\beta} + \left[a(1-b)x^3/3\right]_{-a\beta}^{a\beta} + \left[ax^3/3 - x^4/4\right]_{a\beta}^{a} \right]$$

$$\sigma^2 = \frac{1}{a^2(1-\beta^2)} [\{(-a^4\beta^3/3 + a^4\beta^4/4) - (-a^4/3 + a^4/4)\}$$
$$+ a(1-\beta)\{a^3\beta^3/3 + a^3\beta^3/3\} + \{(a^4/3 - a^4/4) - (a^4\beta^3/3 - a^4\beta^4/4)\}].$$

$$\sigma^2 = \frac{1}{a^2(1-\beta^2)} \left[a^4(1-\beta^4)/6\right]$$

$$\sigma^2 = \frac{a^2(1+\beta^2)}{6} \tag{3.10}$$

Hence standard uncertainty, which is equal to standard deviation, is $a\sqrt{\frac{1+\beta^2}{6}}$.

In many cases for type B evaluation of uncertainty, either of three aforesaid distributions is used. Uncertainty stated by the superior laboratory is taken as range of the either distribution and square root of variance so calculated gives type B evaluation of standard uncertainty.

Hence, Type B uncertainty is

$a/\sqrt{3}$ for rectangular distribution

$a/\sqrt{6}$ for triangular distribution

and

$$a\sqrt{(1+\beta^2)}/\sqrt{6} \text{ for trapezoidal distribution.} \tag{3.11}$$

3.3 Small Sample Distributions

In the normal distribution μ is the mean of very large number of random variable and same is the case for variance. In actual practice, we know only the sample mean and its variance. Size of the sample, in a measurement laboratory is normally 3–5. Hence theory of small sampling will be applicable to most of the measurements carried out in a measurement laboratory.

3.3.1 The Student's t Distribution

One of the special distributions derived from the normal distribution is the Student's t distribution. Student is the pen name of Prof. W. S. Gosset. Let us consider the definite integral of normal probability distribution, in which deviations have been expressed in terms of its population standard deviation.

$$\int_{-k}^{k} \exp(-y^2/2)\mathrm{d}y. \tag{3.12}$$

The integral represents the area covered by the normal distribution between the limits $y = -k$ to $y = k$,

if

$$y = (x - \mu)/\sigma. \tag{3.13}$$

The integral represents the cumulative probability for x lying between the limits

$$\mu - k\sigma \leq x \leq \mu + k\sigma.$$

Here μ is the mean and σ is the standard deviation of the normal distribution. Normally σ the standard deviation is not known. It is the estimate of the standard deviation s_ν from the finite number of observations which is known.

If we replace σ by s_ν in (3.13), we get Student's t variable i.e.

Student's t variable t_ν is defined as

$$t_\nu = \frac{x - \mu}{s_\nu}. \tag{3.14}$$

The symbol ν is the degree of freedom of the t_ν variable.

Another way of defining the Student's t variable is as follows:

If X is a (0, 1) normal variate and Y is a χ^2 variable, then the Student's t_ν is also defined as

$$t_\nu = \frac{X}{\sqrt{Y/\nu}}. \tag{3.15}$$

The proof of their equality is as follows:

By definition

$$X = \frac{(x - \mu)}{\sigma},$$

$$Y = \chi^2 = \sum_{1}^{n} (x - \mu)^2/\sigma^2,$$

giving

$$\chi^2/\nu = \frac{\sum_{1}^{n} (x - \mu)^2/\sigma^2}{n - 1},$$

but

$$s_\nu^2 = \frac{\sum_1^n (x-\mu)^2}{n-1}.$$

Hence

$$\frac{\chi^2}{\nu} = \frac{s_\nu^2}{\sigma^2}. \qquad (3.16)$$

So (3.15) becomes

$$t_\nu = \frac{(x-\mu)/\sigma}{(s_\nu^2/\sigma^2)^{1/2}} = \frac{(x-\mu)}{s_\nu}.$$

Now, if we wish to find out the probability of

$$-k \le t_\nu \le k$$

or its alternate expression, the probability of

$$\mu - ks_\nu \le x \le \mu + kt_\nu,$$

then required cumulative probability is no longer given by (3.12) but by the integral of Student's density function $f(t_\nu)$

$$\int_{-k}^{k} f(t_\nu)dt_\nu = \frac{\Gamma\{(\nu+1)/2\}}{\Gamma(\nu/2)\sqrt{n\pi}} \int_{-k}^{k} \left(1+\frac{t_\nu^2}{\nu}\right)^{-(\nu+1)/2} dt \qquad (3.17)$$

or the probability density function of Student's t is

$$f(t_\nu) = \frac{(1+t^2/\nu)^{-(\nu+1)/2}\Gamma\{(\nu+1)/2\}}{\sqrt{\nu\pi}\{\Gamma(\nu/2)\}} - \infty \le t \le \infty, \qquad (3.18)$$

where

$$\Gamma(n/2) = (n/2-1)(n/2-2)\cdots 3.2.1 \quad \text{for all even values of } n$$

and

$$\Gamma n/2 = (n/2-1)(n/2-2)\cdots 3/2.1/2.\sqrt{\pi} \quad \text{for all odd values of } n.$$

Recurrence formula for gamma function $\Gamma n/2$

$$\Gamma n/2 = (n/2-1)\Gamma(n/2-1).$$

This distribution is useful for small samples. For sample of size n from the normal population having μ as mean and variance σ^2, the sample mean $\bar{x}$ and sample variance $s^2$are calculated in the usual manner and the Student's t function is used to find the probability for the given t or the total probability (covered area) for the range of Student's t. In other words, Student's t distribution is used for small sample in the same way as normal distribution is used for samples of larger size say $n \cong 500$.

3.3.1.1 Mean and Variance of Student's t Function

The mean of the distribution is zero. The standard deviation is $\nu/(\nu - 2)$. The ν is greater than 2. The distribution is symmetric about $t = 0$. Obviously it will tend to the standard normal distribution as ν approaches to a larger number. The Student's t distribution is shown in Fig. 3.4. The values of t statistic for different values of degrees of freedom and probabilities are given in Table A.6.

3.3.1.2 Comparison of Normal and t Distributions

Figure 3.5 gives a plot of $f(t)$ for $n = 5$ i.e. $\nu = 5-1 = 4$ with a plot of normal distributions of equal standard deviation. It is noted that the Student's t distribution has comparatively more probability concentrated in its tails. Like normal distribution it is symmetrical, continuous and bell shaped and with increasing n or ν it rapidly converges to a normal distribution.

3.3.1.3 Applications of t-Statistic

The t-distribution has a large number of applications, some of which useful for a calibration laboratory are enumerated below:

1. To test the sample mean, if it differs significantly from the population mean. In this case population mean is known, which in some cases may be the resultant mean from various laboratories or a moving average of the attribute of an artefact measured at periodical intervals. We assume, in this case, that the attribute of the artefact does not change with time and environmental conditions or known

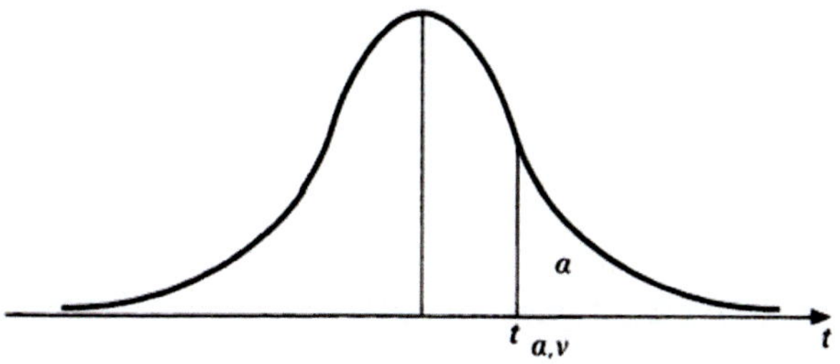

Fig. 3.4 t' distribution

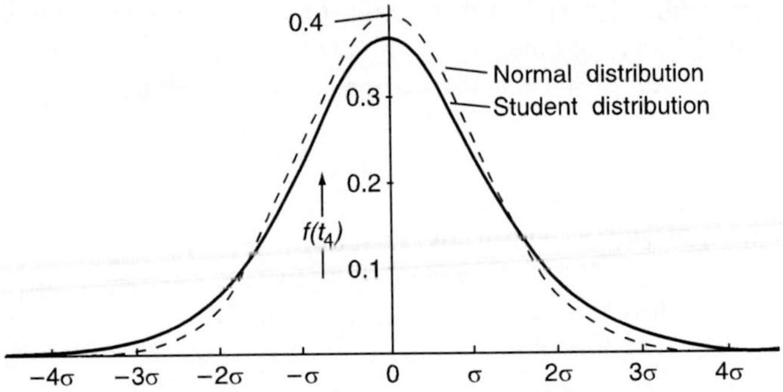

Fig. 3.5 Normal distribution and t distribution for $n = 5$

relations exist between the attribute and time or environmental conditions. For example, the value of the resistance of Ni-Chrome resistor changes with time and mass of a platinum iridium cylinder changes with time after cleaning in steam. Then due corrections are applied to the moving average.

2. To test the difference between the two sample means. This is useful to evaluate the quality of test results obtained by different observers of the same laboratory or results obtained from various laboratories in a round robin test.
3. To test the significance of an observed sample correlation coefficient and sample regression coefficient.
4. To test the significance of the observed partial correlations coefficients.

3.3.1.4 *t*-Test for a Sample Mean

The statistic used for this purpose is

$$t = \frac{\bar{x} - \mu}{s/\sqrt{n}}. \tag{3.19}$$

It may be noted that denominator of t is the standard deviation of the mean of the sample. The same is obtained by calculating

$$s^2 = \frac{1}{n-1}\sum_{p=1}^{p=n}(x_p - \bar{x})^2 \tag{3.20}$$

and dividing is by $\sqrt{n}$.

If calculated value of t is less than the tabulated value of t at a given probability (say 0.05) then the sample mean is not significantly different from the population mean at 5% level.

3.3.1.5 Numerical Example

Measured value of a 5 mm slip gauge at 20°C by a calibrating laboratory is 5.000 mm. The user laboratory measures its length as 5.042 mm at 20°C by repeating the measurement ten times and obtaining the standard deviation of 0.040 mm. We wish to know if the result may be taken as statistically genuine result.

$$\text{Here calculated value of } t \text{ statistic} = \frac{(5.042 - 5000)}{0.04/\sqrt{9}} = 3.15.$$

Conclusion: Tabulated t value for 9 degrees of freedom is 2.2622 for 0.95 cumulative probability. Here calculated value is larger than the tabulated value; hence we may say that sample mean may not be taken as statistically genuine result at 5% level of significance.

For clarity, say another measurement is carried out, which gave mean value of 5.021 mm from 10 repetitions and standard deviation of 0.040.

In this case

$$t = \frac{5.021 - 5.000}{0.04/\sqrt{9}} = 1.57.$$

The calculated value is less than the tabulated value of t at the cumulative probability of 0.95. Hence sample mean may be taken as statistically genuine result at 5% level of significance.

Alternatively had there been the standard deviation of 0.066 mm instead of 0.04 mm in the first example, other figures remain unchanged as in first case, t will then be given as

$$t = \frac{5.042 - 5.000}{0.066} = 1.909.$$

Here also t value is less than the tabulated value; hence the result may be taken as statistically genuine result. Hence it may be seen that the outcome of the conclusion depends both on the measured value and the standard deviation. Smaller value of standard deviation requires measured value closer to the population mean.

3.3.1.6 *t*-Test for Difference of Two Means

Let there be two samples of size n_1 and n_2 and means $\bar{x}_1$ and $\bar{x}_2$, respectively. We wish to test if these samples come from same population with standard deviation σ and common mean μ. The t statistic is

$$t = \frac{(\bar{x}_1 - \bar{x}_2)}{s\sqrt{\frac{1}{n_1} + \frac{1}{n_2}}}. \tag{3.21}$$

Here s is an unbiased estimate of the common population standard deviation and given by

$$s^2 = \frac{1}{n_1 + n_2 - 2}\left\{\sum_{p=1}^{p=n_1}(x_p - \bar{x}_1)^2 + \sum_{q=1}^{q=n_2}(x_q - \bar{x}_2)^2\right\}. \quad (3.22)$$

3.3.1.7 Numerical Example

Let a weight of 10 g was calibrated against a standard of known mass at two occasions with the following results.

Sample 1	Sample 2
10.000025	10.000044
10.000032	10.000034
10.000030	10.000022
10.000034	10.000010
10.000024	10.000047
10.000014	10.000031
10.000032	10.000040
10.000024	10.000030
10.000030	10.000032
10.000031	10.000035
10.000035	10.000018
10.000025	10.000021
	10.000035
	10.000029
	10.000022
$n_1 = 12$	$n_2 = 15$

By calculation, we get

$$\bar{x}_1 = 10.000028\,\text{g} \quad \bar{x}_2 = 10.000030\,\text{g}$$

$$\sum_{p=1}^{p=12}(x_p - \bar{x}_1)^2 = 380 \times 10^{-12}\text{g}^2 \quad \sum_{q=1}^{q=15}(x_q - \bar{x}_2)^2 = 1,410 \times 10^{-12}\text{g}^2$$

$$\begin{aligned} s^2 &= \frac{1}{n_1 + n_2 - 2}\left\{\left\{\sum_{p=1}^{p=n_1}(x_p - \bar{x}_1)^2 + \sum_{q=1}^{q=n_2}(x_q - \bar{x}_2)^2\right\}\right\} \\ &= \frac{(380 + 1410) \times 10^{-12}}{(12 + 15 - 2)} \\ &= 71.6 \times 10^{-12}\,\text{g}^2, \end{aligned}$$

giving

$$t = \frac{(30 - 28) \times 10^{-6}}{\sqrt{71.6(1/12 + 1/15)10^{-12}}} = 0.610.$$

Looking at the t tables for 25 degrees of freedom at probability of 0.05, we get the value of t as 2.06.

Since the calculated value is much smaller than the tabulated value at 0.05, we may say that two means belong to the same population.

3.3.1.8 Assumption Made for Student's t Test

1. The parent population from which the samples are drawn is normal.
2. The sample observations are independent.
3. The population standard deviation is not known.

3.3.1.9 Paired t-Test for Difference of Means

Let us consider the case when two samples are not independent. Size of the samples is same and observations are paired. For example, a weight is calibrated with two positions of the riders so that one observation is supposed to be higher by the mass value equivalent of the rider positions. Naturally, it is expected that difference of the two observations is equivalent to the mass of the two rider positions. In this case, the observations of two samples are paired as (x_i, y_i) for all p taking values from 1 to n, n being the sample size. The statistic t is

$$t = \frac{\bar{d}}{s/\sqrt{n}}, \tag{3.23}$$

where

$$\bar{d} = \frac{\sum_{p=1}^{p=n} (x_p - y_p)}{n} \tag{3.24}$$

and

$$s^2 = \frac{\sum (d_p - \bar{d})^2}{n-1}. \tag{3.25}$$

3.3.1.10 Numerical Example

During the calibration of a 10 g weight, two observations are taken by shifting the rider on rider bar by one notch. After taking into account the change in positions of the rider, the differences of the paired observations are given as

$$5, 2, 8, -1, 3, 0, -2, 1, 2, 5, 0, 4 \text{ and } 6\,\mu\text{g}.$$

Ideally each difference should be zero; we wish to test if the differences are real or due to statistical fluctuations.

Calculations

$$\text{Sum of all differences} = 31,\ \bar{d} = 31/12 = 2.58$$

$$\text{Sum of the squares of differences} = 185,$$

giving

$$s^2 = \frac{1}{n-1}\left\{\sum_{p=1}^{p=12} d_p^2 - \frac{\left(\sum_{p=1}^{p=12} d\right)^2}{n}\right\} = \frac{1}{11}\{185 - (31)^2/12\} = 9.5382$$

$$t = 2.58/(\sqrt{9.5328}/12) = 2.89.$$

Tabulated value of t for 11 degrees of freedom at 0.05 is 1.80 (Table A.6). The calculated value is greater than the tabulated value; hence it is assumed that the differences are not due to statistical fluctuations. The reason, in this particular case, may be "not positioning the rider precisely at the same position".

3.3.2 The χ^2 Distribution

If X is $N(0, 1)$ distribution, i.e. a Gaussian distribution with zero mean and deviations being expressed in terms of σ, then the random variable

$$Y = X^2 \tag{3.26}$$

is said to be χ^2 random variable with one degree of freedom.

In general, if $X_1, X_2, \ldots, X_n$ are n independent variables of $N(0, 1)$ distribution, then

$$Y = \sum_{r=1}^{r=n} X_r^2 \tag{3.27}$$

is said to be a χ^2 variable with n degrees of freedom.

In fact, $X_1, X_2, \ldots, X_n$ are the deviations from the mean which is zero in this case. So Y is sum of squares of deviations from the mean.

Similarly, if $X_1, X_2, X_3, \ldots, X_n$ are the variables of $N(\mu, \sigma^2)$ then

$$Y = \sum_{i=1}^{n} (X_i - \mu)^2/\sigma^2. \tag{3.28}$$

Y is also a χ^2 random variable of n degrees of freedom.

If $x_1, x_2, x_3, \ldots, x_n$ is a random sample of size n from $N(\mu, \sigma^2)$. Here as stated earlier μ is unknown, then the sampling distribution of

$$\sum_{r=1}^{r=n} \{x_r - \bar{x})^2/\sigma^2 = (n-1)s^2/\sigma^2 \text{ is also } \chi^2. \tag{3.29}$$

The χ^2 has $n-1$ degrees of freedom. One degree is reduced as μ has been replaced by $\bar{x}$. The probability distribution of $y = \chi^2$ is given by

$$f(y) = \frac{y^{(\nu-2)/2}e^{-y/2}}{2^{\nu/2}\Gamma\nu/2} \quad \text{for } y \geq 0$$

$$= 0 \quad \text{for all other values of } y. \tag{3.30}$$

Mean value of y is ν

and (3.31)

Standard deviation of y is 2ν.

The curve of χ^2 distribution is continuous and asymmetrical. The mode or maximum frequency of the function is at $y = \nu - 2$. The distributions for different degrees of freedom are shown in Fig. 3.6. It is always skewed to right, but tends to become a normal distribution if ν approaches infinity. Like the Student's distribution it is dependent of degrees of freedom but independent of μ and σ.

A typical χ^2 distribution is shown in Fig. 3.7.

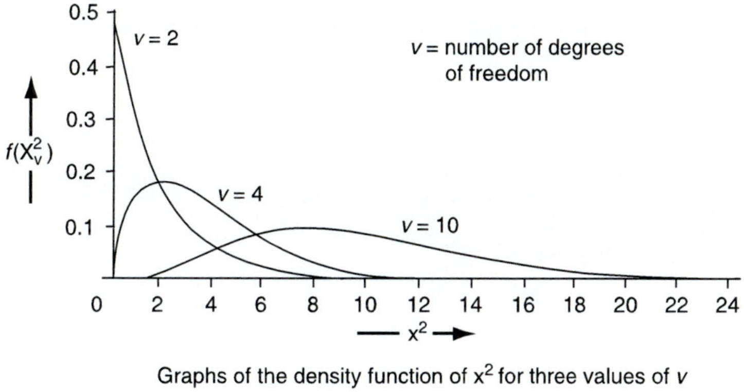

Fig. 3.6 χ^2 distribution of various degrees of freedom

Fig. 3.7 A typical χ^2 curve

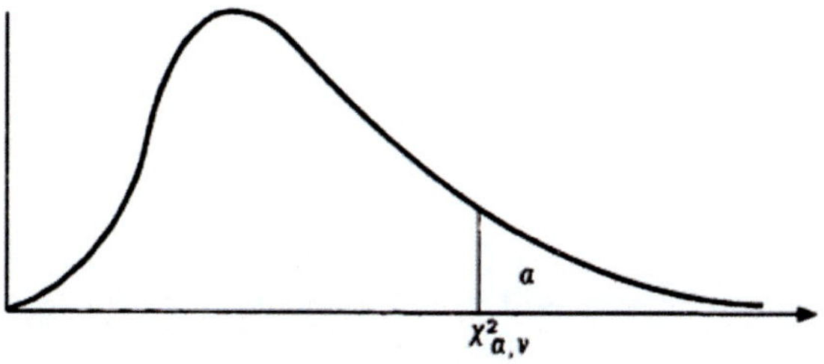

3.3.2.1 Use of χ^2 Distribution to Find a Range of Standard Deviation for Given Probability

By definition (3.28)

$$\sum_{p=1}^{p=n} (x_p - \bar{x})^2 = \chi_\nu^2 \sigma^2 \tag{3.32}$$

also

$$\sum_{p=1}^{p=n} (x_p - \bar{x})^2 = s_\nu^2 (n-1) = s_\nu^2 \times \nu. \tag{3.33}$$

Thus

$$\chi_\nu^2 = \nu \times s_\nu^2 / \sigma^2. \tag{3.34}$$

Here $\nu = n - 1$ for a single variable and ν is number of degrees of freedom giving us

$$\sigma = s_\nu \sqrt{\frac{\nu}{\chi_\nu^2}}, \tag{3.35}$$

$$\sigma = s_\nu k,$$

$$\text{where } k = \sqrt{\frac{\nu}{\chi_\nu^2}}. \tag{3.36}$$

If the value of χ_ν^2 is chosen for a given probability and known degrees of freedom ν, we get the corresponding value of σ.

We can determine the probability of σ exceeding the value found from

$$\int_{\chi 1}^{\infty} f(\chi_\nu^2) \mathrm{d}\chi_\nu^2$$

and less than the value found from

$$\int_0^{\chi 2} f(\chi_\nu^2) \mathrm{d}\chi_\nu^2.$$

The values of χ_ν^2 for six probability levels for the integral $\int_{\chi 1}^{\infty} f(\chi_\nu^2) \mathrm{d}\chi_\nu^2$ are given in Table A.9.

Consider the expression

$$\left\{ 1 - \int_{\chi 1}^{\infty} f(\chi_\nu^2) d\chi_\nu^2 - \int_0^{\chi 2} f(\chi_\nu^2) d\chi_\nu^2 \right\}. \tag{3.37}$$

Here choosing χ_1 and χ_2 in such a way that the two integrals are equal to each other for all values of ν, and also making the sum of the two integrals equal to 0.10, 0.05, 0.02 and 0.01, respectively. Then the probability of χ^2 lying between χ_1 and χ_2

becomes respectively 0.90, 0.95, 0.98 and 0.99. The minimum and maximum values of χ^2 from a χ^2 table for proper degrees of freedom are found out. From (3.36) for a given value of ν – the degrees of freedom, we can find out maximum and minimum values of k corresponding to two values of χ. Hence for given degrees of freedom and probability, we can find out the range of σ for given probability. Combining the two steps, a separate Table A.11 has been constructed giving the values of k_{min} and k_{max} for probabilities of 0.90, 0.95, 0.98 and 0.99. Multiplication of k_{min} and k_{max} with the sample standard deviation gives the range of σ for given probability and degrees of freedom.

$$k_{\min} s_{\nu} \leq \sigma \leq k_{\max} s_{\nu}. \tag{3.38}$$

The values of $k_{\min}$ and $k_{\max}$ for different probabilities and sample size are given in Table A.11.

3.3.3 The F-Distribution

The distribution, which was discovered by R. A. Fisher, is related to χ^2 distribution.

If X_1 and X_2 are independent χ^2 random variables with ν_1 and ν_2 degrees of freedom respectively then the random variable

$$F = \frac{X_1/\nu_1}{X_2/\nu_2} = \frac{\chi_1^2/\nu_1}{\chi_2^2/\nu_2} \tag{3.39}$$

is said to have an F-distribution with ν_1 and ν_2 degrees of freedom.

Let s_1^2 and s_2^2 be the estimates of the variance σ^2 of a normal distribution. Thus

$$s_1^2 = \sigma^2 \chi_1^2/\nu_1$$

$$\text{and} \tag{3.40}$$

$$s_2^2 = \sigma^2 \chi_2^2/\nu_2.$$

From (3.40), we get

$$F = \frac{s_1^2}{s_2^2} = \frac{\chi_1^2/\nu_1}{\chi_2^2/\nu_2}. \tag{3.41}$$

Two random samples of size n_1 and n_2 are taken from a normal distribution with mean μ and variance σ^2. The two sample variances s_1^2 and s_2^2 are calculated in the usual way. Then the sampling distribution of $(n_1 - 1)s_1^2/\sigma^2$ is a χ^2 with $(n_1 - 1)$ degrees of freedom and the sampling distribution of $(n_2 - 1)s_2^2/\sigma^2$ is again a χ^2 variable with $(n_2 - 1)$ degrees of freedom. Thus, the statistic F for the two sample is given by

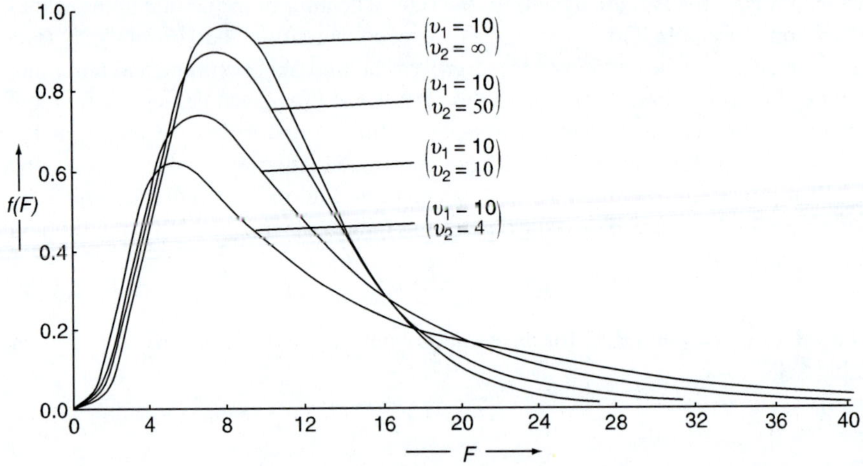

Fig. 3.8 *F* distribution curves with different sets of degrees of freedom

$$F = \frac{(n_1 - 1)s_1^2/\sigma^2}{(n_1 - 1)} \Bigg/ \frac{(n_2 - 1)s_2^2/\sigma^2}{(n_2 - 1)},$$

giving

$$F = s_1^2/s_2^2. \tag{3.42}$$

F has Fisher (Snedecor's) distribution with $(n_1{-}1)$ and $(n_2{-}1)$ degrees of freedom.

So instead of comparing the estimates of variances, we can compare the two χ^2 variables with ν_1 and ν_2 degrees of freedom, respectively.

The variable F is the ratio of positive quantities; hence its range is from 0 to ∞. Fisher showed that F had a probability distribution given as

$$f(F) = \frac{\Gamma\{(\nu_1 + \nu_2)/2\}}{\Gamma(\nu_1/2)\Gamma(\nu_2/2)} \nu_1^{\nu_1/2} \nu_2^{\nu_2/2} \frac{F^{\nu_1/2-1}}{(\nu_2 + \nu_1 F)^{(\nu_1+\nu_2)/2}}. \tag{3.43}$$

Like χ^2 distribution, the F distribution is a continuous asymmetrical distribution with a range from 0 to ∞. A few F distribution curves with different degrees of freedom are shown in Fig. 3.8. Here we see that the curve becomes more symmetrical as the degrees of freedom increase.

3.3.3.1 Parameters of F Distribution

A typical F distribution curve is shown in Fig. 3.9. The curve is continuous for F from 0 to ∞ and is always skewed to right.

Mean of the distribution is

$$\nu_2/(\nu_2 - 2) \quad \text{for } \nu_2 > 2. \tag{3.44}$$

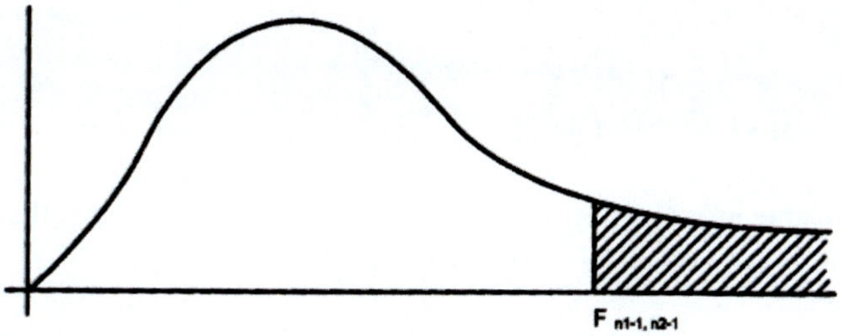

Fig. 3.9 A typical F curve

Here we see that mean approaches 1 for larger values of ν_2.

Variance σ^2 *of the distribution*

$$2\left(\frac{\nu_2}{\nu_2-2}\right)^2\left\{\frac{\nu_1+\nu_2-2}{\nu_1(\nu_2-4)}\right\}. \tag{3.45}$$

3.3.4 Upper and Lower Percentage Points

We know by definition

$$F_{(\nu1,\nu2)} = \frac{s_1^2}{s_2^2}. \tag{3.46}$$

Here ν_1 is the degree of freedom of s_1^2 – the sample variance in the numerator and ν_2 is the degree of freedom of s_2^2 – the sample variance in the denominator.

From (3.46), we get

$$\frac{1}{F_{(\nu1,\nu2)}} = \frac{s_2^2}{s_1^2} = F_{(\nu2,\nu1)},$$

$$P[(F_{\nu1,\nu2)} \geq c] \Rightarrow P\left[1/c \geq \frac{1}{F_{(\nu1,\nu2)}}\right]. \tag{3.47}$$

Let

$$P[F_{(\nu1,\nu2)} \geq c] = \alpha \tag{3.48}$$

then

$$1-\alpha = 1 - P[F_{(\nu1,\nu2)} \geq c], \tag{3.49}$$

$$= 1 - P\left[\frac{1}{F_{(\nu1,\nu2)}} \leq 1/c\right], \tag{3.50}$$

$$\alpha = P\left[\frac{1}{F_{(\nu1,\nu2)}} \leq 1/c\right] = 1 - P\left[F_{(\nu2,\nu1)} \geq 1/c\right].$$

Thus, giving

$$1 - \alpha = P[F_{(\nu 2,\nu 1)} \geq 1/c]. \tag{3.51}$$

Thus $F_{(\nu 2,\nu 1)}$ – the significant point at the probability $(1 - \alpha)$ – is the reciprocal of $F_{(\nu 1,\nu 2)}$ at the probability α.

Hence F value at probability α with ν_1, ν_2 degrees of freedom is the reciprocal of F value at $1 - \alpha$ with ν_2, ν_1 degrees of freedom.

This means

if $F(8,4)$ at probability of 0.05 is 6.04 then $1/6.04 = 0.166$ will be the value of $F(4,8)$ at probability of 0.95.

The following relation relates the upper percentage point of the distribution to the lower percentage point:

$$F_{1-\alpha,\nu_1,\nu_2} = \frac{1}{F_{\alpha,\nu_2,\nu_1}}. \tag{3.52}$$

The values of F for different degrees of freedom at 5% points are given in Table A.7 and the values of F for different degrees of freedom at 1% points are given in Table A.8.

3.3.4.1 Notation

The notation F_{1-a,ν_1,ν_2} means that $1-\alpha$ is the probability of the $F = s_1^2/s_2^2$ degrees of freedom of variances in numerator and denominator being ν_1 and ν_2, respectively.

3.3.5 *Application of F-Test*

3.3.5.1 Testing for Equality of Population Variances

F-distribution is used for testing the homogeneity of two samples. Let there be a sample of size n_1 and variance s_1^2. Similarly let there be another sample of size n_2 and variance s_2^2. The ratio of the two variances is the F statistic with n_1–1 and n_2–1 degrees of freedom; if the samples belong to the same population then s_1^2/s_2^2 must be close to unity. The larger sample variance is taken as the numerator for F statistic so that $F > 1$.

$$F_{n1-1,n2-1} = \frac{s_1^2}{s_2^2}. \tag{3.53}$$

If calculated value of F is less than the tabulated values for given degrees of freedom for a given value of probability then two variances are said to belong to the same population. Normally probability value taken is 0.05; however in some cases, the probability of 0.01 is also chosen.

3.3.5.2 Numerical Example

During evaluation of two 1 kg balances, the sum of squares of deviations from the sample mean of size 8 was found 84.4 $(\mu g)^2$ for one balance, while it was found 102.6 $(\mu g)^2$ for the sample size 11 for the second balance. We wish to know if one balance has a better repeatability than the other (i.e. standard deviations are significantly different from each other or not).

$$\sum_{p=1}^{p=8} (x_p - \bar{x})^2 = 84.4 \text{ giving sample variance } s_p^2 = 84.4/(8-1) = 12.057.$$

$$\sum_{q=1}^{q=10} (x_q - \bar{x})^2 = 102.6 \text{ giving sample variance } s_q^2 = 102.6/9 = 11.4.$$

$$F \text{ statistic} = \frac{s_p^2}{s_q^2} = 12.057/11.4 = 1.057. \tag{3.54}$$

The calculated value is smaller than 3.29 (Table A.8) the tabulated value for 7 and 9 degrees of freedom at probability of 0.05. This suggests that there is no significant difference in the two variances; hence repeatability of two balances is same within statistical fluctuations.

3.3.6 For Equality of Several Means

This test is carried out by the technique of analysis of variance. This plays a very important role in design of experiments. The method has been discussed in Chap. 4.

3.4 Combining of Variances of Variables Following Different Probability Distribution Functions

In a measurement process there are variety of sources contributing to uncertainty, each of which is calculated in terms of variance. Assigning one separate variable to each source gives rise to as many random variables as there are sources. We may not know as which variable follows which probability distribution. Now we wish to find out the combined variance of variables following different probability distribution functions.

Earlier we have proved that the variance of a variable, which is a linear function of variables, each variable following the normal probability distribution, is the weighted sum of the variances. The weight factor of each variance is the square of the coefficient of its variables in the linear function. Mathematically:

If

$$Z = a_1x_1 + a_2x_2 + a_3x_3 + \cdots + a_nx_n = \sum a_r x_r. \quad (3.55)$$

All $x_1, x_2, x_3, \ldots, x_n$ variables are independent and follow normal distributions.

Then

$$\sigma_z^2 = \sum_{r=1}^{r=n} a_r^2 \sigma_r^2. \quad (3.56)$$

It can however be proved that the above relation (3.56) is true even if all the variables do not belong to the normal distribution. Mathematical proof consists of (1) obtaining the combined probability function (pdf) of any two pdfs defined in a finite range, (2) deriving a result for variance for the two linearly related variables with given pdf and (3) finally generalizing the expression of variance of a variable, which is a linear function of n independent variables. All this is too tedious and unnecessary for the user metrologists and hence been omitted.

References

1. S.C. Gupta, V.K. Kapoor, *Fundamentals of Mathematical Statistics* (Sultan Chand, New Delhi, 1994), p. 13.27, p. 14.66
2. C.F. Dietrich, *Uncertainty, Calibration and Probability* (Adam Hilger, New York, 1991), pp. 39–41, χ^2 212, 213, F 215–216S

Chapter 4
Evaluation of Measurement Data

4.1 Introduction

In a metrology laboratory, measurements are carried out assuming that measuring instruments are unbiased; hence all measured values derived from the observations of the measuring instrument belong to a normal distribution. In order to calculate the value of a measurand (output variable), we take observations from the measuring instruments, and apply known corrections from the calibration certificate and any corrections due to environmental parameters such as temperature, pressure and humidity of surrounding air or corrections due to the time elapsed since the calibration of measuring instrument or standard used. To assess the quality of the measured values Dixon test is applied (Sect. 4.2). If there is more than one sample, individual mean of each sample is evaluated for quality (Sect. 4.3). Quality of variances obtained for different samples is assessed (Sect. 4.4 onward). After doing all this drill, one should find out all sources of uncertainty, calculate the contribution from each source in form of variance along with degrees of freedom and combine the variances from all sources. State the results together with uncertainty of measurement.

4.2 Evaluation of Validity of Extreme Values of Measurement Results

So the first thing we do is to take observations of a measuring instrument; these observations make the base for further calculations. We apply certificate and environmental corrections to get the corrected observations. Before making any further calculations we need to ascertain the validity of these corrected observations. Most of the corrected observations form a cluster, but some observations are a little away from this cluster. We may reject these observations, but this violates the principle of physical measurement of not leaving any observation. We may reject a

S.V. Gupta, *Measurement Uncertainties*, DOI 10.1007/978-3-642-20989-5_4,

certain observation if sufficient reason of its rejection is found out. One method of finding out such outlier observations is the Dixon test.

4.2.1 *Outline (Dixon)Test*

Basic assumption of this test is that all good observations belong to a normal distribution. All observations are arranged in ascending order for testing the extreme large value of the observation. For calculating the validity of extreme low value of observation, these are arranged in descending order. The test parameter depends upon the total number of observations (n) and is given in column 2 of Table 4.1 against the number of observations in column 1 of the said table.

It has been assumed that number of repetitions in a set is not more than 25.

If the parameter corresponding to known value of n is more than the value given in column 3 of Table 4.1, then X_n may be taken as an outlier and hence may be rejected. The test may be repeated for next extreme values after rejection of the outlier, but then value of n is to be reduced by 1.

Table 4.1 Critical values for Dixon outlier test

n(observation)	Test parameter	Critical value
3		0.941
4		0.765
5	$(X_n - X_{n-1})/(X_n - X_1)$	0.620
6		0.560
7		0.507
8		0.554
9	$(X_n - X_{n-1})/(X_n - X_2)$	0.512
10		0.477
11		0.576
12	$(X_n - X_{n-2})/(X_n - X_2)$	0.546
13		0.521
14		0.546
15		0.525
16		0.507
17		0.490
18		0.475
19	$(X_n - X_{n-2})/(X_n - X_3)$	0.462
20		0.450
21		0.440
22		0.430
23		0.421
24		0.413
25		0.406

4.3 Evaluation of the Means Obtained from Two Sets of Measurement Results

After verifying the validity of individual measurements we make use of these measurements in calculating the average or arithmetic mean. Having discussed the evaluation and validity of individual measurement, we switch over to mean of the measurements. Let a set of measurements be with sample mean $\bar{x}$ and we wish to know that if it belongs to a population with mean μ and standard deviation (σ). Then test statistic z will belong to a normal distribution N (0, 1). Two cases may arise.

(a) When population standard deviation σ is known then

$$z = \frac{\bar{x} - \mu}{\sigma/\sqrt{n}}. \tag{4.1a}$$

(b) When population standard deviation σ is not known, standard deviation s of the sample is taken as population standard deviation, and test statistic is

$$z = \frac{\bar{x} - \mu}{s/\sqrt{n}}. \tag{4.1b}$$

Here n is the number of measurements carried out to calculate sample mean. The statistic z in (4.1b) will belong to normal distribution for larger sample size; for smaller size sample z will belong to Student's t distribution.

Example 4.1. Let there be population having mean of 1,350 and standard deviation of 150. If there is sample with mean 1,300 from 25 measurements. We wish to know if this sample mean belongs to the population.

$$\begin{aligned} z &= (1{,}350 - 1{,}300)/150/\sqrt{25} \\ &= 50 \times 5/150 = 1.666. \end{aligned}$$

This value of z is less than 2, corresponding to the probability of 0.05. Hence we may conclude that the calculated mean belongs to the same population and should be accepted for further calculations.

In fact we can determine the range of $\bar{x}$ within which it can be assumed to belong to the population mean 1,350 and standard deviation 150. Taking $z = \pm 2$,
we get

$$\begin{aligned} \bar{x} &= 1{,}350 \pm 2 \times 150/\sqrt{25} \\ &= 1{,}350 \pm 60. \end{aligned}$$

In general

$$\bar{x} \le \mu + 2\sigma/\sqrt{n},$$
$$\text{and} \quad \bar{x} \ge \mu - 2\sigma/\sqrt{n}$$
$$\text{or} \quad \mu - 2\sigma/\sqrt{n} \le \bar{x} \le \mu + 2\sigma/\sqrt{n}.$$

Example 4.2 In another set of measurements of resistance of 1Ω, let the mean of very large values be 1.0001350 Ω. If 1.0001250 Ω is the mean value of 16 measurements with standard deviation of 0.0000160 Ω, we wish to know if the sample mean may be taken as belonging to the same population.

The z in this case is given as

$$z = (1.0001350 - 1.0001250)/0.0000160/\sqrt{16} = 2.5.$$

The z value is more than 2; hence mean of measurements may not be taken to belong to the same population at 5% probability. The value z at probability of 0.01 is 2.56; hence the measured mean may be taken as belonging to the same population at 1% probability.

4.3.1 Two Means Coming from the Same Source

Quite often we take a set of measurements on a particular day and calculate the mean, say $\bar{x}_1$, and repeat the same measurements under same environments with same or similar instruments and obtain the arithmetic mean, say $\bar{x}_2$. We wish to know if the two means are significantly different or consistent within statistical fluctuations. In this case, it may be assumed that the two sets of measurements belong to the same population. Before proceeding further let us calculate the population standard deviation from the standard deviations of the two samples.

4.3.1.1 Standard Deviation of the Two Means

Let the two sets of measurements be distinguished by the suffixes 1 and 2.

Let n_1 measurement results be taken on the first day and n_2 be the number of measurement results on the repeat performance, then means and variances of two sets are given as

$$\bar{x}_1 = \frac{\sum_{p=1}^{p=n_1} x_p}{n_1}, \tag{4.2}$$

$$\bar{x}_2 = \frac{\sum_{p=1}^{p=n_2} x_p}{n_2}, \tag{4.3}$$

$$S_1^2 = \sum_{p=1}^{p=n_1} (x_p - \bar{x}_1)^2 = s_1^2 \nu_1, \tag{4.4}$$

$$S_2^2 = \sum_{p=1}^{p=n_2} (x_p - \bar{x}_2)^2 = s_2^2 \nu_2. \tag{4.5}$$

Here n_1, n_2 are number of measurement results, ν_1, ν_2 are degrees of freedom and s_1, s_2 are standard deviations respectively of the two sets.

As stated above, both sets of measurement results have been made; on the same/similar instruments under same environmental conditions; hence it should have a common standard deviation. The best estimate of this standard deviation using the data of both sets is given by

$$s_\nu = \sqrt{\frac{S_1^2 + S_2^2}{n_1 + n_2 - 2}}, \tag{4.6}$$

$$s_\nu = \sqrt{\frac{s_1^2 \nu_1 + s_2^2 \nu_2}{n_1 + n_2 - 2}} \tag{4.7}$$

that is s_ν is the square root of the weighted mean of squares of s_1 and s_2. The estimates of the standard deviations of the two means respectively will be

$$\frac{s_\nu}{\sqrt{n_1}} \text{ and } \frac{s_\nu}{\sqrt{n_2}}. \tag{4.8}$$

Thus, the estimate of the standard deviations of the difference of the two means will be

$$s_{\nu d} = s_\nu \sqrt{\frac{1}{n_1} + \frac{1}{n_2}}. \tag{4.9}$$

4.3.1.2 Test for Two Means of Samples of Smaller Size

The standard deviation of the difference of two means namely $\bar{x}_1 - \bar{x}_2$ is given in (4.9). Dividing the difference of two means by their combined standard deviation we get Student's t statistic for the differences of two means belonging to same population, i.e. having common standard deviation

$$t_{\nu d} = \frac{\bar{x}_1 - \bar{x}_2}{s_{\nu d}} = \frac{\bar{x}_1 - \bar{x}_2}{s_\nu} \sqrt{\frac{n_1 n_2}{n_1 + n_2}}. \tag{4.10}$$

$t_{\nu d}$ is the Student's t statistic and the degree of freedom is

$$n_1 + n_2 - 2. \tag{4.11}$$

From (4.10) t_{vd} is calculated and corresponding probability from the table for t distribution with $(n_1 + n_2 - 2)$ degrees of freedom can be found out. If t_{vd} exceeds the value of t for predetermined value of the probability then it is assumed that the difference between the two means is more than the tolerated. Second method is that we specify the percentage probability of tolerance and find the corresponding value of t_{vd} for the degrees of freedom determined from (4.11), and if the calculated value of t_{vd} from (4.10) is more than the value of t obtained from the table, then the difference between the two means is taken too large to be considered due to normal statistical fluctuations. However if t_{vd} is found to be less than the tabulated value, then the difference between the two means may be taken due to statistical fluctuations only.

Normally tables for Student's t are given for the area covered β_{tp}; hence P_t for a given value of t_{vd} is

$$P_t = \frac{(1 - \beta_{tp})}{2}. \tag{4.12}$$

The chosen value of P_t is normally 0.05. If the value t_{vd} from (4.10) is less than the value of t corresponding to probability of 0.05, then the difference of the two means is taken due to normal statistical fluctuations and the two means are said to be consistent. If it lies between the values of t corresponding to P_t values of 0.05 and 0.01, then the consistency of the two means should be taken with caution. If the calculated value t_{vd} exceeds the value of t corresponding to P_t of 0.01, then two means are said to be inconsistent; i.e. difference between the means is large enough to say that one of the means is not correct. But it is not possible from this statistics to say which mean is not correct. The only alternative is to take another set of measurements and compare its mean with the other two means, and retain the two sets whose means are consistent.

4.3.1.3 SD and Mean Value of Two Means

Once we get the set of consistent means the question comes of the combined mean of the two sets of measurement results and its standard deviation.

Let us distinguish the two sets of measurement results by 1 and 2, giving us

$$\bar{x}_1 = \frac{\sum_{p=1}^{p=n_1} x_{1p}}{n_1}, \tag{4.13}$$

$$\bar{x}_2 = \frac{\sum_{p=1}^{p=n_2} x_{2p}}{n_2}. \tag{4.14}$$

Now if $\bar{x}$ is the combined mean then it is given by

$$\bar{x} = \frac{\sum_{p=1}^{p=n_1} x_{1p} + \sum_{p=1}^{p=n_2} x_{2p}}{n_1 + n_2},$$

$$\bar{x} = \frac{n_1\bar{x}_1 + n_2\bar{x}_2}{n_1 + n_2}. \tag{4.15}$$

Combined variance s_c^2 is given by

$$s_c^2 = \frac{\sum_{p=1}^{p=n_1} (x_{1p} - \bar{x})^2 + \sum_{p=1}^{p=n_2} (x_{2p} - \bar{x})^2}{n_1 + n_2 - 2}. \tag{4.16}$$

There were n_1 independent variables in the first set of measurement results, but by fixing the mean as $\bar{x}_1$, the actual number of independent variable is n_1–1; similar is the situation for second set of measurement results; hence the total degree of freedom remained is $n_1 + n_2 - 2$. Hence sum of the squares of deviation from the combined mean of each variable is divided by $n_1 + n_2$–2.

For simplicity of calculations, consider only the numerator of (4.16), which can be written as

$$\sum_{p=1}^{p=n_1} x_{1p}^2 + n_1\bar{x}^2 - 2\bar{x}\sum_{p=1}^{p=n_1} x_{1p} + \sum_{p=1}^{p=n_2} x_{2p}^2 + n_2\bar{x}^2 - 2\bar{x}\sum_{p=1}^{p=n_2} x_{2p}$$

$$= \sum_{p=1}^{p=n_1} x_p^2 + \sum_{p=1}^{p=n_2} x_p^{\prime 2} + (n_1 + n_2)\bar{x}^2 - 2\bar{x}\left\{\sum_{p=1}^{p=n_1} x_p + \sum_{p=1}^{p=n_2} x_p\right\}$$

$$= \sum_{p=1}^{p=n_1} x_{1p}^2 + \sum_{p=1}^{p=n_2} x_{2p}^2 + (n_1 + n_2)\bar{x}^2 - 2\bar{x}\{(n_1 + n_2)\bar{x}\}$$

$$= \sum_{p=1}^{p=n_1} x_{1p}^2 + \sum_{p=1}^{p=n_2} x_{2p}^2 - (n_1 + n_2)\bar{x}^2. \tag{4.17}$$

But s_1^2 and s_2^2 are given as

$$\nu_1 s_1^2 = \sum_{p=1}^{p=n_1} x_{1p}^2 - n_1\bar{x}_1^2$$

and

$$\nu_2 s_2^2 = \sum_{p=1}^{p=n_2} x_{2p}^2 - n_2 \bar{x}_2^2.$$

Substituting the values of summations in (4.17),

we get

$$\nu_1 s_1^2 + \nu_2 s_2^2 + n_1 \bar{x}_1^2 + n_2 \bar{x}_2^2 - (n_1 + n_2)\bar{x}^2 = \nu_1 s_1^2 + \nu_2 s_2^2 + n_1 \bar{x}_1^2 + n_2 \bar{x}_2^2$$
$$-(n_1 \bar{x}_1 + n_2 \bar{x}_2)^2/(n_1 + n_2)$$
$$= \nu_1 s_1^2 + \nu_2 s_2^2 - n_1 n_2 (\bar{x}_1 - \bar{x}_2)^2/(n_1 + n_2).$$

Substituting this value for the numerator of (4.16), we get

$$s_c^2 = \frac{\nu_1 s_1^2 + \nu_2 s_2^2 - n_1 n_2 (\bar{x}_1 - \bar{x}_2)^2/(n_1 + n_2)}{n_1 + n_2 - 2}. \tag{4.18}$$

4.3.1.4 Numerical Example

Let two means for the mass of a kilogram weight obtained by using two different sets of similar equipment be 1000.000103 and 1000.00011 g, and two estimated standard deviations be 0.00003 and 0.00006 g. The first mean is of 15 measurements and the second is of 10 measurements.

Now s_ν from (4.7) is given by

$$s_\nu = \left\{ \frac{0.03^2 \times 14 + 0.06^2 \times 9}{15 + 10 - 2} \right\}^{1/2}$$
$$= \{(0.0126 + 0.0324)/23\}^{1/2} = 0.04423,$$

giving $s_{\nu d}$ from (4.9)

$$s_{\nu d} = 0.04423\,(1/10 + 1/15)^{1/2} = 0.018056.$$

Therefore

$$t_{\nu d} = \frac{\bar{x}_1 - \bar{x}_2}{s_{\nu d}} = 0.007/0.018056 = 0.388,$$
$$\nu_d = 10 + 15 - 2 = 23.$$

From the Student's t table, for 23 degrees of freedom (take $\nu = 20$ or 25), t for 0.05 probability is 1.7247 for $\nu = 20$ and 1.7081 for $\nu = 25$ (Table A.6). Here we see that $t_{\nu d}$ value is much smaller than either of the two values; hence the two means are consistent. Hence we can take the weighted mean of the two means to represent the estimated value of the mass of the kilogram weight under test.

$$\text{Mass of the kilogram} = (15 \times 1000.000103 + 10 \times 1000.000110)/(15 + 10)$$
$$= 1000.000106\,\text{g}.$$

Here we may notice that standard deviations of the two sets are quite different and are in fact in the ratio of 2:1. It is therefore tempting to apply the Fisher's test for homogeneity of two standard deviations.

$$F \text{ statistic} = (0.00006)^2/(0.00003)^2 = 4.$$

The value of F for degrees of freedom of 9 and 14 is 3.89 for $P_t = 0.01$ probability and 2.59 for $P_t = 0.05$; the calculated value of F is larger than that of for 0.01. Hence the standard deviations do not appear to be homogeneous. This is in contradiction of our statement that means are consistent. The reason being obvious as s_1 and s_2 essentially come in the denominator of the expression for t_{vd}, which reduces the numerical value of t_{vd} for larger value of standard deviations; hence the two means are consistent. Therefore, it is suggested that before we declare the two means to be consistent, it is advisable to apply the F-test for consistency of standard deviations.

4.3.2 Test for Two Means Coming from Different Sources

Quite often different laboratories determine the value of same parameter of an object, for example, mass of a travelling standard weight. Each laboratory will give the mean value and standard deviation of its measurements. Here, it may not be necessary to have same population standard deviation. In this case, mean $\bar{x}$ will be given by

$$\bar{x} = \frac{\dfrac{\bar{x}_1}{s_1^2/n_1} + \dfrac{\bar{x}_2}{s_2^2/n_2}}{\left(\dfrac{1}{s_1^2/n_1} + \dfrac{1}{s_2^2/n_2}\right)}$$

or

$$\bar{x} = \frac{n_1 s_2^2 \bar{x}_1 + n_2 s_1^2 \bar{x}_2}{n_1^2 s_2^2 + n_2^2 s_1^2}. \tag{4.19}$$

It may be noted that $\bar{x}$ is the weighted mean of $\bar{x}_1$ and $\bar{x}_2$ where weight factors are

$$\frac{1}{s_1^2/n_1} \text{ and } \frac{1}{s_2^2/n_2}\text{, respectively.} \tag{4.20}$$

These are the reciprocals of the variances of each mean.

Combined standard deviation is given by

$$s_{vd} = \left(\frac{s_1^2}{n_1} + \frac{s_2^2}{n_2}\right)^{1/2}. \tag{4.21}$$

For the purpose of assessing the closeness of two means, Student's t statistic is taken as

$$t_{vd} = \frac{\bar{x}_1 - \bar{x}_2}{s_{vd}}$$

or

$$t_{vd} = \frac{\bar{x}_1 - \bar{x}_2}{(s_1^2/n_1 + s_2^2/n_2)^{1/2}}. \tag{4.22}$$

Effective degree of freedom in this case is not simply $n_1 + n_2$–2 but is to be calculated from Welch–Satterthwaite [1,2] formula:

$$\frac{1}{\nu_{\text{eff}}} = \sum_{p=1}^{p=n} \frac{s_p^4}{\nu_p} \bigg/ \left(\sum_{r=1}^{r=n} s_p^2\right)^2. \tag{4.23}$$

Taking s_1^2/n_1 and s_2^2/n_2 as the variances of each mean, we get the value of effective degree of freedom as

$$\nu_{\text{eff}} = (n_2 s_1^2 + n_1 s_2^2)^2 \bigg/ \left(\frac{n_2^2 s_1^4}{n_1} + \frac{n_1^2 s_2^4}{n_2}\right). \tag{4.24}$$

Rest of the test will be the same as before; namely if t_{vd} is less than the value of Student's t statistic for probability of 0.05 then we can safely assume that the two means are close enough to be considered consistent. In that case, there combined mean will be given by (4.19) and the estimate of the standard deviation of the weighted mean is given by

$$\frac{1}{s^2} = \frac{n_1}{s_1^2} + \frac{n_2}{s_2^2}. \tag{4.25}$$

4.3.2.1 Numerical Example

A metre bar was measured at two laboratories with different equipment, with following results:

Means	1000.014 mm	1000.054 mm
Standard Deviations	0.032 mm	0.074 mm
No of measurements	11	16

t_{vd} from (4.22)

$$t_{vd} = \frac{1000.014 - 1000.054}{0.032^2/11 + 0.074^2/16} = \frac{0.04}{9.309 \times 10^{-5} + 3.422 \times 10^{-4}} = \frac{0.04}{0.02086}$$
$$= 1.917.$$

The degree of freedom ν_{eff} is

$$\nu_{\text{eff}} = (n_2 s_1^2 + n_1 s_2^2)^2 \Big/ \left(\frac{n_2^2 s_1^4}{n_1} + \frac{n_1^2 s_2^4}{n_2}\right) = \frac{(16 \times 0.032^2 + 11 \times 0.074^2)^2}{16^2 \times 0.032^4 + 11^2 \times 0.074^4}$$
$$= \frac{(16.384 \times 10^{-3} + 60.236 \times 10^{-3})^2}{256 \times 1.049 \times 10^{-6}/10 + 121 \times 29.986 \times 10^{-6}/15}$$
$$= \frac{5870.624}{26.843 + 241.649} = \frac{5870.624}{268.493} = 21.86.$$

Now with 20 degrees of freedom the value of t is 1.7247 for $P_t = 0.05$ and is 2.5280 for $P_t = 0.01$.

The calculated value of t_{vd} is between the values of t for $P_t = 0.05$ and $P_t = 0.01$; hence the two means may be taken as consistent with a caution.

4.4 Comparison of Variances of Two Sets of Measurement Results

We started with validity of single observation, and then went to arithmetic means. Next step naturally should be to deal the variances. For verifying the consistency of two sets of measurement results, similar to the assessing of closeness of means, we can assess the ratio of the variances of two sets. The statistic involved is F, which is the ratio of the variances s_1^2 and s_2^2 of the two sets. These are given by

$$s_1 = \sqrt{\frac{\sum_{r=1}^{r=n_1} (x_{1r} - \bar{x}_1)^2}{n_1 - 1}}, \tag{4.26a}$$

$$s_2 = \sqrt{\frac{\sum_{r=1}^{r=n_1} (x_{2r} - \bar{x}_2)^2}{n_2 - 1}}. \tag{4.26b}$$

Statistic is

$$F = s_1^2/s_2^2.$$

Probabilities of F are indicated in the Tables Table A.7 and A.8 for F greater than unity. Hence the larger of two variances is taken as numerator of F statistic. If the value of F is less than the tabulated value for probability of 0.05 in Table A.7 then the two variances are supposed to be equal and two sets consistent. In case the value of F lies between the tabulated value of F for probabilities of 0.05 and 0.01 then results of the two sets may be taken as consistent with caution. In case F is greater than the tabulated value (Table A.8) for probability 0.01 then the two sets are said to inconsistent.

In tables giving F, n is usually replaced by ν – the degree of freedom. The two values of degree of freedom are given as

$$\nu_1 = n_1 - 1,$$
$$\nu_2 = n_2 - 1.$$

The values of ν_1 in the tables always apply to the variance on the numerator and ν_2 to that of the variance in denominator in the expression of F. The variance taken on numerator is always greater than that of denominator.

4.4.1 *Numerical Example*

Two laboratories reported the following voltage measurement at 10 V level.

	Lab. 1	Lab. 2
Standard deviation	0.2 mV	0.5 mV
No. of readings	16	11

To check if the two standard deviations are consistent
F statistic is given as

$$F = \frac{0.5^2}{0.2^2} = \frac{0.25}{0.04} = 6.25.$$

The value of F for $P_t = 0.01$ for 10 and 15 degrees of freedom is 3.80. The calculated value is much greater than the value of F even for $P_t = 0.01$; hence the two standard deviations are not consistent.

4.5 Measurements Concerning Travelling Standards

An object [3] is circulated within certain number of laboratories and each laboratory makes the measurement of the given parameter and sends the results to a nodal laboratory. The results normally consist of the estimated value of the parameter, number of observations taken to arrive at the estimated value and the standard deviation of each observation. One of the responsibilities of nodal laboratory is to see the quality and validity of the set of measurement of each laboratory. The nodal laboratory first calculates the overall mean. For calculating the mean value of the parameter, there are two methods, namely

(A) The weighted mean of the estimated values of means supplied by each laboratory is taken. The weight factor is $\frac{1}{s^2/n}$.
(B) The weighted mean of all the estimated values of means supplied by each laboratory is taken with unity as weight factor.

Further the nodal laboratory estimates the standard deviations at various steps and to inter-compare them.

In a mathematical sense, there will be m sets of observations one each by m laboratories. Let qth laboratory has taken n_q observations and reported the value of the measurand as $\bar{x}_q$ with σ_q as standard deviation of each observation. Let an observation is denoted by x_{pq}. The laboratory-wise observations are as follows:

$$\begin{array}{llll}
\text{Lab} & \text{Observations} & \text{Mean} & \text{S.D.} \\
1 & x_{11}, x_{12}, x_{13}, x_{14}, \ldots, x_{1n_1} & \bar{x}_1 & \sigma_1 \\
2 & x_{21}, x_{22}, x_{23}, x_{24}, \ldots, x_{2n_2} & \bar{x}_2 & \sigma_2 \\
3 & x_{31}, x_{32}, x_{33}, x_{34}, \ldots, x_{3n_3} & \bar{x}_3 & \sigma_3 \\
\vdots & \vdots & \vdots & \vdots \\
q\text{th} & x_{q1}, x_{q2}, x_{q3}, x_{q4}, \ldots, x_{qn_q} & \bar{x}_q & \sigma_q \\
\vdots & \vdots & \vdots & \vdots \\
\vdots & \vdots & \vdots & \vdots \\
m\text{th} & x_{m1}, x_{m2}, x_{m3}, x_{m4}, \ldots, x_{mn_m} & \bar{x}_m & \sigma_m
\end{array} . \tag{4.27}$$

The standard deviation of each observation is σ_{pq} with $p$taking values from 1 to n_q. Let the most likely value of x be $\bar{x}$; the meaning of it is yet undefined.

Taking rightly that each observation is belonging to a normal distribution, its probability of happening is

$$\frac{\exp -(x_{pq} - \bar{x})^2/2\sigma_{pq}^2}{\sigma_{pq}(2\pi)^{1/2}}. \tag{4.28}$$

Probability of occurrence of the observations x_{pq} [3], where p takes integral values from 1 to n_q and q takes integral values from 1 to m, is given by

$$Q = \frac{\exp\left[-\sum\sum (x_{pq} - \bar{x})^2/2\sigma_{pq}^2\right]}{\left(\prod_{q=1}^{q=m}\prod_{p=1}^{p=n_q} s_{pq}\right)(2\pi)^{N/2}}. \tag{4.29}$$

Here $\sum_{q=1}^{q=m} n_q$ is N

and

$$\prod_{p=1}^{p=n_q} \sigma_{pq} = \sigma_{1q} \times \sigma_{2q} \times \sigma_{3q} \cdots \sigma_{n_q q}.$$

Write

$$w_{pq} = k^2/\sigma_{pq}^2.$$

Here w_{pq} is the weight factor of observation x_{pq}. Substituting σ_{pq}^2 in terms of k and the weight factor, we get

$$Q = \frac{\prod_{q=1}^{q=m} \prod_{p=1}^{p=n_q} (w_{pq})^{1/2} \exp\left\{-\sum_{q=1}^{q=m} \sum_{p=1}^{p=n_q} w_{pq}(x_{pq} - \bar{x})^2/2k^2\right\}}{(k)^N (2\pi)^{1/2}}. \tag{4.30}$$

Taking logarithm of both sides and writing log Q as T, we get

$$T = \frac{1}{2}\sum_{q=1}^{q=m} \sum_{p=1}^{p=n_q} \log(w_{pq}) - N \log k - \sum_{q=1}^{q=m} \sum_{p=1}^{p=n_q} w_{pq}(x_{pq} - \bar{x})^2/2k^2 - \frac{N}{2}\log(2\pi). \tag{4.31}$$

Most likely values for $\bar{x}$ and k are those for which Q is a maximum. This mean T becomes a maximum, for which the necessary conditions are that $\frac{\delta T}{\delta k}$ and $\frac{\delta T}{\delta \bar{x}}$ are separately zero.

Now $\frac{\delta T}{\delta \bar{x}} = \sum_{q=1}^{q=m} \sum_{p=1}^{p=n_q} w_{pq}(x_{pq} - \bar{x})/k^2$.

This gives

$$= \sum_{q=1}^{q=m} \sum_{p=1}^{p=n_q} w_{pq} x_{pq} - \bar{x} \sum_{q=1}^{q=m} \sum_{p=1}^{p=n_q} w_{pq}. \tag{4.32}$$

Equating it to zero, we get

$$\bar{x} = \frac{\sum_{q=1}^{q=m} \sum_{p=1}^{p=n_q} w_{pq} x_{pq}}{\sum_{q=1}^{q=m} \sum_{p=1}^{p=n_q} w_{pq}}. \tag{4.33}$$

Now

$$\frac{\delta T}{\delta k} = 0 = -\frac{N}{k} + \sum_{q=1}^{q=m} \sum_{p=1}^{p=n_q} w_{pq}(x_{pq} - \bar{x})^2/k^3, \tag{4.34}$$

giving us

$$k^2 = \sum_{q=1}^{q=m} \sum_{p=1}^{p=n_q} w_{pq}(x_{pq} - \bar{x})^2 / N. \tag{4.35}$$

Now variance $\sigma_{\bar{x}}^2$ of the mean $\bar{x}$ is given

$$\frac{1}{\sigma_{\bar{x}}^2} = \sum_{q=1}^{q=m} \sum_{p=1}^{p=n_q} \frac{1}{\sigma_{pq}^2} = \sum_{q=1}^{q=m} \sum_{p=1}^{p=n_q} \frac{w_{pq}}{k^2}. \tag{4.36}$$

Substituting the value of k from (4.35), we get

$$\sigma_{\bar{x}}^2 = \frac{k^2}{\sum\limits_{q=1}^{q=m} \sum\limits_{p=1}^{p=n_q} w_{pq}} = \frac{\sum\limits_{q=1}^{q=m} \sum\limits_{p=1}^{p=n_q} w_{pq}(x_{pq} - \bar{x})^2}{N \sum\limits_{q=1}^{q=m} \sum\limits_{p=1}^{p=n_q} w_{pq}}. \tag{4.37}$$

For finite number of observations best estimate of variance $s_{\bar{x}}^2$ is obtained by multiplying the variance $\sigma_{\bar{x}}^2$ for very large number of observations by $N/N-1$ giving us

$$\sigma_{\bar{x}}^2 = \frac{\sum\limits_{q=1}^{q=m} \sum\limits_{p=1}^{p=n_q} w_{pq}(x_{pq} - \bar{x})^2}{(N-1) \sum\limits_{q=1}^{q=m} \sum\limits_{p=1}^{p=n_q} w_{pq}}. \tag{4.38}$$

4.5.1 Mean and Standard Deviation for each Laboratory

Following the similar steps as above for composite mean of all the laboratories, the mean of the qth laboratory is given by

$$\bar{x}_q = \frac{\sum\limits_{p=1}^{p=n_q} w_{pq} x_{pq}}{\sum\limits_{p=1}^{p=n_q} w_{pq}}. \tag{4.39}$$

Similarly the variance of the qth mean is given by

$$\sigma_{\bar{x}_q}^2 = \frac{\sum\limits_{p=1}^{p=n_q} w_{pq}(x_{pq} - \bar{x}_q)^2}{n_q \sum\limits_{p=1}^{p=n_q} w_{pq}}. \tag{4.40}$$

Hence estimated variance for qth laboratory for finite number of observations is

$$s_{\bar{x}_q}^2 = \frac{\sum_{p=1}^{p=n_q} w_{pq}(x_{pq} - \bar{x}_q)^2}{(n_q - 1)\sum_{p=1}^{p=n_q} w_{pq}}. \tag{4.41}$$

Equation (4.41) gives the variance within the laboratory.

We will now show that the value of mean $\bar{x}$ is same whether taken as the weighted mean of the means of all the laboratories or taken as the mean of all the weighted observations.

The weight factor of qth mean will be

$$\sum_{p=1}^{p=q} w_{pq} = w_q. \tag{4.42}$$

So $\bar{x}$ the mean of the means of all laboratories is given by

$$\bar{x} = \frac{\sum_{q=1}^{q=m} w_q \bar{x}_q}{\sum w_q}. \tag{4.43}$$

Substituting the values of w_q and $\bar{x}_q$, we get

$$\bar{x} = \frac{\sum_{q=1}^{q=m}\left\{\sum_{p=1}^{p=n_q} w_{pq}x_{pq} \times \sum_{p=1}^{p=n_q} w_{pq} \Big/ \sum_{p=1}^{p=n_q} w_{pq}\right\}}{\sum_{q=1}^{q=m} w_q},$$

$$\bar{x} = \frac{\sum_{q=1}^{q=m}\sum_{p=1}^{p=n_q} w_{pq}x_{pq}}{\sum_{q=1}^{q=m}\sum_{p=1}^{p=n_q} w_{pq}}. \tag{4.44}$$

Thus the weighted mean of the m laboratories is same as the weighted mean of all the observations taken together. Hence methods 1 and 2 give the same numerical value of the final mean if weight factors are taken into consideration.

4.5.2 Inter-Laboratories Standard Deviation

The variance of the overall mean $\bar{x}$ from m means of the m laboratories is given by

$$_2\sigma_{\bar{x}}^2 = \frac{\sum_{q=1}^{q=m} w_q(\bar{x}_q - \bar{x})^2}{m \sum_{q=1}^{q=m} w_q}. \tag{4.45}$$

Equation (4.45) has been obtained by replacing x_{pq} by $\bar{x}_q$, as the number of variables (means) is m; hence N has been replaced by m.

But $w_q = \sum_{p=1}^{p=n_q} w_{pq}$.

Hence (4.45) becomes

$$_2\sigma_{\bar{x}}^2 = \frac{\sum_{q=1}^{q=m} \sum_{p=1}^{p=n_q} w_{pq}(\bar{x}_q - \bar{x})^2}{m \sum_{q=1}^{q=m} \sum_{p=1}^{p=n_q} w_{pq}}. \tag{4.46}$$

Suffix 2 of $_2\sigma_{\bar{x}}^2$ denotes the standard deviation by external consistency. In other words, this is the standard deviation of the means of several laboratories from the overall mean. Hence it denotes the inter-laboratory standard deviation.

From (4.46) the best estimate of the standard deviation by external consistency (inter-laboratory) is given as

$$s_2^2(\bar{x}) = \frac{\sum_{q=1}^{q=m} \sum_{p=1}^{p=n_q} w_{pq}(\bar{x}_q - \bar{x})^2}{(m-1) \sum_{q=1}^{q=m} \sum_{p=1}^{p=n_q} w_{pq}}. \tag{4.47}$$

Equation (4.47) gives the best estimate of the inter-laboratories variance, from which best estimate of inter-laboratories standard deviation is obtained.

4.5.3 Intra-Laboratory Standard Deviation

Equation (4.40) gives the variance of the qth laboratory, so sum of variances of all laboratories is given by

$$\sigma^2 = \frac{\sum\limits_{q=1}^{q=m}\sum\limits_{p=1}^{p=n_q} w_{pq}(x_{pq} - \bar{x}_q)^2}{N\sum\limits_{q=1}^{q=m}\sum\limits_{p=1}^{p=n_q} w_{pq}}. \tag{4.48}$$

Hence estimate of the variance σ^2 from finite number of observations is obtained by multiplying σ^2 by $N/(N-m)$ giving

$$s_1^2 = \frac{\sum\limits_{q=1}^{q=m}\sum\limits_{p=1}^{p=n_q} w_{pq}(x_{pq} - \bar{x}_q)^2}{(N-m)\sum\limits_{q=1}^{q=m}\sum\limits_{p=1}^{p=n_q} w_{pq}}. \tag{4.49}$$

Here we see m standard deviations have been taken from N observation, so degree of freedom for the expression is $N-m$. This explains the reason for $(N-m)$ in the denominator of the expression in (4.49).

This estimate of the standard deviation of observations from the mean $\bar{x}$ is known as the standard deviation of the mean by internal consistency. In other words, it is within the laboratory (Intra laboratory) standard deviation.

4.6 F-test for Internal and External Consistency

4.6.1 F-test for Inter- and Intra-Laboratory Variances

Besides being used to decide the consistency of two sets of data as done in Sect. 4.4, F-test can be used to decide whether a number of sets observations as provided by the measurement laboratories are part of a larger normal population. If individual sets are biased then the complete set of observations will not be homogeneous and F-test may be used for revealing this inhomogeneity.

Let there be a set of N observations belonging to population having μ as mean and σ as the standard deviation. The N observations have been divided in m groups each having n_q observations. The q takes all integral values from 1 to m. In other words, there are m laboratories each has given the measured value $\bar{x}_q$ of the particular measurand as mean of n_q observations.

Let us consider the χ^2 of degree N of all the observations x_{pq} with population mean μ and standard deviation σ

$$\chi^2 = \sum_{q=1}^{q=m}\sum_{p=1}^{p=n_q} \frac{w_{pq}(x_{pq} - \mu)^2}{\sigma^2}. \tag{4.50}$$

Expressing the term inside the two summations as the sum of squares of deviations from means, we get

$$\chi^2 = \sum_{q=1}^{q=m} \sum_{p=1}^{p=n_q} \frac{w_{pq}\left\{(x_{pq} - \bar{x}_q) + (\bar{x}_q - \bar{x}) + (\bar{x} - \mu)\right\}^2}{\sigma^2}. \tag{4.51}$$

Here $\bar{x}$ is the mean of all observations.

Expanding the term inside two summations will result into six terms, three terms of square of each term inside the curly bracket and three terms of the products of the three terms. Using the properties of the means, summations of the three products will be zero giving us

$$\chi^2 = \sum_{q=1}^{q=m} \sum_{p=1}^{p=n_q} \frac{w_{pq}(x_{pq} - \bar{x}_q)^2}{\sigma^2} + \sum_{q=1}^{q=m} \sum_{p=1}^{p=n_q} \frac{w_{pq}(\bar{x}_q - \bar{x})^2}{\sigma^2} + \sum_{q=1}^{q=m} \sum_{p=1}^{p=n_q} \frac{w_{pq}(\bar{x} - \mu)^2}{\sigma^2}. \tag{4.52}$$

Expressing all deviations in terms of σ, we may interpret the three terms as follows:

The first term on the right-hand side is sum of the square of deviations from m individual means. As m means have been obtained from N observations, the degree of freedom of the first term is $N - m$. It is due to intra-laboratory (within a laboratory) variations. It is due to internal consistency. In other words, it is a χ^2 of $N - m$ degrees of freedom; let us write it as χ_1^2.

The second term of right-hand side is the sum of squares of deviations between m means; hence its degree of freedom is $m - 1$. Finding the overall mean has used up one degree of freedom $\bar{x}$. This term is due to inter-laboratory variations and may be termed as variations due to external consistency. The second term is also χ^2 and is of $m - 1$ degrees of freedom.

The last term is a χ^2 of one degree of freedom.

We further see that first term divided by $m - n$ gives s_1^2 the variance by internal consistency or the intra-laboratory variance. The second term when divided by $m - 1$ gives s_2^2.

F-test may be applied to these two variances giving F value as

$$F = \frac{(m-1) \sum_{q=1}^{q=m} \sum_{p=1}^{p_q} w_{pq}(x_{pq} - \bar{x}_q)^2}{(N-m) \sum \sum w_{pq}(\bar{x}_q - \bar{x})^2}, \tag{4.53}$$

as s_1^2 from (4.49) is given by

$$s_1^2 = \frac{\sum_{q=1}^{q=m} \sum_{p=1}^{p=n_q} w_{pq}(x_{pq} - \bar{x}_q)^2}{(N-m) \sum_{q=1}^{q=m} \sum_{p=1}^{p=n_q} w_{pq}}. \tag{4.54}$$

and s_2^2 from (4.47) is

$$s_2^2(\bar{x}) = \frac{\sum_{q=1}^{q=m} \sum_{p=1}^{p=n_q} w_{pq}(\bar{x}_q - \bar{x})^2}{(m-1) \sum_{q=1}^{q=m} \sum_{p=1}^{p=n_q} w_{pq}}. \quad (4.55)$$

Hence (4.53) is also equal to

$$F = \frac{s_1^2}{s_2^2}. \quad (4.56)$$

4.6.2 Weight Factors

In the above formulations the weight factor w_{pq} is quite general. We may choose the value of it as per requirement of the analysis. We have two choices, namely

All weight factors are taken equal and each is equal to unity

Weight factors are equal to the inverse of square of standard deviation of the observation

4.6.2.1 Case 1 Weight Factor Is Unity

From (4.33), the overall mean $\bar{x}$ is given by

$$\bar{x} = \frac{\sum_{q=1}^{q=m} \sum_{p=1}^{p=n_q} x_{pq}}{N}. \quad (4.57)$$

But from (4.39)

$$\bar{x}_q = \sum_{p=1}^{p=n_q} x_{pq}/n_q. \quad (4.58)$$

So (4.57) may be written as

$$\bar{x} = \frac{\sum_{q=1}^{q=m} n_q \bar{x}_q}{N}. \quad (4.59)$$

From s_1^2 from (4.54)

$$s_1^2 = \frac{\sum_{q=1}^{q=m} \sum_{p=1}^{p=n_q} (x_{pq} - \bar{x}_q)^2}{N(N-m)}. \quad (4.60)$$

From s_2^2 (4.55)

$$s_2^2 = \frac{\sum_{q=1}^{q=m} n_q(\bar{x}_q - \bar{x})^2}{N(m-1)}. \tag{4.61}$$

Equations (4.60) and (4.61) give the value of F as

$$F = \frac{s_1^2}{s_2^2} = \frac{(m-1)\sum_{p=1}^{p=m}\sum_{q=1}^{q=n_q}(x_{pq} - \bar{x}_q)^2}{(N-m)\sum_{q=1}^{q=m} n_q(\bar{x}_q - \bar{x})^2}. \tag{4.62}$$

4.6.2.2 Case 2 Weight Factor Other than 1

It is usually not possible to give an independent weight factor to each observation, but it is possible to give a weight factor to the set of one group with n_q observations. We know the variance of the set with n_q observations is $s^2(x_q)$; hence the weight factor for the qth set of n_q observations is $1/s^2(x_q)$. Substituting it in (4.39), we get

$$\bar{x}_q = \frac{\sum_{p=1}^{p=n_q} x_{pq}/s^2(x_q)}{n_q/s^2(x_q)},$$

$$\bar{x}_q = \frac{\sum_{p=1}^{p=n_q} x_{pq}}{n_q}. \tag{4.63}$$

This means that take simple arithmetic mean of the qth set of observation. For the overall mean $\bar{x}$ take the weighted mean of the m means obtained from (4.63). The weight factor is the inverse of the variance of the corresponding mean. So we get

$$\bar{x} = \frac{\sum_{q=1}^{q=m} \bar{x}_q/s^2(\bar{x}_q)}{\sum_{q=1}^{q=m} 1/s^2(\bar{x}_q)}. \tag{4.64}$$

From (4.54), we get s_1^2 as

$$s_1^2 = \frac{\sum_{q=1}^{q=m}\sum_{p=1}^{p=n_q}(x_{pq} - \bar{x}_q)^2/s^2(x_q)}{(N-m)\sum_{q=1}^{q=m} n_q/s^2(x_q)},$$

but

$$\sum_{p=1}^{p=n_q} (x_{pq} - \bar{x}_q)^2 / s^2(x_q) = n_q - 1,$$

giving s_1^2 as

$$s_1^2 = \frac{\sum_{q=1}^{q=m} (n_q - 1)}{(N - m) \sum_{q=1}^{q=m} n_q / s^2(x_q)}$$

$$= \frac{N - m}{(N - m) \sum_{q=1}^{q=m} n_q / s(x_q)^2} \tag{4.65}$$

$$= \frac{1}{\sum_{q=1}^{q=m} 1 / s^2(\bar{x}_q)},$$

where

$$s^2(\bar{x}_q) = s^2(x_q)/n_q. \tag{4.66}$$

Again from (4.55), we get s_2^2 as

$$s_2^2 = \frac{\sum_{q=1}^{q=m} \sum_{p=1}^{p=n_q} (\bar{x}_q - \bar{x})^2 / s^2(x_q)}{(m - 1) \sum_{q=1}^{q=m} n_q / s^2(x_q)},$$

$$s_2^2 = \frac{\sum_{q=1}^{q=m} n_q (\bar{x}_q - \bar{x})^2 / s^2(x_q)}{(m - 1) \sum_{q=1}^{q=m} n_q / s^2(x_q)},$$

$$s_2^2 = \frac{\sum_{q=1}^{q=m} (\bar{x}_q - \bar{x})^2 / s^2(\bar{x}_q)}{(m - 1) \sum_{q=1}^{q=m} 1 / s^2(\bar{x}_q)}. \tag{4.67}$$

Hence variable F will be

$$F = \frac{s_1^2}{s_2^2} = \frac{(m - 1)}{\sum (\bar{x}_q - \bar{x})^2 / s^2(\bar{x}_q)}. \tag{4.68}$$

The degree of freedom of s_1^2 is $N - m$ and that of s_2^2 is $m - 1$.

4.6.3 F-test for Variances

Now compare the calculated F value with the tabulated value of F corresponding to ν_1 and ν_2 degrees of freedom for probability 0.05. If the calculated value is lesser than the tabulated value then the results are consistent, or in other words, the results reported by the various laboratories are very much the part of the same population and hence acceptable. If the calculated value lies between the values for probabilities of 0.05 and 0.01, then results are acceptable with a caution. If the calculated value is more than the F value for probability of 0.01, then results do not belong to the same population.

If it is so then the results with maximum variance may be isolated and values of s_1^2 and s_2^2 are recalculated from the remaining results and whole procedure stated above is repeated. In such a case, it is advisable to use the inverse of variance of each laboratory as the weight factor for calculating the overall mean result.

4.7 Standard Error of the Overall Mean

From (4.38) the standard error of the overall mean is

$$s_{\bar{x}}^2 = \frac{\sum_{q=1}^{q=m} \sum_{p=1}^{p=n_q} w_{pq}(x_{pq} - \bar{x})^2}{(N-1) \sum_{q=1}^{q=m} \sum_{p=1}^{p=n_q} w_{pq}}.$$

Using the properties of the mean, this can be expressed in terms of s_1^2 and s_2^2 as follows:

$$s^2(\bar{x}) = \frac{(N-m)}{N-1} s_1^2 + \frac{(m-1)}{N-1} s_2^2. \tag{4.69}$$

4.7.1 Results Inconsistent

After going through all the rigorous evaluation of measurement data from several laboratories, and finding that the results are inconsistent, it may not be possible to either reject the whole set of data or reject the data from a certain laboratory. In that case overall mean and variance are to be reported.

The best method to report the mean is to take the mean of all the means by different laboratory without attaching any weight factor to a mean. That is

$$\bar{x} = \sum_{q=1}^{q=m} \bar{x}_q / m. \tag{4.70}$$

The square of the standard error of this mean is given by

$$s^2(\bar{x}) = \frac{1}{m(m-1)} \sum_{q=1}^{q=m} (\bar{x}_q - \bar{x})^2. \tag{4.71}$$

It leads to the same formula if each set of measurements contains equal number of observations and the weight assigned to each measurement is unity.

4.8 Analysis of Variance

4.8.1 One-Way Analysis of Variance

We have discussed methods that are most frequently used to test the equality of two means or two variances. Quite often we need to analyse data of several samples (more than 2). In addition of the method given above (in Sects. 4.4.6 and 4.4.7) we discuss one-way analysis of variance, which is a little simpler method with unity weight factor to each observation.

We start with the hypothesis that

Samples means are same namely

$$H_0: \quad \mu_1 = \mu_2 = \mu_3 = \cdots = \mu_m. \tag{4.72}$$

Alternative Hypothesis H_1: At least two means are not equal.

Notations used are the same as in (4.27). For ready reference, these are repeated below in Table 4.2.

Table 4.2 Observations, means and variances of various laboratories

	Observations				
Laboratories	1	2	3...	m	
Observations	$x_1 1$	$x_1 2$	$x_1 3$	$x_1 m$	
	$x_2 1$	$x_2 2$	$x_2 3$	$x_2 m$	
	$x_3 1$	$x_3 2$	$x_3 3$	$x_3 m$	
	...	...	...	...	
	...	...	...	...	
	...	...	$x_{n3} 3$	...	
	$x_{n1} 1$		...	...	
		$x_{n2} 2$	...	...	
				$x_{nm} m$	
					Sum of the row
Column sum T_q =					T
Column mean $\bar{x}_q$					$\bar{x}$
Column variance s_q^2					
$(n_q - 1)s_q^2$					$\sum_{q=1}^{q=m} (n_q - 1)s_q^2 = \text{SSW}$

We know that
The pth observation of qth laboratory$= x_{pq}$

$$\text{Total of observations of } q\text{th laboratory } T_q = \sum_{p=1}^{p=n_q} x_{pq}. \tag{4.73}$$

$$\text{Mean of observations of } q\text{th laboratory } \bar{x}_q = \frac{T_q}{n_q} = \frac{\sum_{p=1}^{p=n_q} x_{pq}}{n_q}. \tag{4.74}$$

$$\text{Variance of observations of } q\text{th laboratory } s_q^2 = \frac{\sum_{p=1}^{p=n_q} (x_{pq} - \bar{x}_q)^2}{n_q - 1}. \tag{4.75}$$

$$\text{Sum total of all observations } T = \sum_{q=1}^{q=m} T_q = \sum_{q=1}^{q=m} \sum_{p=1}^{p=n_q} x_{pq}. \tag{4.76}$$

$$\text{Total number of observations } N = \sum_{q=1}^{q=m} n_q. \tag{4.77}$$

$$\text{Overall mean (mean of all observations) } \bar{x} = \frac{T}{N} = \frac{\sum_{q=1}^{q=m} \sum_{p=1}^{p=n_q} x_{pq}}{N}. \tag{4.78}$$

Variance s_m^2 of all N observations regarded as single sample

$$s_m^2 = \frac{\sum_{q=1}^{q=m} \sum_{p=1}^{p=n_q} (x_{pq} - \bar{x})^2}{N - 1}. \tag{4.79}$$

Let μ_q and σ_q^2 be the mean and standard deviation of the qth laboratory and q takes values $1, 2, 3, \ldots, m$.

From (4.79), we have

$$\begin{aligned}
(N-1)s_m^2 &= \sum_{q=1}^{q=m} \sum_{p=1}^{p=n_q} (x_{pq} - \bar{x})^2 = \sum_{q=1}^{q=m} \sum_{p=1}^{p=n_q} \left[(x_{pq} - \bar{x}_q) + (\bar{x}_q - \bar{x})\right]^2 \\
&= \sum_{q=1}^{q=m} \sum_{p=1}^{p=n_q} \left[(x_{pq} - \bar{x}_q)^2 + 2(x_{pq} - \bar{x}_q)(\bar{x}_q - \bar{x}) + (\bar{x}_q - \bar{x})^2\right] \qquad (4.80) \\
&= \sum_{q=1}^{q=m} \sum_{p=1}^{p=n_q} (x_{pq} - \bar{x}_q)^2 + 2 \sum_{q=1}^{q=m} \sum_{p=1}^{p=n_q} (x_{pq} - \bar{x}_q)(\bar{x}_q - \bar{x}) + \sum_{q=1}^{q=m} \sum_{p=1}^{p=n_q} (\bar{x}_q - \bar{x})^2.
\end{aligned}$$

Overall mean $\bar{x}$ is the mean of column means $\bar{x}_q$; hence the sum of deviations of the laboratory means $\bar{x}_q$ from their overall mean $\bar{x}$ is zero making the second term in (4.80) zero,

giving us

$$\sum_{q=1}^{q=m}\sum_{p=1}^{p=n_q}(x_{pq}-\bar{x})^2=\sum_{q=1}^{q=m}\sum_{p=1}^{p=n_q}(x_{pq}-\bar{x}_q)^2+\sum_{q=1}^{q=m}\sum_{p=1}^{p=n_q}(\bar{x}_q-\bar{x})^2$$

or (4.81)

$$\mathrm{SST}=\mathrm{SSW}+\mathrm{SSA},$$

where

$$\mathrm{SST}=\sum_{q=1}^{q=m}\sum_{p=1}^{p=n_q}(x_{pq}-\bar{x})^2, \quad (4.82)$$

$$\mathrm{SSW}=\sum_{q=1}^{q=m}\sum_{p=1}^{p=n_q}(x_{pq}-\bar{x}_q)^2=\sum_{q=1}^{q=m}(n_q-1)s_q^2, \quad (4.83)$$

$$\mathrm{SSA}=\sum_{q=1}^{q=m}\sum_{p=1}^{p=n_q}(\bar{x}_q-\bar{x})^2. \quad (4.84)$$

4.8.1.1 Testing the Null Hypothesis

Let us consider SSW, the sum of squares of deviations from their respective means (refer to right-hand side of (4.83). Population variance σ^2 is the average values of variances s_q^2; hence

$$\begin{aligned}\sum_{q=1}^{q=m}(n_q-1)s_q^2&=(n_1-1)s_1^2+(n_2-1)s_2^2+(n_3-1)s_3^2+\cdots+(n_m-1)s_m^2\\&=(n_1-1)s^2+(n_2-1)s^2+(n_3-1)s^2+\cdots+(n_m-1)s^2.\\&=(N-m)\sigma^2.\end{aligned} \quad (4.85)$$

Hence $\mathrm{SSW}/(\mathrm{N}-\mathrm{m})$ is an unbiased estimate of σ^2. (4.86)

Similarly, it can be proved that

$$\mathrm{SSA}=\sum_{q=1}^{q=m}\sum_{p=1}^{p=n_q}(\bar{x}_q-\bar{x})^2=(m-1)\sigma^2+\sum_{q=1}^{q=m}n_q(\mu_q-\mu)^2. \quad (4.87)$$

Here μ is the overall mean and it, therefore, is

$$\mu = \frac{\sum_{q=1}^{q=m} n_q \mu_q}{N}. \tag{4.88}$$

Thus average value of SSA/$(m-1)$ is

$$\mathrm{SSA}/(m-1) = \sigma^2 + \left[\sum_{q=1}^{q=m} n_q(\mu_q - \mu)^2\right] \Big/ (\mu - 1). \tag{4.89}$$

Now if $\mu_1 = \mu_2 = \mu_3 = \ldots = \mu_m = \mu$ so that null hypothesis H_0 is true, the second term in (4.89) must be zero.

Therefore, the quantities in (4.85) and (4.89) must be compatible and their ratio must be nearly 1 within the normal statistical fluctuations; hence Fisher's F-test

$$F = \frac{\mathrm{SSA}/(m-1)}{\mathrm{SSW}/(N-r)}$$

is applied to test the hypothesis that all samples belong to the same population having common mean and variance.

The results are summarized in Table 4.3 as follows:

The expected mean squares (EMS) column deserves further comment. It has been pointed that the average value of

$$\frac{\mathrm{SSW}}{N-q} = \mathrm{MSW} = \sigma^2 \tag{4.90}$$

and the average value of

$$\frac{\mathrm{SSA}}{q-1}\mathrm{MSA} = \sigma^2 + \frac{\sum_{q=1}^{q=m} n_q(\mu_q - \mu)^2}{m-1}. \tag{4.91}$$

Table 4.3 Summary of Results

Source of variance	Sums	Degree of freedom	Mean of sums	Expected mean sum (EMS)	F-test
Between laboratories (Groups)	SSA	$m-1$	SSA/$(q-1)$ = MSA	$\sigma^2 + \sum_{q=1}^{q=m} \frac{n_q(\mu_q - \mu)^2}{m-1}$	MSA/MSW
Within laboratories	SSW	$N-m$	SSW/$(N-q)$ = MSW	σ^2	
Total	SST	$N-1$			

These are expected mean squares and will always be included in a column of the analysis of variance table. For the special case when

$$n_1 = n_2 = n_3 = \ldots = n_m = n. \tag{4.92}$$

From (4.92), we get

$$1/(m-1)\sum_{q=1}^{q=m} n_q(\mu_q - \mu)^2 = (n/m-1)\sum_{q=1}^{q=m}(\mu_q - \mu)^2. \tag{4.93}$$

In more complicated situations, which we are going to discuss later, a glance at the EMS column will indicate which mean square is to be taken for F ratio test.

4.8.1.2 Numerical Example

A high-pressure gauge was calibrated by four laboratories at 10, 15, 20 and 25 bars (1 bar= 10^5 Pa). Each laboratory took 5 observations at each of the four points and reported the results as given in Tables 4.4 and 4.5.

Table 4.4 Means & standard deviations at 4 test points

Points on the scale	Mean	Laboratories				Level
		A	B	C	D	
10 bar	Mean	10.0	10.0	10.1	9.9	10.0
	SD	0.071	0.071	0.122	0.071	
15 bar	Mean	15.0	15.1	15.1	14.8	15.00
	SD	0.158	0.187	0.122	0.122	
20 bar	Mean	19.8	20.02	20.1	19.9	20.0
	SD	0.173	0.122	0.187	0.071	
25 bar	Mean	25.0	24.8	25.4	25.2	25.1
	SD	0.187	0.224	0.212	0.200	
Overall mean		17.45	17.525	17.675	17.45	17.525

Table 4.5 Detailed observations of the four laboratories at 20 bar level

Laboratory	Observations			
	A	B	C	D
	19.7, 20.1, 19.7 19.7, 19.8	20.2, 20.3, 20.2, 20.2, 20.3	20.3, 20.0, 19.9 20.3, 20.0	19.9, 19.8, 19.9, 19.9, 20.0,
Mean	19.8	20.2	20.1	19.9
SD	0.173	0.122	0.187	0.071
Overall mean	20.0			

We wish to calculate
Overall mean $\bar{x}$
Within a laboratory variance σ_w^2
Between the laboratories σ_b^2

Let there be q laboratories that participated in a measurement programme. Each laboratory took $n_1, n_2, n_3, \ldots, n_q$ observations and has reported the means as $\bar{x}_1, \bar{x}_2, \bar{x}_3, \ldots, \bar{x}_q$, and standard deviations $s_1, s_2, s_3, \ldots, s_q$.

Overall mean

$$\bar{x} = \frac{n_1\bar{x}_1 + n_2\bar{x}_2 + n_3\bar{x}_3 + \ldots + n_q\bar{x}_q}{n_1 + n_2 + n_3 + \ldots + n_q}. \tag{4.94}$$

Estimate of overall variance

$$s_m^2 = \frac{\sum_{p=1}^{p=m}\sum_{q=1}^{q=n_q}(x_{pq} - \bar{x})^2}{\sum_{q=1}^{q=n_q}(n_q) - 1}, \tag{4.95}$$

$$s_m^2 = \frac{(n_1 - 1)s_1^2 + (n_2 - 1)s_2^2 + (n_3 - 1)s_3^2 + \ldots + (n_q - 1)s_q^2}{(n_1 - 1) + (n_2 - 1) + (n_3 - 1) + \ldots + (n_q - 1)}.$$

We will calculate sum of squares of deviations from the mean

$$\sum_{q=1}^{q=4}\sum_{p=1}^{p=5}(x_{pq} - \bar{x}_q)^2 = \text{SSW}, \tag{4.96}$$

$$\sum_{q=1}^{q=4} n_q(\bar{x}_q - \bar{x})^2 = \text{SSA}, \tag{4.97}$$

$$\sum_{q=1}^{q=4}\sum_{p=1}^{p=5}(x_{pq} - \bar{x})^2 = \text{SST}. \tag{4.98}$$

SST is the sum of first two summations

Calculation sheet at nominal values at level of 20 bar is given in Table 4.6.

Referring to figures in the calculation sheets Table 4.6, we obtain the figures given in Table 4.7

$$F = 0.1667/0.02125 = 7.84.$$

Tabulated value of F with 3 and 16 degrees of freedom at probability value for $0.05 = 3.25$.

Tabulated value of F with 3 and 16 degrees of freedom at probability value for $0.01 = 5.32$.

This shows that variances between the laboratories and within a laboratory are not equal and difference is too large to be attributed due to statistical fluctuations.

To further investigate the reason of too much difference in standard deviations, we look for the bias in the results of four laboratories. The details are given in Table 4.8.

Table 4.6 Calculation sheet at nominal values at level of 20 bar

Serial No	Laboratory A			Laboratory B			Laboratory C			Laboratory D		
	x	$X - x_q$	$(x - x_q)^2$	x	$x - x_q$	$(x - x_q)^2$	x	$x - x_q$	$(x - x_q)^2$	x	$x - x_q$	$(x - x_q)^2$
1	19.7	−0.1	0.01	20.2	0.0	0.00	20.3	0.2	0.04	19.9	0.0	0.00
2	20.1	0.3	0.09	20.3	0.1	0.01	20.0	−0.1	0.01	19.8	0.1	0.01
3	19.7	−0.1	0.01	20.2	0.0	0.00	19.9	−0.2	0.04	19.9	0.0	0.00
4	19.7	−0.1	0.01	20.0	−0.2	0.04	20.3	0.2	0.04	19.9	0.0	0.00
5	19.8	0.0	0.0	20.3	0.1	0.01	20.0	0.1	0.01	20.0	−0.1	0.01
Sum	–	0.0	0.12	–	0.0	0.06	–	0.0	0.14	–	0.0	0.02
Mean	19.8			20.2			20.1			19.9		
SD			0.173			0.122			0.187			0.071
$\bar{x}$	20.0											
Sum	$0.12 + 0.06 + 0.14 + 0.02 = 0.34 = \text{SSW}$											
$\bar{x} - x_q$	0.2			0.2			0.1			0.1		
$(\bar{x} - x_q)^2$	0.04			0.04			0.01			0.01		
$5\Sigma(\bar{x} - x_q)^2$	$5(0.04 + 0.04 + 0.01 + 0.01) = 0.5 = \text{SSA}$											

Table 4.7 Variances between the laboratories

Sources of error	Sum of squares of deviation	Degree of freedom	Variances
Between Labs.	0.50	3	$0.1667 = \sigma_b^2$
Within a Lab	0.34	16	$0,021225 = \sigma_w^2$
Total sum	0.84	19	–

Table 4.8 Bias estimates (Laboratory mean – overall mean of 4 laboratories) at all levels

Levels	A	B	C	D	Mean	SD
1	0.00	0.00	0.10	−0.1	0.00	0/0816
2	0.00	0.10	0.10	−0.2	0.00	0.1414
3	−0.20	0.20	0.10	−0.10	0.00	0.1826
4	−0.10	−0.3	0.30	0.10	0.00	0.2582
	−0.075	0.000	0.150	−0.075	0.00	

Note: To deal with smaller numbers 19.7 has been subtracted from each observation
Inferences:
Laboratory C appears to have a positive bias of 0.15
Laboratories A and D appears to have negative bias

Uncertainty of overall mean

$$\pm t\sqrt{\left\{\frac{\sigma_b^2}{q} + \frac{\sigma_L^2}{\sum_{q=1}^{q=4} n_q}\right\}}. \tag{4.99}$$

4.9 Tests for Uniformity of Variances

4.9.1 *Bartlett's Test for Uniformity of Many Variances*

Quite often to evaluate measurement data received from various laboratories or data collected over a period of years in a laboratory, we need to ascertain if the variances do belong to the same population. In this case our Null Hypothesis and its alternate are

H0: $\sigma_1^2 = \sigma_2^2 = \sigma_3^2 = \ldots = \sigma_m^2$
H1: At least two variances are different

In Bartlett test only precondition is that sample size is same; here we assume that:

(a) m samples (data received from m laboratories) are drawn from m populations and
(b) Each of the m populations has the normal-Gaussian probability distribution.

The Bartlett statistic denoted as χ is

$$\chi = \frac{2.3026}{C}\left[(N-m)\log s_p^2 - \sum (n_q - 1)\log s_p^2\right]. \tag{4.100}$$

Here C is given as

$$C = 1 + \frac{1}{3(m-1)}\left[\sum_{q=1}^{q=m}\frac{1}{(n_q-1)} - \frac{1}{N-m}\right]. \tag{4.101}$$

And

$$s_p^2 = \sum_{q=1}^{q=m}(n_q-1)s_q^2/(N-m). \tag{4.102}$$

We know that

$$\sum_{q=1}^{q=m} n_q = N.$$

The more the s_q^2 differs from each other, the larger will be the value of χ. If all the s_q^2 are more or less the same then the statistic χ will be small. The statistic in this case will be approximately distributed as Chi-square χ^2 with $m-1$ degrees of freedom.

4.9.1.1 Numerical Example

Let three laboratories calibrated 1 kg weight and sent data of standard uncertainty, i.e. standard deviation along with the mass values assigned to the weight. The standard uncertainty and sample size data were as follows:

$m = 3$, $n_1 = 5$, $n_2 = 8$, $n_3 = 10$, respective standard uncertainty is 14.3 μg, 12.5 μg, 20.4 μg

Giving us

$$s_p^2 = \frac{4\times 204 + 7\times 156 + 9\times 416}{23-3} = 282.6.$$

$$C = 1 + \frac{1}{6}\left[\frac{1}{4} + \frac{1}{7} + \frac{1}{9} - \frac{1}{20}\right] = 1.0757.$$

$$(n_1-1)\log s_1^2 = 9.23852,$$
$$(n_2-1)\log s_2^2 = 15.35187,$$
$$(n_3-1)\log s_3^2 = 23.57184.$$

Giving us

$$\sum_{q=1}^{q=3} (n_q - 1) \log s_q^2 = 48.15871$$

$$(N - q) \log s_p^2 = 20 \log (282.3) = 20 \times 2.45071 = 49.01422.$$

Giving

$$\chi = 2.3026\,(49.01222 - 48.15871)/1.0757 = 1.7662.$$

From the Chi-square table for the value of Chi square at the probability of 0.05 = 5.99. Calculated value is much smaller than the critical value; hence we can assume safely that the variances reported by the laboratories are equal within statistical fluctuations.

4.9.2 *Cochran Test for Homogeneity of Variances*

If size of the samples (number of measurements taken for arriving at the results) is same then a much-simplified Cochran test may be applied for the homogeneity of variances. The statistic is

$R_{n,r}$ = Ratio of largest variance and sum of all the variances

That is

$$R_{n,q} = \frac{\text{largest } s_q^2}{\sum_{q=1}^{q=m} s_q^2}.$$

4.9.2.1 Numerical Example

The values of variances reported by three laboratories in measurement of 10 Ω resistor were 140, 660, and 200 $(\mu\,\Omega)^2$. Number of observation taken to arrive at the result by each laboratory is 10.

$$R_{10,3} = \frac{660}{140 + 200 + 660} = 0.66.$$

The value of $R_{10,3}$ for probability of 0.05 is 0.6167 and for probability of 0.01 is 0.6912. Hence the variances are said to be homogenous at 1% level but nonhomogeneous at 5% level.

For the purpose of establishing degree of equivalence for national measurement standards participating in regional comparison, one may refer to [4].

References

1. F.E. Satterthwaite, Pscychometrika **6**, 309–316 (1941)
2. F.E. Satterthwaite, Biometrics **2**(6), 110–114 (1846)
3. C.F. Dietrich, *Uncertainty, Calibration and Probability* (Adam Hilger, Bristol, 1991), pp. 42–47, 274–276
4. I.A. Khartitonov, A.G. Chunovkina, Evaluation of regional key comparison data-two approaches. Metrologia **43**, 470–476 (2006)

Chapter 5
Propagation of Uncertainty

5.1 Mathematical Modelling

Quite often the quantity under measurement (measurand) Y is not measured directly, but is the result of measurement of several independent quantities X_1, X_2, X_3, ..., X_n. The measurand Y is also called the output quantity and X_1, X_2, X_3, ..., X_n as input quantities. The quantity Y is related to output quantities through some well-defined relation. That is Y is expressed in terms of X_1, X_2, X_3, ..., X_n as

$$Y = f(X_1, X_2, X_3, \ldots, X_n). \tag{5.1}$$

Notations:

- For economy of notation, the same symbol is used for the physical quantity and for the random variable that represent the possible outcome of an observation of that quantity. When it is stated that an input quantity X_p has a particular probability distribution then X_p is a random variable. The physical quantity itself is invariant and has a unique, fixed value.
- In a series of observations, the qth observed value of X_p is denoted as $x_{p,q}$. The estimate of the X_p is denoted by x_p, which in fact is the expected value of X_p. Quantities, in general, are expressed in capital letters, while their numerical values by the corresponding small case letters. The value of the quantity X_n is expressed as x_n for all integral values of n. Hence y the estimated value of the quantity Y is expressed as

$$y = f(x_1, x_2, x_3, \ldots, x_n), \tag{5.2}$$

where x_1, x_2, x_3, ..., x_n are the measured estimates of the physical quantities X_1, X_2, X_3, ..., X_n, respectively.

It is assumed that each input estimate is corrected for all known systematic effects, which are likely to influence significantly.

S.V. Gupta, *Measurement Uncertainties*, DOI 10.1007/978-3-642-20989-5_5,

5.1.1 Mean of Measurand (Dependent Variable)

For the case of repeated measurements, the estimate of y may be obtained in two ways [1].

1. Find the average of all the input quantities and substitute their values in the function. This is expressed as

$$\bar{y} = f(\bar{x}_1, \bar{x}_2, \bar{x}_3, \ldots, \bar{x}_n). \tag{5.3}$$

2. For each set of n measurement values of input quantities, find the estimated value of y and then take the mean of m estimated values of y. m is the number of sets of n measured input quantities. This is expressed as

$$\bar{y} = \frac{1}{m}\sum_{q=1}^{q=m} y_q = \frac{1}{m}\sum_{q=1}^{q=m} f(x_{1,q}, x_{2,q}, x_{3,q}, \ldots, x_{n,q}). \tag{5.4}$$

First method is applicable if the input quantities are independent of each other and are not influenced by the environmental conditions or proper corrections are applied due to change in environmental conditions. Second method is applicable for input quantities, which are dependent and which are affected by influence quantities including environmental conditions. Second method is preferable if the output quantity Y is a non-linear function of input quantities. Two methods are equivalent when Y is a linear function of input quantities.

5.1.2 Functional Relationship and Input Quantities

The input quantities X_1, X_2, X_3, ..., X_n are themselves measured quantities which may further depend upon some other quantities, including corrections and correction factors for systematic effects, thereby leading to a complicated functional relationship, which may rather be difficult if not impossible to write down explicitly. Further, it may not be an algebraically defined function and may be a portion of it is determined experimentally or it exists only as an algorithm that is calculated numerically. Sometimes $\delta f/\delta x_p$ is determined experimentally by measuring the change in Y by incorporating a change in X_p. In this case, knowledge of f is or a portion of it is correspondingly reduced to an empirical first order Taylor's expansion. So the function f may be taken in a broader sense.

If the mathematical model does not satisfy the degree of accuracy desired, then additional input quantities may be included in the function f to eliminate the inadequacy.

For example for ordinary day-to-day weighing in a market place, mass of the commodity is taken as the nominal mass of the weight. For a better degree of accuracy, we take into account the actual mass of the weight. For still better accuracy, we apply air buoyancy correction for which we may take density of air as 1.2 kg/m^3. For still better accuracy, we calculate values of the density of air, and those of weight and commodity by measurements. To improve the accuracy further, we may like to know the actual composition of air or measure the density of air inside the balance at the time of weighing only. In the first case it is a simple relation, a correction due to mass of weight is applied in the second case, air buoyancy is added in the third case, a relationship of air density with environmental conditions is added further and a few measurements of density of weight and commodity are to be taken. To improve further experiment for determination of air density in situ is carried out and added in the relationship.

In another example, power dissipated across a given resistor is given by

$$P = f(V, R) = V^2/R. \tag{5.5}$$

To improve accuracy variation of resistance of R with temperature is to be considered giving

$$P = f(V, R, \alpha, t) = V^2/R_0(1 + \alpha t). \tag{5.6}$$

Relationship becomes more and more complex, if dependence of α with temperature and measurement of temperature are taken into account.

5.1.3 Expansion of a Function

If in (5.1), an input quantity X_p is changed by a small amount ΔX_p, for all integral values of p, there will be a corresponding change in the dependent variable, say by ΔY related as

$$Y + \Delta Y = f(X_1 + \Delta X_1, X_2 + \Delta X_2, \mathrm{K}, X_n + \Delta X_n). \tag{5.7}$$

Using Taylor's expansion, we get

$$Y + \Delta Y = f(X_1, X_2, X_3, \mathrm{K}, X_n) + \sum_{p=1}^{p=n} \frac{\delta f}{\delta X_p} \Delta X_p + R. \tag{5.8}$$

R is the remainder term of the expansion and R is zero if f is a linear function of the input quantities.

Subtracting (5.1) from (5.8), we get

$$\Delta Y = \sum_{p=1}^{p=n} \frac{\delta f}{\delta X_p} \Delta X_p + R. \tag{5.9}$$

R is zero if Y is a linear function X_p for all p.

The condition for Y to be a linear function of input quantities set out for uncertainty calculation by ISO GUM [1] method is necessary to make the remainder R equal to zero.

To make the mathematics simpler, generality will not be lost if we consider only two independent variables, namely X_1 and X_2. Y is the dependent variable of these two independent variables. Further if Y is a linear function of X_1 and X_2, then R is zero; then retaining only the first term of (5.9), we get

$$\Delta Y = \Delta X_1 \delta Y/\delta X_1 + \Delta X_2 \delta Y/\delta X_2. \tag{5.10}$$

Expressing ΔY as $y_p - \bar{y}$ in terms of deviations of the two independent variables, we get

$$(y_p - \bar{y}) = (x_{1p} - \bar{x}_1)\delta Y/\delta X_1 + (x_{2p} - \bar{x}_2)\delta Y/\delta X_2. \tag{5.11}$$

All partial derivatives are calculated at the mean values of the independent variables. Further we know that variance of y is denoted as σ_Y^2 and is given as

$$\sigma_Y^2 = \lim_{n\Rightarrow\infty}\left[\frac{1}{n}\sum_{p=1}^{p=n}(y_p - \bar{y})^2\right]. \tag{5.12}$$

Hence from (5.11) becomes

$$\begin{aligned}\sigma_Y^2 &= \lim_{n\Rightarrow\infty}\left[\frac{1}{n}\sum(x_{1p} - \bar{x}_1)^2(\delta Y/\delta X_1)^2\right]\\ &\quad+\lim_{n\Rightarrow\infty}\left[\frac{1}{n}\sum(x_{2p} - \bar{x}_2)^2(\delta Y/\delta X_2)^2\right]\\ &\quad+2\lim_{n\Rightarrow\infty}\left[\frac{1}{n}\sum(x_{1p} - \bar{x}_1)(x_{2p} - \bar{x})(\delta Y/\delta X_1)(\delta Y/\delta X_2)\right].\end{aligned} \tag{5.13}$$

Similar to variance there is another term covariance, which is defined as follows:

$$\mathrm{Cov}(x_1, x_2) = \lim_{n\Rightarrow\infty}\left[\frac{1}{n}\sum_{p=1}^{p=n}(x_{1p} - \bar{x}_1)(x_{2p} - \bar{x})\right]. \tag{5.14}$$

Expressing $\frac{1}{n}\sum(x_{1p} - \bar{x}_1)^2$ as σ_{x1}^2 in (5.13), we get

$$\sigma_Y^2 = \sigma_{x1}^2(\delta Y/\delta X_1)^2 + \sigma_{x2}^2(\delta Y/\delta X_2)^2 + 2\mathrm{Cov}(X_1, X_2)(\delta Y/\delta X_1)(\delta Y/\delta X_2). \tag{5.15a}$$

Generalizing it for many variables, (5.15) becomes

$$\sigma_Y^2 = \sum_{p=1}^{p=n}\sum_{q=p}^{q=p}\sigma_{xpq}^2(\delta Y/\delta X_p)^2 + 2\sum_{p=1}^{p=n}\sum_{q=p+1}^{q=n}\mathrm{Cov}(x_p, x_q)(\delta Y/\delta X_p)(\delta Y/\delta X_q). \tag{5.16a}$$

It may be noted that $\delta Y/\delta X_p$ for all values of p are calculated at the observed values $x_1, x_2, x_3, \ldots, x_n$. The value of $\delta Y/\delta X_p$ at the input quantities is quite often called as coefficient and is denoted as c_p. If all input quantities X_1, X_2 etc. are independent of each other, then their covariance will be zero. Hence (5.15a) and (5.16a) respectively become

$$\sigma_Y^2 = \sigma_{x1}^2(\delta Y/\delta X_1)^2 + \sigma_{x2}^2(\delta Y/\delta X_2)^2. \tag{5.15b}$$

$$\sigma_Y^2 = \sum_{p=1}^{p=n}\sum_{q=p}^{q=p} \sigma_{xpq}^2(\delta Y/\delta X_p)^2. \tag{5.16b}$$

But σ_Y^2 is the square of the standard uncertainty; hence in (5.16b), uncertainty of the output quantity is related to the uncertainties of the input quantities.

5.1.4 Combination of Arithmetic Means

Let there be N samples of $n_1, n_2, n_3, \ldots, n_N$ items (observations) and $\bar{x}_1, \bar{x}_2, \bar{x}_3, \ldots, \bar{x}_N$ be their respective means, then $\bar{x}$ – combined mean of all the samples – is given by

$$\bar{x} = \frac{n_1\bar{x}_1 + n_2\bar{x}_2 + n_3\bar{x}_3 + \cdots + n_N\bar{x}_N}{n_1 + n_2 + n_3 + \cdots + n_N}. \tag{5.17}$$

5.1.5 Combination of Variances

The law applicable to arithmetic means is applicable to variances also. If $s_1^2, s_2^2, s_3^2, \ldots, s_N^2$ be respective variances of $n_1, n_2, n_3, \ldots, n_N$ observations, then their combined variance $\bar{S}^2$ is given by

$$\bar{S}^2 = \frac{n_1 s_1^2 + n_2 s_2^2 + n_3 s_3^2 + \cdots + n_N s_N^2}{n_1 + n_2 + n_3 + \cdots + n_N}. \tag{5.18}$$

5.1.6 Variance of the Mean

We have seen in the earlier chapter that best estimate of the variance of mean $s^2(\bar{x}_p)$ of a quantity observed n times is

$$s^2(\bar{x}_p) = \frac{s^2(x_p)}{n} = \frac{\sum_{p=1}^{p=n}(x_p - \bar{x}_p)^2}{n(n-1)}. \tag{5.19}$$

In order to obtain a reliable estimate of the expectation μ_p of the random variable X_p it is necessary that quite a large number of observations should be taken. This will also ensure that the estimate of variance $s^2(X_p)$ is close enough to $\sigma^2(X_p)$ of the probability distribution of the random variable X_p. The difference between $s^2(X_p)$ and $\sigma^2(X_p)$ should be taken into account when constructing confidence intervals. If the random variable X_p has a normal distribution and $s^2(X_p)$ is taken for relatively smaller sample than Student's t factor should be used to make it close enough to $\sigma^2(x_p)$.

Though $s^2(X_p)$ is more fundamental quantity and easier to manipulate mathematically, the use of its positive square root – the standard deviation – is more frequent as it has the same dimension and unit of measurement that of the variable.

5.2 Uncertainty

The uncertainty of measurement characterizes the dispersion of the values that could reasonably be attributed to the stated value of the output quantity (measurand). In other words, the uncertainty U is the interval within which the conventional true value of the output quantity is likely to lie. For example, if Y is the calculated value of measurand (output quantity) from the data of measured input quantities and U is the uncertainty, then the conventionally true value is likely to lie between $Y - U$ and $Y + U$.

The uncertainty of the output quantity Y consists of uncertainties in measurement of the input quantities. There are two methods of their evaluation, namely Type A evaluation and Type B evaluation. But in either case the quantities calculated are variances (Type A evaluation) or are in nature similar to the variances (Type B evaluation).

5.2.1 Combined Standard Uncertainty

The estimated standard deviation of the estimate y is termed as combined standard uncertainty. It is denoted as $u_c(y)$. The uncertainty $u_c(y)$ is characterized by the positive square root of sum of the squares of the products of standard deviation and its corresponding partial derivative; similar terms are added for dependent input quantities for their covariances. The $u_c(y)$, in this case, is known as standard uncertainty.

For an independent quantity X_P, there may be more than one source of uncertainty, a quantity similar to standard deviation is determined for each source and their squares are added. The square root of the sum is termed as standard uncertainty of X_P and is denoted as $u(x_p)$. Numerically the standard uncertainty is equal to the combined standard deviation of X_P.

5.2.1.1 Measurand (Output Quantity) Is a Function of Single Input Quantity

When the output quantity (measurand) is a function of single input quantity, then combined standard uncertainty is the positive square root of the sum of squares of the uncertainties evaluated by Type A and Type B methods of the single input quantity. The uncertainty of the dependent quantity Y is the product of the value of the partial derivative at x_p and the uncertainty of the input quantity and is expressed as

$$u_c(Y) = (\delta Y/\delta X) \times u(x_p).$$

5.2.1.2 Measurand (Output Quantity) Is a Function of Several Quantities

When the measurand Y is a linear function of a number of input quantities, the combined standard uncertainty of the result of its measurement is equal to the positive square root of the sum of the weighted variances and covariances of these quantities. The weights associated being the partial derivatives of the measurand function with respect to each input quantity (coefficients). It can be expressed as

$$u_c(y)^2 = \sum_{p=1}^{p=n}\sum_{q=p}^{q=n} u_{xpq}^2(\delta Y/\delta X_p)^2 + 2\sum_{p=1}^{p=n}\sum_{q=p+1}^{q=n} \mathrm{Cov}(x_p, x_q)(\delta Y/\delta X_p)(\delta Y/\delta X_q).$$

In the case, when all input quantities are independent of each other (covariances are zero), i.e. second term in the above equation is zero, the standard uncertainty $u_c(y)$ is given as

$$u_c(y)^2 = c_1^2u_{x1}^2 + c_2^2u_{x2}^2 + c_3^2u_3^2 + \cdots + c_n^2u_n^2. \tag{5.20}$$

Here $c_p = (\delta Y/\delta X_p)$ for all values of p from 1 to n.

Taking the uncertainty equal to the value of the standard deviation has the following advantages:

1. No multiplication factor is necessary, which depends upon the confidence level.
2. Different components of standard uncertainty can be simply combined by the quadrature method.
3. Uncertainties calculated either by Type A or Type B evaluation method may be treated in the same fashion.

5.2.2 *Expanded Uncertainty*

The expanded uncertainty of a measurement process is equal to the product of a coverage factor K and the combined standard uncertainty. The value of the factor K depends upon the level of confidence at which one wish to state the result and the effective degrees of freedom. If the level of confidence is 95.45%, then for infinite degrees of freedom (more than 50) K is 2; if the chosen level is 99.7%, then K is 3. Expanded uncertainty U is the range $\pm K \times u_c$ around the stated value in which the true value of a measurand is likely to lie with the stated level of confidence. The magnitude of the factor K, therefore, depends upon the level of confidence at which one wishes to assign the value to the measurand. So U the semi-range of uncertainty is given as

$$U = K \times u_c.$$

If $\bar{y}$ is the measured value, then $\bar{y} - U$ and $\bar{y} + U$ is the interval in which the true value is expected to lie at the stated level of confidence.

In case of expanded uncertainty the factor K should always be clearly stated. Any detailed report of the uncertainty should consist of a complete list of the uncertainty components; the method used to obtain the value of each uncertainty component should also be specified.

5.3 Type A Evaluation of Uncertainty

Type A evaluation of uncertainty is the estimation of variance of the data obtained by direct measurements. Normal statistical method of finding the mean and square of its deviations from the observation is used to estimate the variance. If the number of observed values of the input quantity X_p is n_p, then the degree of freedom is $n_p - 1$. Each input estimate of x_p and its variance is obtained from a distribution of possible values of the input quantity X_p. This probability distribution is frequency based, that is based on a series of observation $x_{p,q}$ of X_p. The Type A evaluation is based on the frequency distribution, which, in most cases, is the normal (Gaussian) distribution. In case of dependent input quantities, covariances are also estimated.

In most cases, the best available estimate or the expected value of a quantity X_p that varies randomly is the arithmetic mean of several corrected observations of this quantity. That is

$$\bar{x}_p = \sum_{q=1}^{q=np} x_{p,q}. \tag{5.21}$$

Here n_p is the number of corrected observations taken under similar conditions.

Individual observation $x_{p,q}$ may differ in value because of random variations in the influence quantities or any other random effect. The variance of observations, which is the best estimate of σ^2 – the variance of the probability distribution of quantity X_p, is obtained from

$$s^2(x_p) = \frac{1}{n_p - 1} \sum_{q=1}^{q=np} (x_{p,q} - \bar{x}_p)^2. \tag{5.22}$$

The positive square root of $s^2(x_p)$ is the standard uncertainty $u_A(x_p)$ determined by Type A evaluation method of uncertainty. The symbol u_A stands for uncertainty arrived at by Type A evaluation.

Table 5.1 Observations and sum of squares of deviations

Serial No.	Mass value in g	$(x_p - \bar{x}_p)$ µg	$(x_p - \bar{x}_p)^2$ (µg)^2
1	1000.000068	−4	16
2	1000.000083	11	121
3	1000.000079	7	49
4	1000.000064	−8	64
5	1000.000063	−9	81
6	1000.000094	22	484
7	1000.000060	−12	144
8	1000.000068	−4	16
9	1000.000076	4	16
10	1000.000065	−7	49
Sum	10000.000720	0	1,040
Mean	1000.000072		

5.3.1 Numerical Example for Calculation of Type A Evaluation of Standard Uncertainty

Let us consider a case of measurement of mass of a 1 kg mass standard. Observations after applying buoyancy and other corrections are given in Table 5.1

Different parameters are as follows:

1. Mean mass value of the kg $= \bar{x}_p = \sum_{q=1}^{q=10} x_{p,q}/10 = 1000.000072\,\text{g}$

Best estimate of the mass of the kilogram is 1000.000072 g

2. Variance $s(x_p)^2 = \frac{(x_p-\bar{x}_p)^2}{n_p-1}$
3. Degrees of freedom$= n_p - 1 = 10 - 1 = 9$
4. Standard deviation $s(x_p) = \sqrt{(x_p - \bar{x}_p)^2/(n_p - 1)} = \sqrt{1040/9} = 10.75\,\mu\text{g}$
5. Standard deviation of the mean

$$s(\bar{x}_p) = s(x_p)/\sqrt{n} = s(x_p)/\sqrt{10}$$
$$s(\bar{x}_p) = 3.4\,\mu\text{g} \tag{5.23}$$

The information about degrees of freedom is required to estimate the extended uncertainty of the mean with given confidence level for a single input quantity and also for the estimation of effective degrees of freedom ν_{eff} for several input quantities.

Here we see that to have a reasonable estimate of Type A uncertainty, a good number of observations need to be taken. In a calibrating laboratory to take such a large number of observations at one calibration point is not feasible. Normally fewer observations say 3 or maximum 5 are taken. In such case the uncertainty u_{A} is full of statistical fluctuations. The author [2] suggested a method to smooth out these statistical fluctuations. The method is of pooling a larger number of variances of the instrument or of the procedure of measurement. The method is described in the following section.

5.4 Pooled Variance

Pooled variance is the arithmetic mean of a number of variances. Due consideration is being given to their respective degrees of freedom.

If $s_1^2, s_2^2, s_3^2, \ldots, s_k^2$ are estimates of variances and $n_1, n_2, n_3, \ldots, n_k$ their respective degrees of freedom, then pooled variance S^2 is given as

$$S^2 = \frac{n_1 s_1^2 + n_2 s_2^2 + \cdots + n_k s_k^2}{n_1 + n_2 + n_3 + \cdots + n_k}, \tag{5.24}$$

giving

$$\text{Standard uncertainty } u_{\mathrm{A}} = S. \tag{5.25}$$

5.4.1 Validity

The method of pooled variance in relation to instruments is valid when the value of the input supplied to the instrument remains constant for each repetition and variation is mainly due to the instrument in use.

5.4.2 Applicable

The method is applicable to cases of calibration of a very large number of measuring instruments, e.g. calibration of a

- Weighing instrument against standard weights
- Linear dial gauge with the help of slip gauges
- Proving ring through dead weights
- The method may be made applicable to calibration of electrical and other instruments, when the constancy of standard input like current or voltage is better by one order of magnitude

In all these cases, observations mainly vary only due to the instruments under use and there is negligible variation of the standard input quantity.

5.4.3 Uses

- Type A evaluation of standard uncertainty in calibration of instruments
- Testing the authenticity of a fresh variance or u_A

- Maintenance of measuring instruments
- Fixing maximum permissible error of measuring instruments for regulatory purposes
- Rejection of an instrument under calibration

5.4.4 Need

5.4.4.1 Calibration of Measuring Instruments

Most of the instruments are calibrated by observing the reading of the instrument and knowing the value of the standard input supplied to it. Ammeters, voltmeters, dial pressure gauges, linear dial gauges and weighing machines are but a few examples.

Such instruments are calibrated at several points. At each point, in addition to, the correction to be applied, both Type A and Type B methods are used to evaluate uncertainties. For the purpose of calculating standard deviation with lesser statistical fluctuations, a larger number of repetitions should be taken at each point under calibration. However, time and resource constraints do not allow to take fairly good number of repetitions at each point. So the value of standard uncertainty u_{A} calculated by using Type A method will be full of statistical fluctuations and thus may create some confusion.

Firstly, the standard deviation of 3 or 5 observations has little meaning, and secondly, it is more likely that there are apparently wide variations in its values from point to point of the same instrument. So the uncertainty u_{A} reported is more likely to be different not only for two calibrated instruments of the same type for the same user, but may also be different at various points at which the instrument has been calibrated. This will naturally confuse the mind of the user and may also raise an alarm in his mind, especially when he will see several different values of uncertainty in the certificate of one instrument. The method of pooled variance may be used to solve this problem. It is, therefore, suggested that the standard deviation at each point of calibration is replaced by the square root of the pooled variance.

5.4.5 Calculation of Pooled Variance

All instruments for the purpose of calculation of pooled variances may be divided into two categories.

5.4.5.1 Category I: The Variance Is Independent of the Input Quantity

All instruments having a linear relation between the input quantities and reading on its scale fall in this category. In this case, variances at different points of the scale may be pooled together.

Let an instrument be calibrated at n number of points, and at each point, m number of observations have been taken. Mean of m observations at the qth point is R_q. Let the pth observation at the qth point be denoted by r_{qp}. Then

$$R_q = \left[r_{q1} + r_{q2} + r_{q3} + \cdots + r_{qm}\right]/m = \frac{\sum\limits_{p=1}^{p=m} r_{qp}}{m}. \tag{5.26}$$

If the variation in the indications of the instrument is independent of the value of the indication, then to obtain pooled variance S^2, the sum of the squares from the mean value of each indication is added and the sum thus obtained is divided by $n(m-1)$. The same is explained in the following paragraph:

Point of the scale	Observations	Mean	Sum of squares from the mean
L_1	$r_{11}, r_{12}, r_{13}, \ldots, r_{1m}$	R_1	$\sum(r_{1p} - R_1)^2$
L_2	$r_{21}, r_{22}, r_{23}, \ldots, r_{2m}$	R_2	$\sum(r_{2p} - R_2)^2$
L_3	$r_{31}, r_{32}, r_{33}, \ldots, r_{3m}$	R_3	$\sum(r_{3p} - R_3)^2$
...	...	...	...
...	...	...	...
...	...	...	...
L_n	$r_{n1}, r_{n2}, r_{n3}, \ldots, r_{nm}$	R_n	$\sum(r_{qp} - R_n)^2$

(5.27)

The pooled variance is

$$S^2 = \frac{\sum\limits_{q=1}^{q=n}\sum\limits_{p=1}^{p=m}(r_{qp} - R_q)}{n(m-1)}. \tag{5.28}$$

5.4.5.2 Category II: The Variance Depends Upon Input Quantity

All instruments having non-linear scales will fall into this category. In this case, variances for the same input quantity only can be pooled together. So the pooling is to be done on different instruments of same kind for each input quantity, thus requiring larger effort and number of instruments before u_{A} is obtained.

In this case variances for similar kinds of instruments are pooled together. The observations are taken for the same magnitude of the input quantity.

If s_q^2 were the variance of n_q number of observations, then pooled variance S^2 for N such sets for a particular point of similar instruments is given by

$$S^2 = \frac{\sum\limits_{q=1}^{q=N}(n_q - 1)s_q^2}{\sum\limits_{q=1}^{q=N}(n_q - 1)}. \tag{5.29}$$

5.4.6 *Uses of Pooled Variance*

5.4.6.1 Estimation of Type A Uncertainty of an Instrument

In the following paragraphs, an example has been cited for justification and calculation of pooled variance from variances taken at different points of calibration of same instrument and also to pool it for several instruments of same kind. The square root of the pooled variance will then be taken as the standard uncertainty from Type A evaluation at all points for the instrument under calibration.

Calibration data of five proving rings have been analyzed. Three proving rings designated as A, B and C are of the same type, each having a dial gauge as an indicating device, while each of the other two proving rings, designated as D and E, is fitted with a six-digit electronic indicating counter, thus having a much better readability in comparison to that of proving rings designated as A, B and C.

Each proving ring was tested at ten points and observations for each point were repeated three times. The variances of three observations at each of the ten points of proving rings are given in Table 5.2.

Abbreviations used in Table 5.3 are as follows:

Mean of the variances = s_{mean}
Maximum variance = s_{max}
Minimum variance = s_{min}

Mean value of the variance s_{mean}, the values of the ratio between the maximum and minimum variances s_{max}/s_{min} and the ratio of the maximum variance with the mean variance s_{max}/s_{mean} have been respectively tabulated in the 1st, 2nd and 3rd

Table 5.2 Variances of five proving rings

S. No.	A	B	C	D	E
1	0.0067	0.0267	0.0600	4.6667	20.6667
2	0.0800	0.0200	0.0867	2.0000	44.6667
3	0.0600	0.0000	0.0867	4.6667	98.0000
4	0.2600	0.1400	0.1400	4.6667	186.000
5	0.4867	0.2067	0.0267	4.6667	254.000
6	0.0600	0.0600	0.1800	56.0000	416.666
7	0.4067	0.0600	0.0867	14.0000	542.000
8	0.3467	0.0467	0.0267	98.6667	772.667
9	0.1800	0.0200	0.0067	42.0000	872.000
10	0.2867	0.3267	0.0200	28.6667	1148.000

Table 5.3 Mean variances s_{mean} and ratios s_{max}/s_{min} and s_{max}/s_{mean}

Proving ring	A	B	C	D	E
s_{mean}	0.2174	0.0906	0.0720	26.000	435.534
s_{max}/s_{min}	72.642	16.335	26.8656	49.3333	55.5807
s_{max}/s_{mean}	2.2387	3.6059	2.5000	3.7948	2.6358

rows of the Table 5.3. Greater variance has always been taken as numerator so that all ratios become suitable for Fisher's F test [3].

Fisher's F test [3] has been applied to test the hypothesis that variances under test belong to the same population. The degrees of freedom for variances at each of ten points are only 2, while degrees of freedom for the mean variance are 20.

The values of variances show a wide range though the application of Fisher's F test indicates that none of the variances in any of the five proving rings is outside the limit of 1% level of significance. Limiting value of F at 1% level of significance for 2 degrees of freedom (Table A.8) for each variance is 99.0. Further Limiting value of F at 5% level of significance for 2 degrees of freedom (Table A.7) is 19.0.

Comparing the ratios of maximum and the mean values of variances in the set indicates that none of the maximum variance is outside the limits even at 5% level of significance for rings A, C and E and at 1% level for rings B and D, as the value of F for 2 and 20 degrees of freedom is 3.49 at 5% and 5.82 at 1%. The ratios of mean variances for A, B and C proving rings show that all mean variances belong to the same population at 1% level of significance.

It may be noticed that ratio of mean variances of proving rings E and D is 16.75, which is much larger than the F value of 2.94 for 2/20 degrees of freedom. Hence the two mean variances cannot be said to belong to the same population even at 1% level of significance.

As the data analyzed have only two degrees of freedom and the last digit in the data is only an eye estimate, a stray case of wide variation may be neglected.

From the above discussions, it may be safely concluded that

To a layman, the variances are apparently different, giving different uncertainties at different points of the same proving ring, thus confusing the user.

To a statistician all variances, in the set, may be pooled and its square root may be taken as the Type A standard uncertainty.

It is also reasonable to take the average of the variances for similar proving rings to form a pooled variance with larger degrees of freedom. When such a data are accumulated for a year or two, a reliable value of the pooled variance may be established. Square root of this pooled variance may thus be taken as Type A standard uncertainty for future use.

Similar process when used for many similar instruments of category II gives a reliable value of pooled variance.

The pooled variance technique is used for a measuring instrument, like balances. In addition of sensitivity figure of the balance, a pooled variance and standard deviation should periodically be calculated and a record should be maintained. Similar exercise should be carried out of all measuring instruments and records of the moving average of its variance should be maintained.

5.4.6.2 Testing the Authenticity of Observations

Standard deviation (S), the square root of pooled variance, may also be used for testing the authenticity of the observations taken in future on a similar instrument.

A criterion may be formulated that the difference between any two observations should not be greater than the standard deviation (S) or its multiple.

For example, assuming the normal distribution for all observations with S as its standard deviation, then no difference between any two observations should be greater than twice the value of 1.96 times the value of S(95% level of confidence). That is semi-range of variations in observations is $\pm 1.96\,S$.

The idea of pooled variance is being used, for routine calibration of weights in Mass Standards activity of National Physical Laboratory, New Delhi, India, for the past so many years.

In a precision balance for a given specific range, the variation in the observations is almost independent of denomination of weights. The standard deviation S from the pooled variance of balance for weights of specific denominations is found out and used as permissible limits for the two mass values obtained for weight of same denomination by two observers under similar environmental conditions. If the difference between the two values is less than $2\,S$, the mean of the two values is taken and reported as the mass value of the weight piece with S as one component of standard uncertainty evaluated by Type A method. Otherwise the mass value of the weight is re-determined. This is a stricter criterion than suggested in the above paragraph (level of confidence is 66%).

5.4.6.3 Maintenance of Laboratory Instruments

For ensuring the good working of a measuring instrument, not only its calibration but also its repeatability should also remain unaltered or within a specified range. The standard deviation square root of variance is a good measure of repeatability of an instrument. Hence continuous monitoring of its variance is to be carried out. Variances, preferably of same number of observations, for ease of calculations, at prescribed regular intervals of time, are found out. The progressive mean of the variances, i.e. the pooled variance, is calculated. For a good working instrument Fisher's F test is applied. The ratio of the new variance with its pooled variance should not exceed the value of F tabulated for appropriate degrees of freedom at a prescribed level of significance (say 5%). Degree of freedom of new variance would be one less than the number of observations for which the variance has been calculated; i.e. for pooled variance degree of freedom is

$$\sum_{q=1}^{q=N} (n_q - 1).$$

Further a graph of all previous values of standard deviations should also be drawn if it shows an upward trend, and if the new value exceeds a certain specified value then instrument may either be discarded or downgraded for less precise work.

5.4.6.4 Fixing Maximum Permissible Error of an Instrument

Having known the pooled variance of an instrument, its maximum permissible error may be fixed. For an instrument, the maximum permissible error should in no case be lower than its standard deviation. An instrument having closer maximum permissible error than its repeatability is of no practical value. Say for example, twice the value of the standard deviation may be a reasonable value of the maximum permissible error of the instrument and vice versa. That is standard deviation of any instrument should not be more than half of its maximum permissible error.

If the pooled variance is proportional to the input value then maximum permissible limits may be in the same proportion. This is the one reason as to why maximum permissible errors are given in percentage rather than in absolute units.

5.4.6.5 Rejection of an Instrument Received for Calibration

If the pooled variance of certain type of instruments is already known, then the particular instrument received for calibration may be rejected if its sample variance is more than the prescribed limits. For the purpose one may think of Fisher's test again; say for example the ratio of the sample variance to the pooled variance of the group of instruments may not exceed the tabulated value for the known degrees of freedom at 5% level of significance. Attention is drawn to the ratio of the mean variances of proving rings D and E, which are 5:1. Had there been already some fixed criterion existed it should have been possible to reject the proving ring on the basis of its variance. For example, according to criterion proposed above, the proving ring E could be rejected.

5.4.7 Concluding Remarks

The concept of pooled variance is relatively new to the scientists engaged in calibration work. So its potential uses have not been fully exploited. It is hoped that the above paragraphs should serve as a step to realize the importance of this concept and full exploitation of its applications. Use of pooled variance for fixing maximum permissible error (MPE) and rejection of an instrument must also help in preparing quality manuals. National Laboratories, custodians of National standards of measurements, should play an active role in this area. As, usually, such laboratories have the capability and resources to generate sufficient data in terms of pooled variances for common measuring instruments and to discuss and finalize the criteria for (a) maintenance and (b) fixing maximum permissible errors of measuring instruments and (c) rejection of measuring instruments received for calibration.

5.5 Type B Evaluation of Standard Uncertainty

5.5.1 Type B Evaluation of Uncertainty

The Type B evaluation of uncertainty involves estimating the quantity u_j^2, which may be taken as an approximation to the corresponding variance, the existence of which is assumed. The quantity u_j^2 is treated like variance and the quantity u_j like standard deviation, and where appropriate the covariances should be treated in a similar way.

In case the probability distribution of x_p is subjective on a predetermined distribution, a quantity of the nature of variance is determined from the knowledge of its distribution. Hence Type B evaluations are based on the predetermined distributions as well.

It may be emphasized that in both cases the distributions are models that are used to represent the state of our knowledge. Uncertainties due to any input quantity X_p should be evaluated by both Type A and Type B evaluation methods.

Prior to BIPM directive, in 1980s, uncertainty consisted of two components: one used to come from random errors and the other due to systematic errors. The two components of uncertainty used to be named as random uncertainty and systematic uncertainty. There is not always a simple correspondence between the classification into Types "A" and "B" and the previously used classification into "random" and "systematic" uncertainties. The term "systematic uncertainty" can be misleading and should be avoided.

In case the measurand – the output quantity – is a function of several input quantities, all variances and covariance should be combined by quadrature method. Variances obtained by Type A evaluation method are not distinguished from those obtained by Type B evaluations. So all variances and covariances should, therefore, be treated as variances or covariances in strict statistical sense.

As mentioned earlier, for those uncertainties, whose estimates have not been obtained by independent repeated observations, Type "B" evaluation method is used. This type of uncertainty is calculated by judgement using all relevant information on the variability of the uncertainty.

For example:

- Previous measurement data
- Experience and general knowledge of the behaviour and properties of relevant materials and instruments
- Manufacturer's specification
- Data provided in the calibration and other certificates
- Uncertainty assigned to reference data taken from handbooks

Uncertainties may creep in a measurement process due to the use of:

- Standards
- Measuring instruments

- Inherent characteristic of the instrument under calibration
- Various physical constants
- Values of physical properties of the material used in standards and measuring instruments
- Operating conditions

5.5.2 Common Uncertainties Evaluated by Type B

1. Uncertainty as reported in the calibration certificates of the standard or the instrument used.
2. Uncertainty due to interpolation between the calibration points of the standard used in the measurement.
3. Uncertainty due to the change in environmental conditions, such as temperature, pressure and relative humidity of air.
4. Uncertainty due to ability to reset, repeatability and threshold discrimination of the instruments used.
5. Uncertainty due to the value taken for some physical constants, or properties of the materials used in the process of measurement, such as values of density of water, acceleration due to gravity and expansion coefficients.
6. Uncertainty in applied corrections based on measurements or the data obtained from standard handbooks.

5.6 Variance and Uncertainty Range

5.6.1 Normal Distribution

Type B uncertainty is also of the nature of standard deviation of the estimated value of x_p. For the estimation of standard deviation from the given range of uncertainty (U), the range U is to be divided or multiplied by a certain factor, whose value will depend upon the confidence level at which the result was stated. For example if the result is stated at a confidence level of 95%, then the range U is to be divided by 1.96 for infinite number of degree of freedom or by the "Student's 't' factor" for the given degrees of freedom.

In older literature, uncertainty figures are given at the confidence level of 50%; in that case one has to multiply it by a factor of 1.48 to obtain the standard deviation.

5.6.2 Rectangular Distribution

If a result has been indicated with a range of a_+ and a_t and it is assumed that it is equally likely for the estimated value to lie anywhere within the given range, i.e. the result has a rectangular probability distribution, then corresponding variance $u^2(x_p)$ will be given as

$$u^2(x_p) = (a_+ - a_-)^2 /12, \tag{5.30}$$

giving standard deviation $u(x_p)$ as

$$u(x_p) = a/\sqrt{3}. \tag{5.31}$$

Here it is assumed that

$$a_+ = a \text{ and } a_- = -a,$$

thus giving

$$(a_+ - a_-) = 2a.$$

5.6.3 *Triangular Distribution*

In many cases, it is more realistic to expect that the chance of reported value lying near the bounds is less than that lying near the midpoints of the range. The probability of occurrence of the result at the extreme boundary points is zero and increases linearly and becomes a maximum at the midpoint of the range. That is the reported value under consideration follows a triangular distribution.

Then $u(x_p)$ is given as

$$u(x_p) = a/\sqrt{6}. \tag{5.32}$$

5.6.4 *Trapezoidal Distribution*

It is more reasonable to assume that the result under consideration has a maximum probability within a range of $\pm a\beta$ about the midpoint of the range and decreases linearly to zero at its ends. That is the input quantity has a symmetric trapezoidal probability distribution having equal sloping sides with a base of width of $2a$ and top of width $2a\beta$. Here β is a proper fraction and may take any value between 0 and 1. When β is 0, the probability distribution becomes triangular distribution and it becomes rectangular distribution for $\beta = 1$. The standard deviation $u(x_p)$, in this case, is given as

$$u(x_p) = a\sqrt{\{(1 + \beta^2)/6\}}. \tag{5.33}$$

Rectangular distribution should be used only when no data are available. Otherwise, logically it will be better to use triangular distribution. Firstly this is similar to normal distribution and secondly more logical. While stating the range of uncertainty, the measurements are carried out, which follow normal distribution. When extended uncertainty is stated, a multiplying factor to standard deviation is given. So when using the uncertainty range to calculate back the standard deviation, one should assume that the reported result is following normal distribution unless contrary is stated or otherwise evident. Type B evaluation of uncertainty should be carried out keeping in view of the hierarchy of standards and laboratories. In practice, for example mass measurement, National prototype kilograms are calibrated by International Bureau of Weights and measure. The calibration certificate contains, besides other data, the mass value with semi-range of uncertainty normally equal to 2 times the standard deviation. So for a national metrology laboratory (NPL in the case of India), the standard deviation should be semi-range divided by 2. It is not justified to assume that reported result is following the rectangular distribution and obtaining standard deviation by dividing semi-range by the square root of 3. All other laboratories should follow a similar method for Type B evaluation of uncertainty. It is emphasized that this method should be used only for applying the mass value of standard and uncertainty associated with the mass value.

In some cases, standards and measuring instruments are calibrated, but no specific value of the standard input versus scale observations with the corresponding uncertainty or correction to the specific points of the scale of the instrument is given. The calibration only ensures that the instrument will perform within certain specified limits. In that case, rectangular distributions or its modified versions may be used for Type B evaluation of uncertainty.

The proper use of available information for Type B evaluation of standard uncertainty of measurement needs greater insight based on experience and general knowledge. It is the skill that can be learned with practical experience and deep study of the mathematical statistics. A well-based Type B evaluation of standard uncertainty can be as reliable as Type A evaluation of standard uncertainty. Type B evaluation of uncertainty assumes greater importance in those cases where direct observed measurement data are small for Type A evaluation of uncertainty. There are many valid reasons for not able to take larger number of observations.

When only single value is known of the quantity X_p, for example a single measured value, a resultant value of the previous measurement, a reference value from the literature or a correction value, this value will be used as x_p. The standard uncertainty $u(x_p)$ associated with x_p is adopted where it is given. Otherwise it has to be calculated from unequivocal uncertainty data. If data of this kind are not available, the uncertainty has to be evaluated on the basis of the experience taken as it may have been stated (often in terms of an interval corresponding to expanded uncertainty).

When probability distribution can be assumed for the quantity X_p, based on theory or experience, then the appropriate expectation or expected value (mean value) and the standard deviation of this distribution have to be taken as the estimate of x_p and the associated standard uncertainty $u(x_p)$, respectively.

References

1. BIPM, IEC, IFCC, ISO, IUPAC, IUPAP and OIML, *Guide to the Expression of Uncertainty in Measurement* (GUM) (ISO, Switzerland, 1993)
2. S.V. Gupta, Pooled variance and its applications with specific reference to Type A Uncertainty in the calibration of measuring instruments. OIML Bull. **XXXVIII**(3), 31–35 (1997)
3. W.C. Guenther, *Analysis of Variance* (Prentice Hall, Englewood Cliffs, NJ, 1964)

Chapter 6
Uncertainty and Calibration of Instruments

6.1 Introduction

Quite often measuring instruments are received for calibration. Scale of the measuring instrument is calibrated at few points only. The correction or the value of standard input is assigned at those selected points of its scale and uncertainty of measurement is also stated at those points only. In most cases, the values of the standard input versus scale readings are given at the selected points. When an instrument is used in the field, the scale readings are recorded, which in general, may not be the same points at which the instrument was calibrated. The correct value is obtained from the corrections at the two nearest calibrated points just by linear manipulation. In this method only small interval containing the observed reading is considered which may not be always justified. However, it is advisable to consider all the points at which calibration is carried out. It is, therefore, necessary that a mathematical relation between the scale reading and standard input is given. So that the user can substitute the value of observed scale reading in the relation and get the value of the input to the instrument. For example an ammeter with range of 100 A and with 100 divisions on the scale is calibrated normally at four points, say at 25 A, 50 A, 75 A, and 100 A graduation marks, but in practice the instrument may read 60 A; then naturally the user would like to know as to what will be the real value of the current passing through it, when the instrument is reading 60 A. This chapter is mainly based on my research paper [1] published in MAPAN – Journal of Metrology Society of India, in 1999.

Another set of instruments are transducer type, which have an arbitrary scale, which is driven by one quantity but depicts a totally different quantity, for example electronic weighing instruments, in which indication will depend upon the electric current through the circuit, but the scale will depict the weight of the body. Proving rings are used to measure force. Force induces linear changes in the diameter of the proving ring, which is measured, but the instrument associated with it indicates force. Same is the case of voltmeter and other electrical measuring instruments etc.

S.V. Gupta, *Measurement Uncertainties*, DOI 10.1007/978-3-642-20989-5_6,

In calibration of hydrometers, the author has observed that, sometimes, the correction at the top of the scale was very large and reduced to almost zero at the bottom of the scale. Such results indicate that the length of the scale is too small or too large. In this case there will be a linear relation between correction and indication of the hydrometer. If the gradient of the linear relation is very small, it suggests that corrections are independent of the scale reading. Similarly the correction assigned to a mercury-in glass thermometer will be a quadratic function of the indication on the thermometer, if diameter of the capillary is uniformly changing.

Normally the number of points at which the instrument is calibrated is very much less than the total number of graduations on it. For finding out the value of input at other points we may like to have some sort of algebraic relation, so that by choosing any numerical value of the independent variable (indication on the scale) the value of the input is obtained. This will enable us to calculate the value of input (dependent variable) for any chosen value of the independent variable – graduation on the scale.

Firstly, we may like to find out if there exists a relation between the corrected input quantities and scale readings. Then we try to establish a graph or an equation of a graph, or an empirical relation between the known inputs and readings taken on the scale of the instrument is required to be given.

Such relations may be a polynomial including a linear relation, a power function, or an exponential function in one variable. A mathematical function becomes specific relation if the values of the constants involved in defining the function are given.

To specify the mathematical relation we use the least square method which gives the best estimates of the constants involved in defining the function. We will find the standard deviation by taking the square root of the average of the squares of the residual errors. This standard deviation is used in calculating the uncertainty of the estimated input for the given scale reading.

Usually a mathematical relation including the values of constants is given after the calibration of an instrument. But in this chapter, we wish to go one step further. In addition of the values of constants involved in the function, we find out the uncertainty in assigning the value of the dependent variable (input quantity), by choosing from the function, any value of the independent variable (indication on the scale of the instrument).

First simple case is that in which scale reading indicated by x bears a linear relationship with the standard input quantity.

6.2 Linear Relation

Simplest function is a linear relation. Let there be n pairs of values of $(x_1, y_1), (x_2, y_2), (x_3, y_3) \dots (x_n, y_n)$, which are to be fitted in a following linear relation expressed as

$$y = c_1 + c_2 x. \tag{6.1}$$

Here c_1 and c_2 are the constants. These are determined from n linear relations obtained by the method of least squares. Considering that (6.1) represents a straight line then c_1 is the intercept on y axis and c_2 is slope of the line.

A very small value of c_2 will indicate independence of y on x. For example corrections at various points of the scale of an unbiased instrument are independent of the magnitude of x.

The values of c_1 and c_2, and their uncertainties are determined by the method of least squares. The method consists of forming linear relation by substituting n pairs of values of (x, y) and finding the best estimates of c_1 and c_2 by minimizing the sum of squares of residual errors E.

$$E^2 = \sum_{p=1}^{p=n} \{y_p - (c_1 + c_2 x_p)\}^2. \tag{6.2}$$

The variables to be adjusted are c_1 and c_2. E is to be minimized, so $\delta E/\delta c_1$ and $\delta E/\delta c_2$ each must be zero.

That is the conditions are as follows:

$$2E\delta E/\delta_{c1} = 0$$

gives us

$$\sum_{p=1}^{p=n} 2\{y_p - (c_1 + c_2 x)\} = 0 \Rightarrow \sum_{p=1}^{p=n} y_p = nc_1 + c_2 \sum_{p=1}^{p=n} x_p \tag{6.3}$$

and $2E\delta E/\delta c_2 = 0$ gives

$$\sum_{p=1}^{p=n} 2\{y_p - (c_1 + c_2 x_p)\} x_p = 0 \Rightarrow \sum_{p=1}^{p=n} x_p y_p = c_1 \sum_{p=1}^{p=n} x_p + c_2 \sum_{p=1}^{p=n} x_p^2. \tag{6.4}$$

Equations (6.3) and (6.4) are two normal equations. These equations when solved for c_1 and c_2 give best estimates of c_1 and c_2. There are two methods of solving them: the classical method and the matrix method.

6.2.1 *The Classical Method*

In the classical method first c_2 is obtained by eliminating c_1 from (6.3) and (6.4). Solution of (6.3) and (6.4) gives best estimates of c_1 and c_2. As their values depend upon the set of n pairs of observed values, these must have some variance and covariance.

Solving (6.3), we get

$$c_1 = [\Sigma x^2 \Sigma y - \Sigma x \Sigma xy]/[n\Sigma x^2 - (\Sigma x)^2], \tag{6.5}$$

$$c_2 = [n\Sigma xy - \Sigma x \Sigma y] / [n\Sigma x^2 - (\Sigma x)^2]. \tag{6.6}$$

6.2.1.1 Variances of c_1 and c_2

In all above derivations x is taken as independent variable that means these are arbitrarily taken values and hence have no uncertainty. The uncertainty is only in the dependent variable y. Hence while calculating variance of c_1 consider component uncertainty due to pth value of the variable y and use in the following relations:

$$V(c_1) = \sigma_y^2 \sum_{p=1}^{p=n} \left\{ \frac{\delta c_1}{\delta y_p} \right\}^2,$$

$$V(c_2) = \sigma_y^2 \sum_{p=1}^{p=2} \left\{ \frac{\delta c_2}{\delta y_p} \right\}^2.$$

Here σ_y^2 is the minimum value of sum of squares of residual errors divided by the number of degrees of freedom. This is obtained by substituting the calculated values of c_1, c_2, and values of n pairs of (x_p, y_p) and corresponding values of y_p in (6.1) and is given as

$$\sigma_y^2 = \Sigma \left[y_p - \{c_1 + c_2 x_p\}^2 \right] / (n-2).$$

Writing

$$D = n\Sigma x^2 - (\Sigma x)^2,$$

we get

$$c_1 = [\Sigma x^2 \Sigma y - \Sigma x \Sigma xy]/D.$$

Giving $\delta c_1/\delta y_p$ the contribution of uncertainty due to pth value of y as

$$\delta c_1/\delta y_p = \frac{\sum x^2 - x_p \sum x}{D},$$

variance of c_1 is the sum of the squares of these deviations multiplied by σ_y^2, giving us

$$V(c_1) = \sigma_y^2 \frac{\sum_{p=1}^{p=n} \left\{ \left(\sum x^2\right)^2 - 2x_p \sum x \sum x^2 + \left(x_p^2\right) \left(\sum x\right)^2 \right\}}{D^2}$$

$$= \sigma_y^2 \frac{n\left(\sum x^2\right)^2 - 2\sum\limits_{p=1}^{p=n} x_p \sum x \sum x^2 + \sum\limits_{p=1}^{p=n} x_p^2 \left(\sum x\right)^2}{D^2}$$

$$= \sigma_y^2 \frac{\sum x^2 \left(n\sum x^2 - \left(\sum x\right)^2\right)}{D^2} = \sigma_y^2 \frac{\left(\sum x^2\right) D}{D^2}$$

and

$$V(c_1) = \sigma_y^2 \frac{\sum x^2}{n\sum x^2 - \left(\sum x\right)^2}. \tag{6.7}$$

Similarly differentiating partially (6.6) with respect to y_p, we get $\frac{\delta c_2}{\delta y_p}$ as

$$\frac{\delta c_2}{\delta y_p} = \frac{n x_p - \sum x}{D},$$

$$\begin{aligned} V(c_2) &= \sigma_y^2 \frac{\sum\limits_{p=1}^{p=n} \left[n x_p - \sum x\right]^2}{D^2} = \sigma_y^2 \frac{\sum\limits_{p=1}^{p=n} \left[n^2 x_p^2 - 2n x_p \sum x + \left(\sum x\right)^2\right]}{D^2} \\ &= \sigma_y^2 \frac{n^2 \sum x_p^2 - 2n \sum\limits_{p=1}^{p=n} x_p \sum x + n\left(\sum x^2\right)^2}{\left\{n\sum x^2 - \left(\sum x\right)^2\right\}^2} \\ &= \sigma_y^2 \frac{n\left(n\sum x_p^2 - \sum x^2\right)}{\left\{n\sum x^2 - \left(\sum x\right)^2\right\}^2}. \end{aligned}$$

Hence variance of c_2–$V(c_2)$ is given as

$$V(c_2) = n \frac{\sigma_y^2}{n\sum x^2 - \left(\sum x\right)^2}. \tag{6.8}$$

6.2.2 *Matrix Method*

The second approach is to use matrix method, which simultaneously gives the solution as well as the variance and covariance of c_1 and c_2.

For brevity Σ stands for $\sum\limits_{p=1}^{p=n}$. The solution of (6.3) and (6.4) in matrix form can be written as

$$\begin{vmatrix} c_1 \\ c_2 \end{vmatrix} = \frac{1}{D}\begin{vmatrix} \Sigma x^2 & -\Sigma x \\ -\Sigma x & n \end{vmatrix}\begin{vmatrix} \Sigma y \\ \Sigma xy \end{vmatrix}. \tag{6.9}$$

The matrix

$$\frac{1}{D}\begin{vmatrix} \Sigma x^2 & -\Sigma x \\ -\Sigma x & n \end{vmatrix} \tag{6.10}$$

is known as variance covariance matrix of c_1 and c_2.D is the determinant of the matrix and is given as

$$D = n\Sigma x^2 - (\Sigma x)^2. \tag{6.11}$$

From (6.9), the values of c_1 and c_2 are obtained from (6.9) and expressed as

$$c_1 = [\Sigma x^2 \Sigma y - \Sigma x \Sigma xy]/[n\Sigma x^2 - (\Sigma x)^2], \tag{6.12}$$

$$c_2 = [n\Sigma xy - \Sigma x \Sigma y]/[n\Sigma x^2 - (\Sigma x)^2]. \tag{6.13}$$

Comparing (6.5) and (6.6) with (6.12) and (6.13), we see that values of c_1 and c_2 obtained by either method are the same.

Further from variance covariance matrix from (6.10), we get

$$\sum_{p=1}^{p=n} x_p^2/D \text{is the variance factor of } c_1.$$

n/D is the variance factor of c_2 and

$$-\sum_{p=1}^{p=n} x_p/D \text{ is the covariance of } c_1 \text{ and } c_2.$$

variance of c_1 *i.e.* $V(c_1)$ is given as

$$V(c_1) = \left[\sigma_y^2\, \Sigma x^2\right]/\left[n\Sigma x^2 - (\Sigma x)^2\right]. \tag{6.14}$$

These are same as obtained from first principle in Sect. 6.2.1.

Variance of $c_2 V(c_2)$ is given as

$$V(c_2) = \left[n\sigma_y^2/[n\Sigma x^2 - (\Sigma x)^2]\right]. \tag{6.15}$$

and covariance of c_1, c_2 Cov (c_1, c_2) is given as

$$\text{Cov}\,(c_1, c_2) = -\sigma_y^2\, \Sigma x/[n\Sigma x^2 - (\Sigma x)^2]. \tag{6.16}$$

Correlation coefficient between c_1 and c_2 is given as

$$r(c_1, c_2) = \text{Cov}\,(c_1, c_2)/\left\{[V(c_1) \cdot V(c_2)]^{1/2}\right\}$$
$$= -\Sigma x/\{[n\Sigma x^2]\}. \tag{6.17}$$

Hence from the above discussions, it is evident that the two methods are identical.

So the calibration laboratory should not only give the values of the best estimate of c_1 and c_2 but their variances and covariances.

In the following paragraph, we discuss the method of calculating the value of uncertainty in the dependent variable "y".

6.3 Uncertainty

The uncertainty $U(y)$ in the estimation of y due to uncertainties associated with c_1 and c_2 for a given value of x from the linear equation

$$y = c_1 + c_2 x \tag{6.1}$$

is given by

$$U^2(y) = [\delta y/\delta c_1 \times U(c_1)]^2 + [\delta y/\delta c_2 \times U(c_2)]^2 + 2[\delta y/\delta c_1 \times \delta y/\delta c_2 \times U(c_1) \times U(c_2) \times r(c_1, c_2)]. \tag{6.18}$$

From (6.1)

$$\delta y/\delta c_1 = 1 \text{ and } \delta y/\delta c_2 = x.$$

Taking uncertainty equal to standard deviation (square root of its variance) and substituting the values of $\delta y/\delta c_1$ and $\delta y/\delta c_2$ in (6.18), it can be written as

$$U^2(y) = V(c_1) + V(c_2) \times x^2 + 2x\,\{V(c_1) \times V(c_2)\}^{1/2} \times r(c_1, c_2). \tag{6.19}$$

We can see that it is a quadratic expression in x; therefore $U^2(y)$ will be a minimum or a maximum for the values of x for which its first differential coefficient with respect to x is zero,

giving us

$$\frac{\mathrm{d}U^2(y)}{\mathrm{d}x} = 2xV(c_2) + \{V(c_1)V(c_2)\}^{1/2}\, r(c_1, c_2) = 0. \tag{6.20}$$

Differentiating again, we get

$$\frac{\mathrm{d}^2U^2(y)}{\mathrm{d}x^2} = V(c_2). \tag{6.21}$$

We see that (6.21), being square of deviations, is always positive ($V(c_2)$).

Hence there will be a minimum of $U(y)$ at x given by

$$x = -r(c_1, c_2)\frac{\sqrt{V(c_1)}}{\sqrt{V(c_2)}}. \quad (6.22)$$

Further x, being the observation on an instrument, is always positive; hence $r(c_1, c_2)$ should be negative for even a minimum to exist.

6.4 Numerical Example

6.4.1 *Calibration of a Proving Ring*

Several sets of data about the force applied and indication, obtained from Force Standards Section of National Physical Laboratory of India, have been fitted into linear equations in two ways:

(a) Force applied is expressed in terms of indications; i.e. indication is dependent variable (y) and force is independent variable (x).
(b) Indications in terms of applied force. In this case Force is dependent variable and indication is independent variable.

Though second equation is more useful from the user's point of view, but during calibration, indicator is observed when a known force is applied; hence we read indications in terms of force applied.

Hence we will give both the equations.

Equation for the uncertainty of dependent variable has been given in each case. It has been observed that uncertainty is a minimum for the same set of values of force and indication irrespective of the fact whether force or indication has been taken as dependent variable. Data used for illustrations are given in Table 6.1.

Table 6.1 Observed and Calculated Values of Indications

Force in kN x	Observed indications y	Calculated y'
2	75.8	75.4
4	151.7	151.7667
6	229.2	229.1333
8	304.1	304.5
10	380.3	380.8667
12	456.0	457.2333
14	533.6	533.6
16	609.	609.9667
18	686.5	686.3334
20	763.6	762.7

Table 6.2 Calculation Sheet of Data in Table 6.1

S. No	x	x^2	y	xy	y^2
1	2	4	75.8	151.6	5,745.64
2	4	16	151.7	606.8	23,012.89
3	6	36	229.2	1,385.2	52,532.64
4	8	64	304.1	2,432.8	92,476.81
5	10	100	380.3	3,803.0	144,628.09
6	12	144	456.0	5,472.0	207,936
7	14	196	533.0	7,462	284,089
8	16	256	609.7	9,755.2	371,734.09
9	18	324	686.5	12,357.0	471,282.25
10	20	400	763.6	15,272.0	583,084.96
Sum	$\Sigma x_p = 110$	$\Sigma x_p^2 = 1,540$	$\Sigma y_p = 4,190$	$\Sigma x_p y_p = 58,696$	$\Sigma y_p^2 = 2,237,162$

In order to calculate the values of c_1 and c_2, we need the following:

$$\sum_{p=1}^{p=n} x_p, \sum_{p=1}^{p=n} x_p^2, \sum_{p=1}^{p=n} y_p \text{ and } \sum_{p=1}^{p=n} \mathrm{x_p y_p}.$$

These can be computed with computer in no time, but in the absence of computers, these can be calculated as given in the tabular form in Table 6.2.

But

$$c_1 = \frac{\sum_{p=1}^{p=n} x_p^2 \sum_{p=1}^{p=n} y_p - \sum_{p=1}^{p=n} x_p \sum_{p=1}^{p=n} x_p y_p}{n \sum_{p=1}^{p=n} x_p^2 - \left(\sum_{p=1}^{p=n} x \right)^2}$$

and

$$c_2 = \frac{n \sum_{p=1}^{p=n} x_p y_p - \sum_{p=1}^{p=n} x \sum_{p=1}^{p=n} y}{n \sum_{p=1}^{p=n} x_p^2 - \left(\sum_{p=1}^{p=n} x \right)^2}.$$

Substituting the values, we get

$$c_1 = -0.966666,$$

$$c_2 = 38.1833.$$

Giving equation of best fit with indications as y and force as x

$$y = -0.9666666 + 38.18333x \tag{6.23}$$

and σ_y^2 is calculated from the data of columns 2 and 3 of Table 6.1

$$\sigma_y^2 = 3.612 \times 10^{-4}.$$

$V(c_1)$ and $V(c_2)$ are again from the values obtained from Table 6.2

$$V(c_1) = 1.680 \times 10^{-4},$$
$$V(c_2) = 7.508 \times 10^{-10},$$
$$\text{and } r(c_1, c_2) = -0.886, \tag{6.24}$$

giving the uncertainty equation as

$$U^2(y) = 1.680 \times 10^{-4} - 6.29 \times 10^{-7}x + 7.508 \times 10^{-10}x^2. \tag{6.25}$$

Equation (6.25) gives indication in terms of force.

If we are required to express force in terms of indications, values of x and y are reversed as given in Table 6.3.

Equation of best fit with force (y) in terms of indications (x)

$$y = 2.5410 \times 10^{-2} + 2.6189 \times 10^{-2}x$$
$$y = 0.0254 + 0.0262x \tag{6.26}$$

and

$$\sigma_y^2 = 3.612 \times 10^{-4}\text{kN}^2,$$
$$V(c_1) = 1.680 \times 10^{-4}\text{kN}^2,$$
$$V(c_2) = 7.508 \times 10^{-10}\text{kN}^2,$$
$$r(c_1, c_2) = -0.886. \tag{6.27}$$

Table 6.3 Applied and Calculated Values of Force

Indications x	Force applied y	Force applied y' calculated
75.8	2	2.010553
151.7	4	3.998314
229.2	6	6.027978
304.1	8	7.989550
380.3	10	9.985168
456.0	12	11.967690
533.6	14	13.999970
609.7	16	15.992970
686.5	18	18.004300
763.6	20	20.023490

Uncertainty equation

$$U^2(y) = 1.680 \times 10^{-4} - 6.29 \times 10^{-7}x + 7.508 \times 10^{-10}x^2. \tag{6.28}$$

Uncertainty is minimum at $x = 419.5$ and

$$y = 11.0\,\text{kN and is equal to } 0.0546\%. \tag{6.29}$$

6.4.2 *Calibration of a Glass Scale*

Let us consider another example of calibration of a glass scale. The observations are taken aligning as much as possible on the line of the scale. Standard values are taken on the microscope graticule. So here indication X is an independent variable and observed corrected value Y is the dependent variable Y. The data for calibration of a glass scale are given in Table 6.4.

Equation for the known values of Y, in mm, against observed indications as X in mm.

$$Y = -1.5151 \times 10^{-4} + 1.000 \times 001X, \tag{6.30}$$

Table 6.4 Calculation Sheet of Data in Table 6.3

Indication on the scale in mm	Value of the standard in mm	Corrections in μm
X	Y 1st Equa.	Y correction (2nd Equa)
10	10.0000	0.0
20	49.9998	−0.2
50.1	50.1000	0.0
50.2	50.2000	0.0
50.3	50.3000	0.0
50.4	50.4000	0.0
50.5	50.4996	−0.4
51.0	51.0000	0.0
51.5	51.4998	−0.2
52.0	52.9998	−0.2
52.5	52.5000	0.0
53.0	53.0002	+0.2
54.0	53.9998	−0.2
54.5	54.4996	−0.4
55.0	55.0002	0.2
100.0	100.0002	0.2
150.0	150.0000	0.0
200.0	200.0000	0.0
250.0	250.0004	0.4
300.0	300.0000	0.0

$$\text{Variance}\left(\sigma_y^2\right) = 3.3559 \times 10^{-8}\ \text{mm}^2,$$
$$V(c_1) = 3.7480 \times 10^{-9}\ \text{mm}^2, \quad (6.31)$$
$$V(c_2) = 2.9640 \times 10^{-13}\ \text{mm}^2,$$
$$r(c_1, c_2) = -0.7573715.$$

Uncertainty equation is given as

$$U^2(Y) = 3.7480 \times 10^{-9} + 2.9640 \times 10^{-13} X^2 - 5.0486 \times 10^{-11} X. \quad (6.32)$$

Similarly corrections (Y) in μm against observed readings on the scale (X) are represented by

$$Y = -5.4236 \times 10^{-2} + 7.4865 \times 10^{-4} X. \quad (6.33)$$

Please note a very small value of c_2, gradient of the line in (6.32), indicates the independence of corrections on the indications of the scale.

$$\text{Variance}(\sigma_y^2) = 3.656 \times 10^{-2},$$
$$V(c_1) = 4.0830 \times 10^{-3}, \quad (6.34)$$
$$V(c_2) = 3.2289 \times 10^{-7},$$
$$r(c1, c2) = -0.7574.$$

Uncertainty equation is given as

$$U^2(Y) = 4.083 \times 10^{-3} + 3.229 \times 10^{-7} X^2 - 7.247 \times 10^{-}5X. \quad (6.35)$$

Minimum value of the uncertainty in each case is observed at reading 85.1667 mm and is equal to 0.04 μ m.

6.5 Other Functions

In many situations the set of ordered pairs $(x_1, y_1), (x_2, y_2), (x_3, y_3), \ldots, (x_n, y_n)$ may not best fit into a linear relation, especially when y's are not corrections but values of the parameter against indications of the instrument attached with the main measuring device. The exponential and power functions are quite common. Similar to linear function, each of these two functions is fully defined by two constants. Taking logarithms of both sides of any of these functions transforms into a linear relation.

6.5.1 Exponential Function

Let the ordered pairs $(x_1, y_1), (x_2, y_2), (x_3, y_3), \ldots, (x_n, y_n)$ be fitted into the function

$$y = c_1 \times \mathrm{Exp}(c_2 x). \tag{6.36}$$

Taking log of both sides, we get

$$\log(y) = \log(c_1) + c_2 x. \tag{6.37}$$

This is a linear relation between $\log(y)$ and x.

Writing $Y = \log(y)$ and $\log(c_1) = B_1$, (6.37) becomes

$$Y = B_1 + c_2 x.$$

With the help of (6.5) and (6.6), we get

$$B_1 = \left[\Sigma x^2 \times \Sigma \log(y) - \Sigma x \times \Sigma x \log(y)\right] / \left[n \Sigma x^2 - (\Sigma x)^2\right], \tag{6.38}$$

$$c_2 = [n \Sigma \{x \log(y)\} - \Sigma x \times \Sigma \log(y)] / \left[n \Sigma x^2 - (\Sigma x)^2\right]. \tag{6.39}$$

Here $Y = \log y$ (base of the logarithm is e and not 10)

$$\text{Variance of } B1 = V(B1) = \left[s^2 \Sigma x^2\right] / \left[n \Sigma x^2 - (\Sigma x)^2\right], \tag{6.40}$$

$$\text{variance of } c_2 = V(c_2) = [n s^2] / [n \Sigma x^2 - (\Sigma x)^2], \tag{6.41}$$

$$\text{covariance of } B_1, c_2 = \mathrm{Cov}(B_1, c_2) = -\sigma_y^2 \, \Sigma x / [n \Sigma x^2 - (\Sigma x)^2], \tag{6.42}$$

where

$$\sigma_y^2 = \Sigma \, [\log(y) - \{B_1 + c_2 x\}]^2 / (n - 2). \tag{6.43}$$

Correlation coefficient between $B1, c2$

$$r(B_1, c_2) = \mathrm{Cov}(B_1, c_2) / \{[V(B_1) \times V(c_2)] = -\Sigma x / \{[n \Sigma x^2]. \tag{6.44}$$

6.5.1.1 Uncertainty

Considering the equation of the exponential function

$$y = \mathrm{Exp}(B1 + c_2 x), \tag{6.45}$$

$$\delta y / \delta B_1 = \mathrm{Exp}(B_1 + c_2 x) = y, \tag{6.46}$$

$$\delta y / \delta c_2 = [Exp(B_1 + c_2 \times x)] \, x = y \times x. \tag{6.47}$$

$$\text{so } U^2(y) = [\delta y / \delta B_1 U(B_1)]^2 / [\delta y / \delta c_2 U(c_2)]^2,$$

$$2[\delta y/\delta B_1 \times \delta y/\delta c_2 \times U(B_1) \times U(c_2) \times r(B_1, c_2)]. \quad (6.48)$$

$$= [yU(B_1)]^2 + [xyU(c_2)]^2 + 2[yU(B_1) \cdot xy \cdot U(c_2) \cdot r(B_1, c_2]. \quad (6.49)$$

Taking uncertainty equal to the standard deviation, we get

$$[U(y)/y]^2 = V(B_1) + V(c_2)x^2 + 2x \times r(B_1, c_2)[V(B_1) \times V(c_2)]^{1/2}. \quad (6.50)$$

But from linear relation $\log(y) = B_1 + c_2 \times x$.

$$U^2\{\log(y)\} = V(B_1) + x^2 \times V(c_2) + 2x \times r(B_1, c_2)[V(B_1) \times V(c_2)] = [U(y)/y]^2. \quad (6.51)$$

Hence the value of uncertainty in y will be y times the value of uncertainty calculated from the linear relation taking $\log(y)$ as a single dependent variable.

6.5.1.2 Numerical Example

Data have been used to illustrate the determination of two constants, variances, and their uncertainty components. The value of variance obtained from the residual errors has been calculated by substituting the values of c_1, c_2, x's and corresponding values of log (y) in (6.37) and the relative variance is obtained from

$$[y_1 - c_1 \times \text{Exp}(c_2 x)]/y_1. \quad (6.52)$$

The two came out, as expected equal, showing thereby that uncertainty taking $\log(y)$ as dependent variable is same as the relative uncertainty of y.

Equation of the exponential function best to the data in Table 6.5 is

$$y = 59.26989 \times \text{Exp}(0.0236853x). \quad (6.53)$$

Table 6.5 Data for Exponential function

Value of x	Value of y	Calculated value of y
5	67	66.7215
10	75	75.11002
15	85	84.5319
20	95	95.1836
25	107	107.1505
30	120	120.6219
35	135	135.7871
40	152	152.8588
45	173	172.077
50	195	193.7112

The uncertainty equation is

$$U^2(y) = 1.2707 \times 10^{-5} + 1.3205 \times 10^{-8}x^2 - 7.2614 \times 10^{-6}x. \tag{6.54}$$

Minimum uncertainty is at $x = 31.02$, y=123.57 and is equal to 0.00165%.

Variance of $\log(y)$ obtained from the square of residual errors is 2.723027×10^{-5} while from

$$\sum \left[[y_p - \{59.26989 \times \mathrm{Exp}(0.0236853 \times x_p)\}]/y_p\right]^2/(n-2) = 2.724717{\times}10^{-5}.$$

This shows that fractional variance is same as the variance of $\log(y)$.

6.6 Power Function

Any power function may be defined as

$$y = c_1 x^{c2}. \tag{6.55}$$

Taking logarithms of both sides, we get

$$\log(y) = \log(c_1) + c_2 \times \log(x). \tag{6.56}$$

Writing $\log(y) = Y$, $\log(x) = X$, and $\log(c_1) = B_1$ we get

$$Y = B_1 + c_2 X. \tag{6.57}$$

Equation (6.57) represents a linear relation. Using (6.5) and (6.6), we get

$$B_1 = \frac{\sum\{\log(x)\}^2 \sum \log(y) - \sum \log(x) \sum \{\log(x) \times \log(y)\}}{n \sum \{\log(x)\}^2 - \{\sum \log(x)\}^2}, \tag{6.58}$$

$$c_2 = \frac{n \sum \log(x) \times \log(y) - \sum \log(x) \times \sum \log(y)}{n \sum \{\log(x)\}^2 - \{\sum \log(x)\}^2}, \tag{6.59}$$

$$\text{Variance of } B_1 = V(B_1) = \frac{s^2 \sum\{\log(x)\}^2}{n \sum \{\log(x)\}^2 - \{\sum \log(x)\}^2}, \tag{6.60}$$

$$\text{Variance of } c_2 = V(c_2) = [ns^2]/[n\Sigma\{\log(x)\}^2 - \{\Sigma \log(x)\}^2], \tag{6.61}$$

$$\text{Covariance of } B_1,\ c_2 = \mathrm{Cov}(B_1, c_2), \tag{6.62}$$

$$= -s^2 \Sigma \log(x)/[n\Sigma\{\log(x)\}^2 - \{\Sigma \log(x)\}^2]. \tag{6.63}$$

Here

$$s^2 = \sum [\log(y) - \{B_1 + c_2 \log(x)\}]^2/(n-2).$$

Correlation coefficient between B_1, c_2

$$r(B_1, c_2) = \mathrm{Cov}(B_1, c_2)/\{[V(B_1) \times V(c_2)]\}^{1/2} \quad (6.64)$$
$$= -\Sigma \log(x) \Big/ \left\{\left[n\Sigma \{\log(x)\}^2\right]\right..$$

6.6.1 Uncertainty

Taking

$$y = \mathrm{Exp}(B_1) \times x^{c2}, \quad (6.65)$$
$$\delta y/\delta B_1 = \mathrm{Exp}(B_1) \times x^{c2} = y, \quad (6.66)$$
$$\delta y/\delta c_2 = \mathrm{Exp}(B_1) \times x^{c2} \times \log(x) = y \times \log, \quad (6.67)$$
$$\text{so } U^2(y) = [\delta y/\delta B_1 U(B_1)]^2/[\delta y/\delta c_2 U(c_2)]^2.$$
$$2[\delta y/\delta B_1 \times \delta y/\delta c_2 \times U(B_1) \times U(c_2) \times r(B_1, c_2] \quad (6.68)$$
$$= [y \times U(B_1)]^2 + [y \times \log(x) U(c_2)]^2 + 2[y \times U(B_1) \times y \times \log(x) \times U(c_2)$$
$$\times r(B_1, c_2]. \quad (6.69)$$

Taking uncertainty equal to standard deviation, we get

$$[U(y)/y]^2 = V(B_1) + \{\log(x)\}^2 \times V(c_2) + 2\log(x) \times r(B_1, c_2)\{V(B_1) \times V(c_2)\}^{1/2}. \quad (6.70)$$

Considering the linear relation

$$Y = B_1 + c_2 X.$$

The uncertainty in Y shall be

$$U^2(Y) = V(B_1) + X^2 \times V(c_2) + 2X \times r(B_1, c_2)\{V(B_1) \times V(c_2)\}^{1/2}.$$

Or

$$U^2\{\log(y)\} = V(B_1) + \{\log(x)\}^2 \times V(c_2) + 2\log(x) r(B_1, c_2)$$
$$\times \{V(B_1) \times V(c_2)\}^{1/2}$$
$$= [U(y)/y]^2. \quad (6.71)$$

Hence in this case also the value of uncertainty in y calculated from the given data will be y times the value of uncertainty calculated from the linear relation taking $\log(y)$ and $\log(x)$ as variables.

6.6.2 Numerical Example

Fictitious data given in the first two columns of Table 6.6 have been fitted to a power function.

Equation of the best fit is

$$y = 22.85071 \times x^{0.3763576}. \tag{6.72}$$

Expression for uncertainty is

$$U^2(\log(y)) = 6.3502 \times 10^{-4} + 6.2151 \times 10^{-5} \{\log(x)\}^2 - 3.8781 \times 10^{-4} \{\log(x)\}. \tag{6.73}$$

The uncertainty is minimum at $x = 24.445$, $y = 73.93393$ and is equal to 0.00548.

6.6.3 Same Data Fitted to Two Functions

Let us consider the data given in Table 6.7.

When the data of a proving ring, given in first two columns of Table 6.7, are fitted to the power function, then equation of best fit is

$$Y = 15.74364 \times X^{0.9996233}, \tag{6.74}$$

which approximates to

$$Y = 0.0 + 15.74 \times X. \tag{6.75}$$

Table 6.6 Data for Power function

Values of x	Values of y	Calculated value of y
5	42.5	41.87572
10	54.4	54.35722
15	62.8	63.31849
20	69.7	70.55896
25	75.3	76.74056
30	80.0	82.19122
35	89.0	87.10065
40	93.0	91.58980
45	96.5	95.74117
50	100.0	99.61389

Table 6.7 Data of Proving ring B

Force in kN x	Indication y	y' Calculated value of y from the power relation	y'' Calculated value of y from linear relation	$(y' - y)^2$	$(y'' - y)^2$
5	79.1	78.67	77.73		
10	156.9	157.30	156.58		
15	235.0	235.91	235.44		
20	313.7	314.52	314.29		
25	392.4	393.11	393.14		
30	471.1	471.70	472.00		
35	550.1	550.29	550.85		
40	630.0	628.87	629.70		
45	709.2	707.45	708.56		
50	788.2	786.02	787.41		
Sum					5.56

If the above data are fitted to a linear relation, then the equation of the best fit is given by

$$Y = -1.1268 + 15.77 \times X. \tag{6.76}$$

Uncertainty equation for the two methods, when analyzed, showed that the minimum percentage uncertainty is 0.1% for power function, while the same is 0.0609% for linear relation. We see that the uncertainty for linear relation is less; i.e. the data fit better to the linear relation. In case the same data are fitted to the two relations, the relation with lower uncertainty should be retained.

The calculated values of indications are also shown in the 3rd and 4th columns of Table 6.7.

In many cases, the relation between input and output quantities is none of the aforesaid functions. For example in a resistance thermometer change in resistance is not linearly related to temperature but is a quadratic function of temperature. Heat produced in a resistor is also a quadratic function of current flowing through it. So now we will consider a general case when output quantity, i.e. the quantity indicated by the instrument, is a polynomial function of the input quantity.

6.7 Method of Least Squares

Let the indication y be related to the input quantity x through a polynomial of degree m and be represented as

$$y = c_0 + c_1 x + c_2 x^2 + \cdots + c_m x^m = \sum_{p=1}^{p=m+1} c_{p-1} x^{p-1}. \tag{6.77}$$

There are n points $y_1, y_2, y_3, \ldots, y_n$, at which the instrument has been checked, and corresponding standard values of input quantities are $x_1, x_2, \ldots, x_n$. So there are n paired values of $(x_1, y_1), (x_2, y_2), (x_3, y_3), \ldots, (x_n, y_n)$, which are to be substituted in (6.77) to determine the values of $m+1$ coefficients $(c_0, c_1, c_2, \ldots, c_m)$. There will be n equations in m variables.

$$\begin{aligned} y_1 &= c_0 + c_1 x_1 + c_2 x_1^2 + \cdots + c_m x_1^m \\ y_2 &= c_0 + c_1 x_2 + c_2 x_2^2 + \cdots + c_m x_2^m \\ y_3 &= c_0 + c_1 x_3 + c_2 x_3^2 + \cdots + c_m x_3^m \\ &\ldots \\ &\ldots \\ &\ldots \\ y_n &= c_0 + c_1 x_n + c_2 x_n^2 + \cdots + c_m x_n^m \end{aligned} \quad . \tag{6.78}$$

These equations can be represented in a compact form as

$$\sum_{q=1}^{q=n} y_q = \sum_{q=1}^{q=n} \sum_{p=1}^{p=m+1} c_{p-1} x_q^{p-1}. \tag{6.79}$$

Once the values of all c_{p-1} are determined from n equations of (6.79), (6.77) will be uniquely defined. But corresponding to one value of x of (6.77) there will be, in general, two values of y. One value of y' is the value of left-hand side of (6.77) obtained after substituting values of coefficients and x. The other value of y is the observed value corresponding to x and obtained from the ordered pair (x, y). The difference in y and y' is called residual error. Let the sum of squares of all residual errors be χ^2. The method of least squares consists of minimizing χ^2. Only variables are the coefficients c_{p-1}; hence differential coefficient of χ^2 with respect to each of the coefficients c_{p-1} will give $m+1$ equations, which are called normal equations. Solution of these equations will give the values of best estimates of the coefficients.

Solution of (6.77) is possible only if $m+1$ is equal to or less than n. If there is no measurement error in y_n and x_n and $m+1$ is equal to n then there will be a unique solution. As y is a measured quantity, there are bound to be errors; hence to estimate the coefficients with certain degree of certainty, $m+1$ should be less than n.

Writing mathematically

$$\begin{aligned} \chi^2 &= \sum_{q=1}^{q=n} (y_q - y_q') \\ &= \sum_{q=1}^{q=n} \left\{y_q - (c_0 + c_1 x + c_2 x^2 + c_3 x^2 + \cdots + c_m x^m)\right\}^2. \end{aligned} \tag{6.80}$$

The minimum value of χ^2 divided by $n - (m+1)$ gives σ_y^2.

The conditions for χ^2 to be a minimum are that each of partial derivatives of χ^2 with respect to c_{p-1}, for all values of p, is zero, giving the following conditions:

$$\begin{aligned}
&\sum_{q=1}^{q=n} y_q = nc_0 + c_1 \sum_{q=1}^{q=n} x_q + c_2 \sum_{q=1}^{q=n} x_q^2 + c_3 \sum_{q=1}^{q=n} x_q^3 + c_4 \sum_{q=1}^{q=n} x_q^4 \cdots c_m \sum_{q=1}^{q=n} x_q^m \\
&\sum_{q=1}^{q=n} y_q x_q = c_0 \sum_{q=1}^{q=n} x_q^1 + c_1 \sum_{q=1}^{q=n} x_q^2 + c_2 \sum_{q=1}^{q=n} x_q^3 + c_3 \sum_{q=1}^{q=n} x_q^4 \cdots c_m \sum_{q=1}^{q=n} x_q^{m+1} \\
&\sum_{q=1}^{q=n} y_q x_q^2 = c_0 \sum_{q=1}^{q=n} x_q^2 + c_1 \sum_{q=1}^{q=n} x_q^3 + c_2 \sum_{q=1}^{q=n} x_q^4 + c_3 \sum_{q=1}^{q=n} x_q^5 \cdots c_m \sum_{q=1}^{q=n} x_q^{m+2} . \\
&\ldots \\
&\ldots \\
&\ldots \\
&\sum_{q=1}^{q=n} y_q x_q^m = c_0 \sum_{q=1}^{q=n} x_q^m + c_1 \sum_{q=1}^{q=n} x_q^{m+1} + c_2 \sum_{q=1}^{q=n} x_q^{m+2} + c_3 \sum_{q=1}^{q=n} x_q^{m+3} \cdots c_m \sum_{q=1}^{q=n} x_q^{2m}
\end{aligned} \tag{6.81}$$

There are $m + 1$ equations in (6.81) and are called normal equations from these normal equations $m + 1$ coefficients of the polynomial equation of (6.78) are found out. The solution may be obtained either by algebraic method or using matrix inversion method.

Using matrix notations, equations in (6.78) are written as

$$[X_{q,p-1}][C_{p-1}] = [Y_q]. \tag{6.82}$$

Here

$[X_{q,p-1}]$ is a n by $m + 1$ matrix, given as

$$[X_{q,p-1}] = \begin{matrix} 1 \; x_1 \; x_1^2 \; \ldots\ldots\ldots \; x_1^m \\ 1 \; x_2 \; x_2^2 \; \ldots\ldots\ldots \; x_2^m \\ 1 \; x_3 \; x_3^2 \; \ldots\ldots\ldots \; x_3^m \\ 1 \; \ldots\ldots\ldots\ldots\ldots\ldots \\ 1 \; \ldots\ldots\ldots\ldots\ldots\ldots \\ 1 \; \ldots\ldots\ldots\ldots\ldots\ldots \\ 1 \; x_n \; x_n^2 \; \ldots\ldots\ldots \; x_n^m \end{matrix} \tag{6.83}$$

$[C_{p-1}]$ is a column matrix of $m + 1$ elements, namely $c_0, c_1, c_2, \ldots, c_m$, and $[Y_q]$ is also a column matrix but of n elements, namely $y_1, y_2, y_3, \ldots, y_n$.

Pre-multiplying both sides by the transpose of the matrix $[X_{q,p-1}]$, which is written as $[X_{q,p-1}]^T$, we get

$$[X_{q,p-1}]^T [X_{q,p-1}][C_{p-1}] = [X_{q,p-1}]^T [Y_q]. \tag{6.84}$$

Equation (6.84) is equivalent to set of normal equations given in (6.81).

Pre-multiplying both sides by inverse of $[X_{q,p-1}]^T[X_{q,p-1}]$ written as $[[X_{q,p-1}]^T \times [X_{q,p-1}]^{-1}$ we get the solution equation for all c's which is given as

$$[\mathrm{C}_{\mathrm{p}-1}] = [[X_{q,p-1}]^T \cdot [X_{q,p-1}]]^{-1}\ [\mathrm{X}_{\mathrm{q,p}-1}]^{\mathrm{T}}[\mathrm{Y}_{\mathrm{q}}]. \tag{6.85}$$

The matrix $[[X_{q,p-1}]^T \cdot [X_{q,p-1}]^{-1}$ is known as variance covariance matrix. The diagonal elements of this matrix will give respective variances of c_{p-1} and other elements will give the covariances.

Computer programmes are easily available to solve (6.85). Substitution of values of c's in (6.80) gives the sum of residual errors and σ_y^2 is obtained by dividing the sum by $n - (m + 1)$. To best fit the data in a polynomial of unknown degree, start with m equal to one (linear relation), increase it in steps of one, and find the value of m for which σ_y^2 is minimum. Once m is known, the calculated values of c's are substituted in (6.77) to give the complete expression of the polynomial. Minimum value of uncertainty due to use of the polynomial and the corresponding value of x is also calculated.

Referece

1. S.V. Gupta, Type A uncertainty in curve fitting. Mapan **14**, 15–20 (1999)

Chapter 7
Calculation of Uncertainty

7.1 Importance of Correct Measurement

We may explain the meaning and importance of uncertainty in measurement in several ways. It is widely recognized that the value of a measured quantity is determined within a certain range. The range depends upon instruments, quality of measurements taken and the confidence level at which the final result is to be stated. Leaving aside the formal definition, half of this range may be called uncertainty of measurement. The uncertainty in a measurement result will depend upon all the three aforesaid elements. Therefore, quantifying a measurable quantity through any measurement process is meaningful only if the value of the quantity measured is given with a proper unit of measurement and is accompanied by an overall uncertainty in measurement.

The quality of a measurement may also be characterized by the semi-range in which the measured value is expected to lie. Incidentally, the word measurement should be understood to mean both a process and the output of that process. The measurements are carried out at different levels. The measurements in industry have assumed greater significance in view of the fact that measurements provide the very basis of all control actions.

Importance of accurate measurement in science may also be illustrated by the following examples.

7.1.1 Discovery of Inert Gases

The density of nitrogen gas was measured, taking samples of nitrogen from air and the chemical reaction in which pure nitrogen was produced. The density of nitrogen sample taken from atmosphere after removing oxygen and CO_2 was found to be more than that of nitrogen through a chemical reaction. The persistence and significant difference of the two values of density of nitrogen made us reach the conclusion on the existence of inert gases such as Helium, and Argon present in atmosphere.

S.V. Gupta, *Measurement Uncertainties*, DOI 10.1007/978-3-642-20989-5_7,

7.1.2 Correction in Composition of Air

More recently, the composition of air has been revised as a result of precise measurements of air density. The density of air used to be calculated by using the CIPM formula [1, 2] expressing density of moist air in terms of pressure, temperature, humidity and involving the composition of air and its molar mass. The density of moist air is calculated by CIPM formula by measurements of pressure, temperature and humidity with reasonable small uncertainty. The density of air has also been measured by gravimetric (artefacts) method [3]. The values obtained by the two methods, though agreed very well within any one of the methods, did not agree with each other. The relative discrepancy was 6.4×10^{-5} [4]. Density obtained by gravimetric method was found to be more than that obtained by CIPM formula. Independent analysis of air samples through spectroscopic means [5] suggested the change in molar fraction of Argon. The CIPM in 2008 [6] subsequently changed the molar fraction of Argon from 0.0917 to 0.09332. This is an example of the benefits of high precision measurements.

7.1.3 Meaning of Quantity Being Exact

The value of velocity of light in vacuum is taken as exact by the international agreement. However, it does not mean that there was no uncertainty in its measurement but by assigning a specific value to the velocity of light in vacuum, we have assigned a new value to the metre. Similar is the case of the value of permeability of free space or magnetic constant, which is taken as $4\pi \times 10^{-7}\,\mathrm{N/A^2}$. This value comes from the definition of the unit of electric current – the ampere, through a specific theoretical formula. The force F acting per unit length on the two current-carrying parallel wires is given as

$$F = \frac{\mu}{4\pi}\frac{I_1 \times I_2}{r}.$$

7.1.4 International Agreement with Uncertainty

Having understood, in 1970s, the benefits of precise measurements along with uncertainty, all metrology laboratories recognized the fact that each measurement result is to be associated with an uncertainty declaration. As a result of which each laboratory started giving the result along with an uncertainty. But there was no uniformity in either achieving or expressing the uncertainty.

With the initiation of globalization, more and more national metrology laboratories started sharing their results of measurements. For better and easy understanding, the results and in assigning some mean value to the results of various laboratories, it was necessary that all laboratories express the measurement results in a uniform way.

7.1.5 *Initiation by BIPM*

In 1978, the International Committee on Weights and Measures (CIPM) – world's highest authority in metrology, requested the International Bureau of Weights and Measures (BIPM), in conjunction with a few national metrology laboratories and other international bodies interested in metrology to look into this problem.

The BIPM prepared a detailed questionnaire covering the issues involved and distributed that to 32 national metrology laboratories and to five international organizations known to have interest in the subject. Almost all agreed that it was necessary to arrive at an internationally accepted procedure for expressing measurement uncertainty and also a uniform method of combining uncertainty components into a single total uncertainty; however, a consensus was not apparent on the method of combining several components. Eleven national laboratories send their experts to the meeting convened by the BIPM. This working group developed a recommendation on statement of uncertainty in 1980 [7,8], which CIPM adopted in 1981 [9] and reaffirmed in 1986 [10]. Before coming to the final conclusion of the efforts of the BIPM along with several other international organizations, we may discuss the procedure followed prior to it.

7.2 Classical Procedure for Uncertainty Calculations

Before 1978, there were various ways of defining the uncertainty. Uncertainty in fact was some function of different sources of errors. The errors were classified into random errors and systematic errors. Those used to be defined as follows.

7.2.1 *Random Error*

An error varies in an unpredictable manner both in magnitude and in sign, when a large number of measurements of the same quantity are made under essentially the same conditions. These errors follow the Gaussian (normal) distribution with zero mean. However, for small sample (smaller number of observations), statistical results which are based on normal distribution are corrected by means of Student's $\sim t'$ factor. These errors may be due to uncontrollable environmental conditions, personal judgement of the observer and inherent instability of the measuring instrument or any other cause of random nature.

7.2.2 Systematic Error

An error is due to the system (including the standards used for the measurement) and cannot be reduced by taking larger number of observations if the equipment and conditions of measurement remain unchanged. These errors may be due to the inability in detection of the measuring system, constant bias, error in the value of the standard, a physical constant and property of the medium or conversion factor used. The value and the sign of this error do not change with the given measuring system. Systematic errors can be broadly classified into (1) constant and (2) variable. Constant systematic errors are those which do not change with respect to time but sometimes, may vary with the magnitude of the measured quantity. Zero setting error in an instrument is a constant systematic error while inaccuracy in the calibration scale may depend upon the magnitude of the quantity measured. Variable systematic errors do depend upon the time; say value of a resistor, which may vary with time because of ageing effect. These may also occur due to insufficient control of environmental conditions.

7.2.3 Calculation of Random Uncertainty (u_r)

The best estimate of the expected value of a random variable of n independent observations $x_1, x_2, x_3, \ldots, x_n$ obtained under same conditions of measurement is the arithmetic mean of n observations.

The mean is given as

$$\bar{x} = \sum_{p=1}^{p=n} x_p/n. \tag{7.1}$$

The measure of dispersion is variance. The best estimate of the population variance from the sample of size n is given

$$s^2 = \frac{\sum_{p=1}^{p=n} (x_p - \bar{x})^2}{n-1}. \tag{7.2}$$

Standard deviation – the positive square root of variance is given by

$$s = \left\{ \frac{\sum_{p=1}^{p=n} (x_p - \bar{x})^2}{n-1} \right\}$$

Standard deviation of the mean $\bar{x}$ is $s(\bar{x})$ and is given by

$$s(\bar{x}) = \left\{ \frac{\sum_{p=1}^{p=n} (x_p - \bar{x})^2}{n(n-1)} \right\}^{1/2}. \tag{7.3}$$

From the standard deviation of the mean $s(\bar{x})$ of the sample of size n, population standard deviation was calculated by multiplying it by the student t factor. The value of student t for chosen level of confidence is taken from the student t Table A.5 by taking $n - 1$ as the degree of freedom. The random standard uncertainty u_r due to single input quantity is given as

$$u_r = t \left\{ \frac{\sum_{p=1}^{p=n} (x_p - \bar{x})^2}{n(n-1)} \right\}^{1/2}. \tag{7.4}$$

The above calculations are based upon the assumption that measured value of the input variable follows the Gaussian (Normal) distribution and $f(x)$ is represented as

$$f(x) = (l/\sigma\sqrt{\pi})\exp[-(x-\mu)^2/2\sigma^2]. \tag{7.5}$$

7.2.4 Combination of Random Uncertainties (u_r)

Let

$$Y = f(X_1, X_2, \ldots, X_n). \tag{7.6}$$

This is a well-defined function of n variables.

The variables are measured to arrive at the measured value of a physical quantity Y. Means and standard deviations of all measured variables are calculated. Standard deviation of the each mean is then multiplied by the student t factor. The student t factor depends upon the chosen level of confidence and upon the degrees of freedom, which is one less than the observations taken.

In order to calculate the contribution to random uncertainty due to variable X_p, its random uncertainty is multiplied by its coefficient. The coefficient C_i is $\delta Y/\delta X_p$, which is the partial differential coefficient of Y with respect to quantity X_p and at the values of independent variable $x_1, x_2, \ldots, x_n$.

Details of calculations are as follows:

Standard deviation s_p is estimated from the sample of size m of the pth variable. The corresponding random uncertainty u_r is determined by multiplying s_p by a student t factor. Student t factor depends upon (a) degrees of freedom $= m - 1$ and (b) the stated level of confidence.

Then the contribution due to random uncertainty of all variables will be for independent variables $X_1, X_2, \ldots, X_n$, uncertainty U_r is given by

$$U_r = \sqrt{\sum_{p=1}^{p=n} u_r \left(\frac{\delta Y}{\delta X_p}\right)^2} = \sqrt{\sum_{p=1}^{p=n} t_p s_p \left(\frac{\delta Y}{\delta X_p}\right)^2}. \qquad (7.7a)$$

If all the measured quantities are not independent of each other, then standard uncertainty will have additional component due to the interdependence of variables in the form of covariance s_{pq}.

Then total random component of standard uncertainty U_r is given as:

$$U_r = \sqrt{\sum_{p=1}^{p=n} (t_p s_p \delta Y/\delta X_p)^2 + 2\sum_{q=1}^{q=n}\sum_{p=q+1}^{p=n} t_{p,q} s_{pq} (\delta Y/\delta X_p)(\delta Y/\delta X_q)}, \qquad (7.7b)$$

where

$$s_{pq} = \frac{\sum_{r=1}^{r=n} \left[\{(x_p)_r - \bar{x}_p)\}\{(x_q)_r - \bar{x}_q\}\right]}{n} = cov(p,q). \qquad (7.8)$$

7.3 Sources of Systematic Uncertainty (U_s)

In determining the magnitude of systematic uncertainty, the contributions due to (a) measuring instrument (b) operating conditions and (c) inherent characteristic of the instrument under test are taken into consideration. Some common types of systematic uncertainties encountered in mechanical measurements may be listed as follows:

- Uncertainty as reported in the calibration certificate of the standard and the instrument used.
- Uncertainty due to the interpolation between the calibration points of the standards used in the measurement.
- Uncertainty due to the change in environmental conditions.
- Uncertainty due to lack of ability to reset, detect and repeat of the instrument under test.
- Uncertainty due to the values taken of some physical constants or properties of the materials used in the measurement process. For example, uncertainty in the values of density of water, acceleration due to gravity, coefficients of linear expansion, temperature, pressure and relative humidity, etc.

It is assumed that the variable of systematic uncertainty follows rectangular distribution, giving the variance as one third of the square of its semi-range and infinite degrees of freedom.

$$\sigma^2 = a^2/3. \qquad (7.9)$$

7.4 Combination of Systematic Uncertainty

Here, standard deviations due to various sources of errors are obtained from the accumulated knowledge about the distribution, the variable follows. Normally, the uncertainties due to such variables are expressed in the form of semi-ranges such as $\alpha_1,\ \alpha_2,\ \alpha_3,\ldots,\alpha_n$. If each variable causing systematic uncertainty is assumed to follow rectangular distribution, then systematic uncertainty due to variable σ_p is given by

$$\sigma_p^2 = \alpha_p^2/3. \tag{7.10}$$

This gives the variance due to total systematic uncertainty as:

$$\sigma_{\mathrm{s}}^2 = \sum_{p=1}^{p=n} (\alpha_p^2/3)(\delta Y/\delta X_p)^2. \tag{7.11}$$

Giving systematic uncertainty as

$$U_{\mathrm{s}} = K\sigma_{\mathrm{s}}. \tag{7.12}$$

K is the value of student t, for $n = \infty$ at a selected level of confidence.

In case, the variances s_p^2 for all values of p are known we may write combined variance as

$$\sigma_{\mathrm{s}}^2 = \sum_{p=1}^{p=n} s_p^2(\delta Y/\delta X_p)^2. \tag{7.13}$$

Hence, combined uncertainty U_{s} is given by

$$U_{\mathrm{s}} = K\sigma_s = K\left[\sum_{p=1}^{p=n} s_p^2(\delta Y/\delta X_p)^2\right]^{1/2}. \tag{7.14}$$

7.5 Dominant Term

Sometimes a source of systematic uncertainty is very prominent such that its component is outstanding at a glance. In that case, sum of absolute components of systematic uncertainties will be less than u_{s} as calculated above, i.e.

$$\sum\left|\frac{\alpha_p}{\sqrt{3}}(\delta Y/\delta X_p)\right| < U_{\mathrm{s}}. \tag{7.15}$$

Then the dominant term – largest of $\left(\alpha_p/\sqrt{3}\right)(\delta Y/\delta X_p)$ say equal to a_{d} is taken out, and systematic uncertainty due to all other sources is calculated by quadrature method as described above. Let it be as U_{s}'. Then total systematic uncertainty U_{s} is given as

$$U_{\mathrm{s}} = a_{\mathrm{d}} + U_{\mathrm{s}}'. \tag{7.16}$$

7.6 Total Overall Uncertainty U

There was no universal agreement in combining the systematic and random uncertainties. One view was to add the two, another was to use quadrature method, while the third was to report them separately.

Arguments in favour of reporting the two uncertainties separately were as follows:

- The concept of confidence level is not applicable to systematic uncertainty, unless the probability distribution of population is known.
- The relative values of two uncertainties may help in deciding the future course of measurement. If random uncertainty is more, then it can, perhaps, be reduced by further experiments, while to reduce systematic uncertainty, equipment and method of measurement are required to be changed.

However, countries participating in The Asia Pacific Program were using the quadrature method. Overall, uncertainty U following quadrature method was

$$U = (U_r^2 + U_s^2)^{1/2}, \text{ if no dominant term is present in } U_s$$

or

$$U = a_d + (U_r^2 + U_s^2)^{1/2} \text{ if dominant term is present in } U_s.$$

To illustrate the above method of calculating the uncertainty, an example is given below.

7.7 Objections to the Above Method

1. Names systematic and random assigned to uncertainties are confusing. The nature of uncertainty due to different sources would not change. Uncertainty due to different sources may be obtained by different methods. So method of evaluation may be different and not the names of uncertainties.
2. In systematic uncertainty, all variance of each of the components is taken as one third of the square of the semi-range, which is not justifiable.
3. Existence of dominant term and methodology of overcoming is arbitrary in nature.
4. There was no universal method of combining the two uncertainties.
5. Coverage factor in terms of student t has been applied separately. Student t takes care of the degree of freedom and the level of confidence. This implies that one should be careful while choosing the value of student t factor.

To avoid these problems and to harmonize the procedure for expressing uncertainty, the CIPM issued recommendations on the two uncertainties. The recommendations are reproduced below.

7.8 The BIPM Recommendations 1980 Basis of ISO Guide (GUM)

The basis of ISO GUM is the Recommendation INC-1, 1980 [7,8]. The uncertainty in the result of a measurement generally consists of several components, which may be grouped into two categories according to the way in which their numerical values are estimated.

(A) Those which are evaluated by statistical method (In this case, the standard deviation is calculated by taking square root of the mean of sum of squares of deviations from the mean). This is the case of repeated observations.
(B) Those which are evaluated by other means.

There is not always a simple correspondence between the classification into categories A or B and the previously used classification into "random" and "systematic" uncertainties. The term "systematic uncertainty" can be misleading and should be avoided.

Any detailed report of the uncertainty should consist of a complete list of the components, specifying for each, the method used to obtain its numerical value.

The components in category A are characterized by the estimated variances s_p^2 (or the estimated standard deviations s_p) and the number of degrees of freedom. When appropriate, the estimated covariance should be given.

The components in Type B should be characterized by quantities u_j^2, which may be considered as approximations to the corresponding variances, the existence of which is assumed. The quantities u_j^2 may be treated like variances and the quantities u_j like standard deviations. Where appropriate, the co-variances should be treated in a similar way.

The combined uncertainty should be characterized by the numerical value obtained by applying the usual method for the combinations of variances. The combined uncertainty and its components should be expressed in the form of "standard deviations".

If, for particular applications, it is necessary to multiply the combined uncertainty by a factor to obtain an overall uncertainty, then the multiplying factor used must always be stated.

As a follow-up action, the international Organization for Standardization (ISO), the International Electro-technical Commission (ICE), the International Federation of Clinical Chemistry and Laboratory Medicine, the International Union of Applied and Pure Chemistry (IUPAC), the International Union of Pure and Applied Physics (IUPAP), the International Organization of Legal Metrology (OIML) and the International Bureau of Weights and Measures (BIPM) joined hands to form a Joint Committee for Guides in Metrology (JCGM) to produce a document. The document "Guide to the Expression of Uncertainty in Measurement (GUM)" was first published in 1993 and reprinted in 1995 with minor corrections [11].

7.9 ISO GUM-Step-by-Step Procedure for Calculation of Uncertainty

1. Particulars of instrument under test (IUT), standards used and the quantity under test Y and its functional relationships with input quantities is established.
2. Express the quantity Y in terms of input quantities $X_1, X_2, X_3, \ldots, X_n$.
3. Find expressions for partial derivative of Y with respect to each input quantities. For $Y = f(X_1, X_2, X_3, \ldots, X_n)$. The partial derivative $\delta Y/\delta X_p$ is determined. It is called as sensitivity coefficient C_p and is calculated at mean measured values of input quantities $x_1, x_2, x_3, \ldots, x_n$.
4. Every uncertainty component is the standard uncertainty, hence replace every s_p^2 by u_p^2.
5. Uncertainty components of each input quantity X_p measured with an instrument are determined by using both Type A and Type B evaluation methods and are combined by using quadrature method and is denoted by u_p for all integral values of p from 1 to n.
6. Express combined standard uncertainty u_c of Y in terms of the uncertainties of each input quantity.
7. If all the input quantities are linearly related to Y and are uncorrelated, then the combined uncertainty u_c

$$u_c^2 = \left(\frac{\delta f}{\delta X_1}\right)^2 \times u_1^2 + \left(\frac{\delta f}{\delta X_2}\right)^2 \times u_2^2 + \left(\frac{\delta f}{\delta X_3}\right)^2 \times u_3^2 + \cdots + \left(\frac{\delta f}{\delta X_n}\right)^2 \times u_n^2$$

$$= \sum_{p=1}^{p=n} \left(\frac{\delta f}{\delta X_p}\right)^2 u_p^2 = \sum C_p^2 u_p^2. \tag{7.17}$$

The above relation is arrived at by the mean value theorem, i.e. following the law of variances $\left(s_p^2\right)$ of linearly related quantities and replacing s_p^2 by u_p^2. This relation is true if all input quantities are independent of each other (un-correlated).

In case of excessive non-linearity of function f, the second-order term in the Taylor's expansion may also be included. Giving

$$u_c^2 = \sum C_p^2 u_p^2 + \sum_{p=1}^{p=n} \sum_{q=1}^{q=n} \left[\frac{1}{2} \left\{ \frac{\delta^2 f}{\delta X_p \delta X_p} \right\}^2 + \frac{\delta f}{\delta X_p} \cdot \frac{\delta^3 f}{\delta X_p \delta X_q^2} \right] u_p^2 u_q^2. \tag{7.18}$$

8. In case, the input quantities are correlated, the expression for combined uncertainty u_c is given by

$$u_c = \sqrt{\sum_{p=1}^{p=n} (s_p \delta Y/\delta X_p)^2 + 2 \sum_{q=1}^{q=n} \sum_{p=q+1}^{p=n} s_{pq} (\delta Y/\delta X_p)(\delta Y/\delta X_q)}. \tag{7.19}$$

Here, s_{pq} is the co-variance $\mathrm{Cov}(x_p, x_q)$ given as

$$\mathrm{Cov}(x_p, x_q) = \frac{\sum_{r=1}^{r=n} \left[\{(x_p)_r - \bar{x}_p)\}\{(x_q)_r - \bar{x}_q\}\right]}{n}.$$

If $r(x_p, x_q)$ is the correlation coefficient, then it is related to covariance as follows:

$$r(x_p, x_q) = \frac{\mathrm{Cov}(x_p, x_q)}{s_p \times s_q} = \frac{\mathrm{Cov}(x_p, x_q)}{u_p \times u_q}.$$

Hence, (7.19) may be expressed as

$$u_c = \sqrt{\sum_{p=1}^{p=n} (u_p \delta Y/\delta X_p)^2 + 2\sum_{q=1}^{q=n} \sum_{p=q+1}^{p=n} r(x_{p,} x_q) \times u_p \times u_q \times (\delta Y/\delta X_p)(\delta Y/\delta X_q)}. \tag{7.20}$$

For a very special case where all the input estimates are correlated with correlation coefficients $r(x_p, x_q) = +1$, then (7.20) becomes

$$u_c = \sqrt{\sum_{p=1}^{p=n} (u_p C_p)^2 + 2\sum_{q=1}^{q=n} \sum_{p=q+1}^{p=n} C_p u_p \times C_q u_q} = \sum_{p=1}^{p=n} C_p u_p. \tag{7.21}$$

9. Find the effective degree of freedom ν_{eff} from Welch–Satterthwaite formula, which given as

$$\nu_{\mathrm{eff}} = \frac{u_c^4(y)}{\sum_{p=1}^{p=n} \frac{u_p^4(y)}{\nu_p}}. \tag{7.22}$$

10. Alternative to the finding of degrees of freedom is to use the Baye's method. For every component of uncertainty through Type A evaluation $u_{\mathrm{A}}(x_p)$, we use the following relation:

$$u_{\mathrm{Abayes}} = \sqrt{\frac{n_p - 1}{n_p - 3}} u_{\mathrm{A}}(x_p). \tag{7.23}$$

All such n components are combined by quadrature method to give

$$u_{\mathrm{Abayes}} = \sqrt{\sum_{p=1}^{p=n} C_p^2 \frac{n_p - 1}{n_p - 3} u_{\mathrm{A}}^2(x_p)}.$$

Then combined uncertainty

$$u_c = \sqrt{u^2_{Abayes} + \sum_{p=1}^{p=n} C_p^2 u_B^2(x_p)}. \quad (7.24)$$

For further details, one may consult [12–14].

11. It may be noted that

$$\nu_{\text{eff}} \leq \sum_{p=1}^{p=n} \nu_p. \quad (7.25)$$

12. Once ν_{eff} is known and level of confidence is decided, then the value k of the coverage factor can be determined from t distribution table.
13. If the output Y is a function of input quantities expressed as their products or quotients, then it is easier to determine the relative uncertainty. Take the logarithm of the function. The partial derivatives $\delta f/\delta X$ will be in the form $1/X_i$. The square of the combined relative uncertainty will then be the sum of squares of relative uncertainties of all input quantities.
14. When an instrument is used to measure, then there will be uncertainty in observed values to be calculated by Type A method. The uncertainties to be calculated by Type B method are due to (a) its calibration (b) its resolution and other properties.

7.10 Calculation of Uncertainty

7.10.1 Procedure for Calculation

1. Identify all sources of uncertainty including uncertainty in the applied corrections, if necessary.
2. Type A evaluation of uncertainty: If the output quantity Y is a function of more than one input quantities, the variance of each set of observations for each input quantity is taken. This is carried out by normal statistical method of finding out standard deviation. This standard deviation is equal to the standard uncertainty of that particular quantity by Type A evaluation method. Calculate uncertainty components for each input quantity. Sometimes in addition of making observations for the determination of Y, observations are also taken of influence quantities such as environmental parameters of pressure, temperature. Uncertainty for each observed influence quantity is calculated by Type A evaluation.
3. Find out variances from the uncertainty values given in the calibration certificates, by Type B evaluation method, for each instrument used. Also, estimate the variances of all physical constants used in the mathematical modelling, their value being taken from the literature. Here, experience and other relevant

information come into play. Variance is calculated depending upon the distribution which the value of quantity likely to follow. May be normal, rectangular or any other probability distribution. Uncertainty for each observed influence quantity is also calculated by Type B evaluation.

4. Find the combined standard uncertainty of Y, using (7.17), (7.18), (7.20) or (7.24) which ever is applicable.
5. For extended uncertainty, effective degrees of freedom are calculated, level of confidence is chosen and the coverage factor k is determined from the Students t factors table. Alternatively, Bayes equation (7.24) may be used using k values from the normal distribution table for given confidence level (probability).

7.10.2 Relation Between Range and Standard Uncertainty

If the probability distribution of the quantity whose semi-range is given, then standard deviation will be calculated from the probability distribution, which the quantity under question follows. For example, let the semi-range in mass measurement of the kilogram is α at 95% confidence level. The measured value will follow normal distribution hence α is 1.96 time the standard deviation. The relevant standard deviation σ is $\alpha/1.96$.

In the absence of knowledge of probability distribution of a quantity, there are four possibilities that the quantity may follow either

1. *Rectangular distribution*, this means that from our experience we know that the true value may lie anywhere within the specific range with equal probability, and standard deviation σ will be $\alpha/\sqrt{3}$

   or
2. *Triangular distribution*, this means the quantity has maximum frequency in centre of the range and then tapers off to zero at the extreme ends of the range standard deviation σ will be $\alpha/\sqrt{6}$
3. *Trapezium distribution*, this means the quantity may maximum uniform probability with a range of $\pm\beta\alpha$ and tapers off uniformly to zero at the extreme end of the range. Although this distribution appears to be most reasonable, we need one more parameter, which is difficult to decide. The standard deviation σ in this case will be $\alpha/\sqrt{\{(1+\beta^2)/6\}}$
4. In case we are taking values of some physical constant from an older literature (say before 1950), then most likely the semi-range α is given for the probable error (50% confidence interval) in that case α is to be multiplied by 1.48 to get the standard deviation.

We discussed two methods for calculations of uncertainty in this chapter arbitrarily named as the Classical and ISO GUM. The GUM is the most internationally accepted document. In either of the methods, we estimate the expectation of Y i.e. mean value and variance of Y. None of the methods estimates the probability distribution function (pdf) of Y.

7.10.3 *Applicability of ISO GUM*

7.10.3.1 For Linear Models

1. No condition is necessary for the valid application of the law of propagation of uncertainty to linear models (models that are linear in X_i).
2. A coverage interval can be determined, in terms of the information provided in the GUM, under the following conditions:

   (a) The Welch–Satterthwaite formula is adequate for calculating the effective degrees of freedom associated with $u(y)$, when one or more of the $u(x_i)$ has an associated degrees of freedom that is finite; this condition is required in order that Y can be characterized by an appropriate scaled and shifted t-distribution.

   (b) The X_i are independent when the degrees of freedom associated with the $u(x_i)$ are finite; the condition is required because the GUM does not treat X_i that are not independent in conjunction with finite degrees of freedom.

   (c) The PDF for Y can adequately be approximated by a Gaussian distribution or a scaled and shifted t-distribution; the condition is satisfied when each X_i is assigned a Gaussian distribution. It is also satisfied when the central limit theorem is applicable to mathematical modeling. However, the GUM uncertainty method may not be validly applicable when there is an X_i whose assigned distribution is non-Gaussian and the corresponding contribution to $u(y)$ is dominant.

3. When the conditions in (2) hold, the results from the application of the GUM uncertainty framework can be expected to be valid for linear models. These conditions apply in many circumstances.

7.10.3.2 For Non-Linear Models

1. The law of propagation of uncertainty can validly be applied for non-linear models under the following conditions:

   (a) f is continuously differentiable with respect to the elements X_i of X in the neighbourhood of the best estimates x_i of the X_i; The condition is necessary for the expansion of the function $f(X)$ by Taylor series upto first-order approximation when the non-linearity of f is insignificant.

   (b) Condition (a) applies for all derivatives up to the appropriate order; the condition is necessary for the application of the law of propagation of uncertainty based on a higher-order Taylor series approximation to $f(X)$. An expression for the most important terms of next highest order to be included is given in (7.19) above.

   (c) The X_i involved in significant higher-order terms of a Taylor series approximation to $f(X)$ is independent; The condition relates to significant model non-linearity in the case of independent X_i. The ISO GUM does not consider X_i that is not independent in this context.

(d) The PDFs assigned to X_i involved in higher-order terms of a Taylor series approximation to $f(X)$ are Gaussian.

(e) Higher-order terms that are not included in the Taylor series approximation to $f(X)$ are negligible. The condition constitutes a correction that the version of the law of propagation of uncertainty using higher-order terms is based on the symmetry of the PDFs for the X_i.

If the analytical determination of the higher derivatives, required when the non-linearity of the model is significant, is difficult or error-prone, suitable software for automatic differentiation can be used.

2. A coverage interval can be determined, in terms of the information provided in the GUM, when conditions (a), (b) and (c) in (2) of Sect. 7.10.3.1 apply, with the exception that the content of (c) in that sub-clause is replaced by "Condition (c) is required in order that coverage intervals can be determined from these distributions".
3. When the conditions in Sects. 7.10.3.1 and 7.10.3.2 hold, the results from the application of the GUM uncertainty approach can be expected to be valid for non-linear models. These conditions apply in many circumstances.

7.11 Propagation of Probability Density Function

It may be noticed that using the ISO Gum method in general propagates the mean and the variance of the input quantities with conditions discussed in Sect. 7.10.3. In the case, the output quantity is linearly related to its input quantities, the probability density function (PDF) of the output quantity in general is a t-distribution with the effective degrees of freedom calculated by Welch–Satterthwaite formula. However in other cases, the PDF of output quantity is derived by using Monte–Carlo method (MCM) [15].

MCM provides a general approach to obtain an approximate representation of Cumulative distribution $G_Y(\eta)$ for Y. The basic of MCM is repeated sampling from the PDFs for the input quantities X_i and evaluation of the model each time. Larger is the number M of samples more accurate results are expected from $G_Y(\eta)$. Step-by-step procedure for MCM is given below.

7.11.1 Step-by-Step Procedure for Monte–Carlo Method

1. Select the number M – the number of Monte Carlo trials. To get coverage of 95%, M should be 10^6.

2. Generate M vectors, by sampling from the assigned PDFs as realizations of the set of N input quantities X_i.
3. Each vector yields a model value of Y say y_r, r takes values from 1, 2, 3, ..., M.
4. Arrange these values in strictly increasing order, these sorted of values give G.
5. Use G to estimate y– the expectation of Y and standard uncertainty $u(y)$ the standard deviation of G. Arithmetic mean of all y_r is an approximation of expectation $E(Y)$. Variance of Y is inversely proportional to M,; hence, closeness of agreement between the averages and $E(Y)$ is inversely proportional to $M^{1/2}$.
6. Use G to form appropriate coverage interval for Y, for a stipulated coverage probability.

7.11.2 Two-Stage Bootstrap Procedure

The procedure enumerated above is valid when PDFs of all input quantities are known. The case of unknown PDFs of input quantities has been discussed by S. V. Crowder and R.D. Moyer [16]. They suggested two-stage bootstrap procedure. The step-by-step procedure is as follows:

1. Estimate the parameters of each input distribution using the observed data. Sample size of 10 is sufficient.
2. Generate a large number, B, of bootstrap samples by simulating data from the distributions estimated in step 1. Each bootstrap sample must be of same size as the original for each input quantity. The authors recommended B to be 10,000.
3. For each bootstrap sample, re-estimate the parameters of the input distributions used in step 1. The variation of these estimated parameters from bootstrap sample to bootstrap sample now represents the uncertainty due to finite size of samples.
4. For each bootstrap sample, generate a large number, M, of Monte Carlo samples from the estimated distributions in step 3 and evaluate measurement equation for each sample. The author found $M = 10,000$ to be sufficient. Evaluate $y = f(x_1, x_2, x_3, \ldots, x_N)$ for each of the Monte Carlo samples. The average is the estimate of $E(y)$ associated with that particular bootstrap sample. Similar estimates of each bootstrap sample provide a distribution of estimates of $E(y)$. The percentiles of this estimated distribution are used to construct a coverage interval.
5. Construct a histogram of the estimate of $E(y)$ from step 4. The interval formed by 2.5% and 97.5% percentiles of this distribution is a 95% ($k = 2$) uncertainty interval for $E(y)$.

Note: For further reading, one may like to go through papers [16–18].

7.12 Bayesian Statistics

Suppose the information about the input quantity X consists of a series of indications regarded as realization of independent, identically distributed random variables characterized by a specific PDF, but with unknown expectation and variance [19]. Bayes Theorem is used to calculate a PDF for X, where X is taken to be equal to the unknown average of these random variables. It is carried out in two steps. First, non-informative joint prior (pre-data) PDF is assigned to the unknown expectation and variance. Using Bayes Theorem, this joint prior PDF is then updated, based on the information supplied by the series of indications, to yield a joint posterior PDF for two unknown parameters. The desired posterior PDF for the unknown average is then calculated as a marginal PDF by integrating over the possible values of the unknown variance.

With the use of Bayes Theorem, the updating is carried out by forming the product of a likelihood function and the prior PDF [20, 21]. The likelihood function in the case of indications obtained independently is the product of the functions, one function for each indication and indication to form e.g. to Gaussian PDF. The posterior PDF is then determined by integrating the product of prior PDF and likelihood over all possible values of the variance. Final expression is obtained after the normalizing resulting expression.

If the indications are characterized by a PDF with only one parameter, a non-informative prior PDF is assigned to the unknown expectation of the random variables and the posterior distribution for X is given directly by Bayes Theorem, without the need for marginalization.

For further details, one may like to consult [22–24].

7.13 Example for Calculations of Uncertainty

In order to elucidate the difference in the Classical and ISO Gum methods, an example of determination of specific resistance of material in the form of wire by measuring its resistance and dimensions is given.

We know, resistance R of a wire of length L and diameter $2r$ is related to specific resistance S by the following relation:

$$S = \pi r^2 R/L. \tag{7.26}$$

Partial derivatives of S with respect of independent variable R, r and L are given below:

$$\delta S/\delta r = 2\pi r R/L = 2S/r, \tag{7.27}$$

$$\delta S/\delta Rr = \pi r^2/L = S/R, \tag{7.28}$$

$$\delta S/\delta L = -\pi r^2 R/L^2 = -S/L. \tag{7.29}$$

This step is same in the two methods.

7.13.1 Calculation of Random Uncertainty

The uncertainty of measured inputs is called random uncertainty in classical method and is evaluated by Type A method in the ISO Gum. The uncertainty of the measuring instrument is considered as systematic uncertainty and calculated separately in classical method. But in ISO Gum method, uncertainty due to observations and in the instruments are considered together, only the methods of evaluation may be different.

7.13.1.1 Measurement of Length

Observations for length of the wire with a scale graduated in mm

$$L = 100.01,\ 99.98,\ 99.99,\ 100.02,\ 100.00cm.$$

The second decimal place of cm in length measurement is obtained by eye estimation

$$\bar{x}_{\mathrm{L}} = 100.00\,\mathrm{cm}. \tag{7.30}$$

The standard deviation of sample in length measurements S_{L} is given as:

$$s_{\mathrm{L}} = (1+4+1+4+0)/4)^{1/2}10^{-2}\mathrm{cm} = \sqrt{2.5}\times 10^{-2}\,\mathrm{cm}. \tag{7.31}$$

Standard deviation of the mean

$$= s(\bar{x}_{\mathrm{L}}) = s_{\mathrm{L}}/\sqrt{n} = \sqrt{2.5}\times 10^{-2}/\sqrt{5} = 0.707\times 10^{-2}\,\mathrm{cm}. \tag{7.32a}$$

Student t factor for 4 degree of freedom at 95% Confidence Level is taken from the table is 2.78.

Hence, random uncertainty in length measurements e_{L}

$$e_{\mathrm{L}} = t\times S_{\mathrm{L}}/\sqrt{n}, \tag{7.32b}$$

$$e_{\mathrm{L}} = 2.78\times\sqrt{2.5}\times 10^{-2}/\sqrt{5}\,\mathrm{cm} = 1.96\times 10^{-2}\,\mathrm{cm}.$$

In the ISO Gum, student t factor is not determined for each standard deviation of the mean.

7.13.1.2 Measurement of Resistance

Resistance was measured and measurements results of 10 repetitions are given

$$R = 100.04, 100.06, 100.05, 100.05, 100.03, 100.02, 100.07, 100.05, 100.05, 100.08\ \text{ohms}$$

Giving the mean value as

$$\bar{x}_R = 100.05\text{ ohms.}$$

estimate of standard deviation of population

$$s_R = \sqrt{\{(1+1+0+0+4+9+4+0+0+9)/9\}} \times 10^{-2} = \sqrt{28/9} \times 10^{-2}\text{ ohm.}$$

Standard deviation of the mean

$$s(\bar{x}_R) = \sqrt{(28/9 \times 10)}\, 10^{-2} = 0.558 \times 10^{-2}\text{ ohm.} \tag{7.33a}$$

Student t factor for 9 degree of freedom at 95% Confidence Level is taken from the table of Student t factor is 2.26.

Uncertainty in resistance measurement

$$t\frac{s_R}{\sqrt{n}} = 2.26 \times \sqrt{(28/90)} \times 10^{-2} = 1.26 \times 10^{-2}\text{ ohm.} \tag{7.33b}$$

In the ISO Gum, student t factor is not determined for each standard deviation of the mean.

7.13.1.3 Measurement of Diameter of the Wire

Diameter of the wire was measured with a micrometer giving 10 value of radius of the wire as follows:

$$r = 1.998,\ 2.001,\ 2.000, 2.001, 1.999, 2.002, 2.000, 1.999, 1.998, 2.001\text{ mm.}$$

Giving mean radius $\bar{x}_d$ as

$$\bar{x}_d = 2.000\text{ mm.}$$

$$s_d = \sqrt{\{(4+1+0+4+1+4+0+1+4+1)9\}} = \sqrt{20/9} \times 10^{-3}\text{mm.}$$

Standard deviation of the mean

$$s(\bar{x}_d) = s_d/\sqrt{n} = \sqrt{(20/9 \times 10)} \times 10^{-3}\text{ mm} = 0.47 \times 10^{-3}\text{ mm.} \tag{7.34a}$$

Student t factor for 9 degree of freedom at 95% Confidence Level is taken from the table of Student t factor is 2.26.

Random uncertainty in diameter measurement

$$e_L = 2.26 \times (\sqrt{2/3}) \times 10^{-3}\text{mm} = 1.065 \times 10^{-3}\text{mm.} \tag{7.34b}$$

In the new method, student t factor is not determined for each standard deviation of the mean.

From the mean values of length, resistance and diameter of the wire, the sensitivity components are

$$\begin{aligned} \delta S/\delta L &= S/L = S/100 \\ \delta S/\delta R &= S/R = S/100. \\ \delta S/\delta r &= S/r = S/1 \end{aligned} \tag{7.35}$$

Hence, component of random uncertainties as per Classical method is:

In measurement of specific resistance due to length measure $= S \times 1.96 \times 10^{-4}$ cm

In measurement of specific resistance due to resistance measurement $S \times 1.26 \times 10^{-4}$

In measurement of specific resistance due to radius measurement $S \times 1.065 \times 10^{-3}$

7.13.1.4 Combined Random Component of Uncertainty

$$U_r = \sqrt{\Sigma\{u_p^2/(\delta X/\delta x_p)^2\}}.$$

As all measurements are independent of each other, so random uncertainty is

$$\begin{aligned} U_r &= S\sqrt{\{(1.96 \times 10^{-4})^2 + (1.26 \times 10^{-4})^2 + (1.065 \times 10^{-3})^2\}/100} \\ &= S\sqrt{118.85} \times 10^{-4} = S \times 10.90 \times 10^{-4}. \end{aligned}$$

In the ISO Gum method, the standard uncertainty, which is equivalent to the standard deviation, is the square root of the sum of the squares of the product of sensitivity coefficients C_i and estimated standard deviation of the mean $\bar{s}_p$.

7.13.1.5 Relative Random Uncertainty

$$U_r/S = 1.09 \times 10^{-4}.$$

7.13.2 *Systematic Uncertainty*

Let the semi-range of uncertainty of scale used in measurement of length $\alpha_L = 0.01$ cm and in accuracy of dial micrometer $\alpha_r = 0.001$ mm. Uncertainty given in the calibration certificate of 100 ohm resistance was 0.0001 ohm at 99% confidence level.

$$U_s = K\sqrt{\Sigma\{\alpha_p^2/3(\delta X/\delta x_p)^2\}}$$

$$= 1.96 \times S\sqrt{[\{(0.0001)^2 + (0.001)^2 + (0.0001/100)^2\}/3]/100}$$

$$= 11.32 \times 10^{-4}\, S.$$

Combined uncertainty $U = \sqrt{\{U_s{}^2 + U_r{}^2\}} = S\sqrt{\{(1.13 \times 10^{-3})^2 + (1.09 \times 10^{-3})^2\}}$ $= S\,1.57 \times 10^{-3}$.

Or relative uncertainty at 95% level of confidence is 0.157%.

There is nothing like systematic and random uncertainties in the ISO Gum method. The uncertainties are evaluated by two methods namely Type A and Type B. Uncertainty component evaluated by Type B method is combined with the uncertainty evaluated by Type A method by quadrature method for each variable. Foe extended uncertainty the effective degrees of freedom are calculated by (7.20). Student's t factor is looked up at the chosen level of confidence and multiplied to standard uncertainty so calculated.

To summarize the ISO Gum method, which has a wider International acceptance, we again take the above example.

1. Due to measurement of wire length

$$\text{Standard deviation} s_\text{L} \text{ from} = \sqrt{2.5} \times 10^{-2}\,\text{cm}$$

Giving
Standard deviation of the mean from (7.32a)= $\sqrt{(2.5/10)} \times 10^{-2}$ cm
Corresponding sensitivity coefficient $c_L = S/100$
So uncertainty through Type A evaluation = $S \times \sqrt{(2.5/10)} \times 10^{-4}$
From the data given above semi-range of uncertainty in vernier caliper = 0.01 cm
Uncertainty through Type B evaluation = $S \times 10^{-4}/\sqrt{3}$
Giving uncertainty component due to measurement wire length

$$u_\text{L} = S \times 10^{-4}\sqrt{0.25 + 1/3} = 0.764 \times S \times 10^{-4}.$$

2. Due to measurement of wire diameter

Standard deviation of the mean diameter (7.34a) = 0.471×10^{-3} mm
Corresponding sensitivity coefficient $c_\text{d} = S/1$
uncertainty through Type A evaluation = $S \times 0.471 \times 10^{-3}$
semi-range of dial micrometer = 0.001 mm
Uncertainty through type B evaluation = $S \times 10^{-3}/\sqrt{3}$ mm
Giving uncertainty component due to diameter measurement

$$U_\text{d} = S \times \sqrt{(0.471)^2 + 1/3} = 0.745 \times S \times 10^{-3}.$$

Due to measurement of resistance
Standard deviation of the mean (7.33a)= $\sqrt{0.25} \times 10^{-2}$ ohm
Corresponding sensitivity coefficient $c_R = S/100$
Uncertainty through Type A evaluation = $S \times \sqrt{0.25} \times 10^{-4}$
Uncertainty in calibration of the resistance at 99% confidence level = 0.0001 ohm
Uncertainty through Type B evaluation = $S \times 10^{-2} \times 0.0001/2.58$ (2.58 is t factor at 99% CL).
Giving uncertainty component due to resistance

$$u_R = S \times 10^{-2}\sqrt{0.25 \times 10^{-4} + 10^{-8}/2.58^2}$$
$$= 0.5 \times S \times 10^{-4}.$$

$$\text{Combined standard uncertainty} = \sqrt{u_R^2 + u_d^2 + u_L^2}$$
$$= S \times 10^{-4}\sqrt{0.5^2 + 7.45^2 + 0.764^2}$$
$$= S \times 7.506 \times 10^{-4}.$$

It may be noted that it is only standard uncertainty (combined standard deviation). For extended uncertainty, we have to find effective degrees of freedom; (7.20) is used for this purpose giving

$$\frac{(S \times 7.506 \times 10^{-4})^4 \cdot 9}{(S \times 10^{-4})^4 \{0.5\}^4 + (4.71)^4 + (0.5)^4} \approx 58.$$

Hence, Student t factor for 95% confidence level is 1.96, giving extend uncertainty $1.96 \times 7.506 \times \mathrm{S} \times 10^{-4} = 0.147 \times \mathrm{S} \times 10^{-2}$ or 0.147%.

Here, we see that relative uncertainty is almost same as obtained by older method.

7.14 Merits and Limitations of ISO Gum Method

7.14.1 Merits of ISO GUM

The primary expression of uncertainty in ISO Gum method is the standard uncertainty. A standard uncertainty is both internally consistent and transferable. In this sense, standard uncertainty is a fundamental expression in measurement.

When the mathematical modeling (measurement equation) is linear, the estimate y and standard uncertainty $u(y)$ are determined by ISO GUM are correct values for all state of knowledge probability distributions for the input variables $X_1, X_2, X_3, \ldots, X_n$ that have the specified expected value x_p, standard deviations,

$u(x_p)$ and correlation coefficients $r(x_p, x_q)$. In this sense, y and $u(y)$ are robust estimate and standard uncertainty for Y.

The estimate y and standard uncertainty $u(y)$ determined by the ISO-Gum method may be reasonable when all non-linear input quantities have small uncertainties.

7.14.2 Limitations of ISO GUM

When the mathematical model is non-linear function and one or more input quantities have large uncertainties, the standard uncertainty $u(y)$ is a poor approximation for the standard deviation of $S(Y)$ for Y.

An uncertainty interval is a secondary expression in ISO Gum-method. Since the ISO-Gum propagates the estimates and standard uncertainties rather than probability distributions for the input quantities, it does not yield an uncertainty interval with specific coverage probability.

References

1. P. Giacomo, Equation for the determination of density of moist air (1981). Metrologia **18**, 33–40 (1982)
2. R.S. Davis, Equation for the determination of density of moist air(1981/1991). Metrologia **29**, 67–70 (1992)
3. A. Picard, H. Fang, Three methods of determining the density of moist air during mass comparisons. Metrologia **39**, 31–40 (2002)
4. A. Picard, H. Fang, M. Glaser, Discrepancy in air density determination between thermodynamic formula and gravimetric method: evidence for a new value of mole fraction of argon in air. Metrologia **41**, 396–400 (2004)
5. S.Y. Park et al., A re-determination of the argon content of air for buoyancy correction in mass standard comparisons. Metrologia **41**, 387–395 (2004)
6. CIPM, Revised formula for density of air (CIPM-2007). Metrologoia **45**, 149–155 (2008)
7. R. Kaarls, NIPM Proc. Verb.com.Int. Poise et Mesure **49**, A1–A12 (1981)
8. P. Giacomo, Metrologia **17**, 73–74 (1980)
9. P. Giacomo, Metrologia **18**, 43–44 (1982)
10. P. Giacomo, Metrologia **24**, 49–50 (1987)
11. ISO, IEC, OIML and BIPM, *Guide to the Expression of Uncertainty in Measurement*. (ISO/TAG 4/WG 3, 1995)
12. K. Raghu, Bayesian alternative to ISO GUMs use of the Welch-Saitterthwaite formula. Metrologia **43**, 1–11 (2006)
13. W. Birch, L. Callegaro, F. Pennechhi, Non-linear models and best estimator in Gum. Metrologia **43**, S196–S199 (2006)
14. K. Raghu, B. Toman, D. Huang, Comparison of ISO GUM, draft GUM Supplement and Bayesian Statistics using simple linear calibration. Metrologia **43**, S167–S177 (2006)
15. M.G. Cox, M.R.L. Siebert, The use of a Monte Carlo method for evaluating uncertainty and expanded uncertainty. Metrologia **43**, S178–S188 (2006)

16. S.V. Crowder, R.D. Moyer, A two stage Monte Carlo approach to the expression of Uncertainty with non-Linear measurement equation with small size. Metrologia **43**, 34–41 (2006)
17. R. Willink, On using Monte Carlo method to calculate uncertainty intervals. Metrologia **43**, L39–L42 (2006)
18. W. Bich, L. Callegaro, F. Pennecchi, Non-linear models and best estimator in the GUM. Metrologia **43**, S161–S166
19. JCGM, Supplement 1 to the Guide to the expression of uncertainty in measurement – Propagation of distribution functions, page 18, (2008)
20. A. Gelman, J.B. Carlin, H.S. Stern, D.B. Rubin, *Bayesian Data Analysis* (Chapman and Hall, London, 2004)
21. P.M. Lee, *Bayesian Statistics* (Oxford University Press, New York, 1997)
22. I. Lira, W. Woger, Comparison between conventional Bayesian approaches to evaluate measurement data. Metrologia **43**, S249–S259 (2006)
23. I. Lira, Bayesian evaluation of comparison data. Metrologia **43**, S231–S234 (2006)
24. A. Balsamo, G. Mana, F. Pennecchi, The expression of uncertainty in non-linear parameter estimates. Metrologia **43**, 396–402 (2006)

Chapter 8
Uncertainty in Calibration of a Surface Plate (Fitting a Plane)

8.1 Introduction

A metrology laboratory, quite often, receives surface plates for calibration. It is required to determine a plane that will best fit the empirically measured points. When such a plane is determined it is possible to calculate the departures from flatness of individual points on the surface of the plate and to determine the overall error. The method of testing given in the following paragraphs is based on Dietrich book [1]

We will first derive the formula for the determination of the heights of set of points which contribute to the main plane. The surface to be measured is usually surveyed by measuring along certain lines. A plan of points to be measured is shown in Fig. 8.1. The dots represent points at which height of the test point is measured. The number of points chosen will depend upon the size of the plates. Size of a surface plate may vary from 300 mm by 120 mm to 2,000 mm by 1,200 mm. Number of points chosen along any line would depend upon the size of the plate and may be between 6 and 20.

The observations are the heights of the individual points taken along the predetermined lines. One of the methods in use is of an auto-collimator and a plane mirror mounted on a three-point suspension table. The suspension table is schematically shown in Fig. 8.2. The height is the product of the angular deflection of the reflected light from the plane mirror and the distance d, which is equal to the distance of the third foot from the other two feet of the plane mirror mount (Fig. 8.3). At each move, the mirror mount is moved by a fixed distance, which is equal to the length of the chosen line by the number of observations. The number of observations along any line is always an even number. Normally the lines chosen are boundary lines of the rectangular grid, its two diagonals, and at least two lines bisecting the boundary lines, one along the length and another along the width of the rectangular grid.

S.V. Gupta, *Measurement Uncertainties*, DOI 10.1007/978-3-642-20989-5_8,

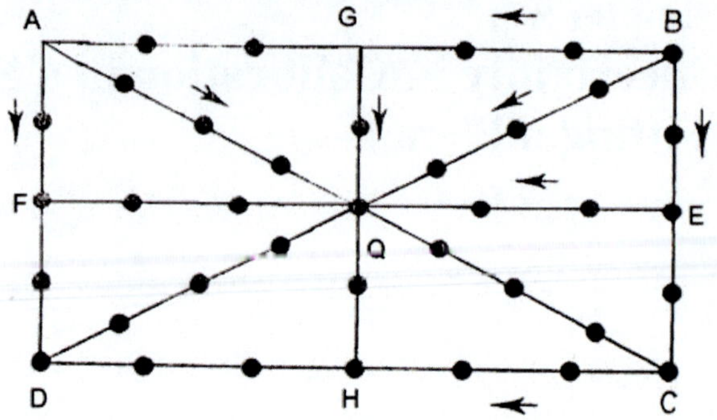

Fig. 8.1 Measurement Plan

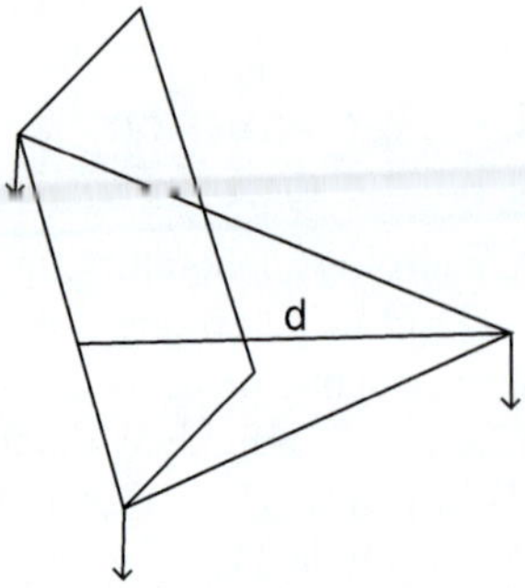

Fig. 8.2 Mirror on its stand

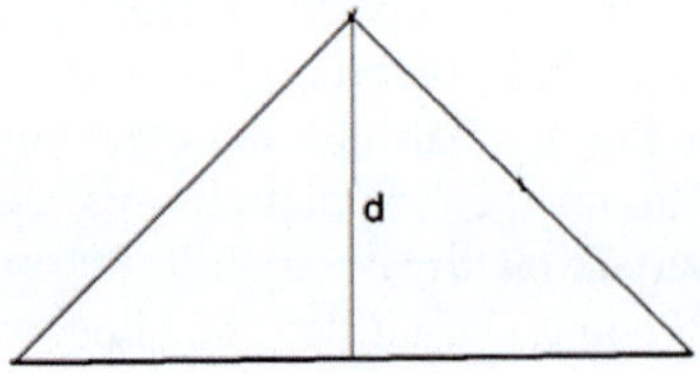

Fig. 8.3 Base of mirror stand

8.2 Procedure

Step-by-step procedure of measurement of heights of various points along the designated line is as follows:

1. Observations are taken along one diagonal say AC, starting with point A.
2. Height at each point is deduced and added to the sum of the heights previously taken. The progressive sum of heights is plotted with distance of the point as abscissa.
3. Join extremity of one ordinate to that of next, till the point C, giving the curve AC (Fig. 8.4).
4. The straight line through AC is drawn Fig. 8.5.
5. The height of the middle point of the line AC is obtained.
6. Ordinate of this point meets the curve at Q Fig. 8.6.

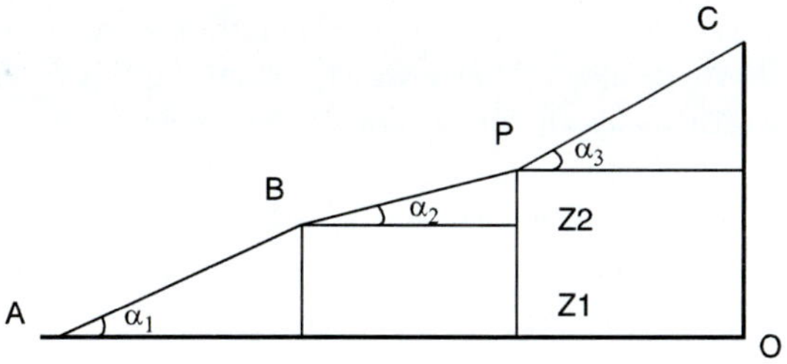

Fig. 8.4 Cumulative heights of various points"

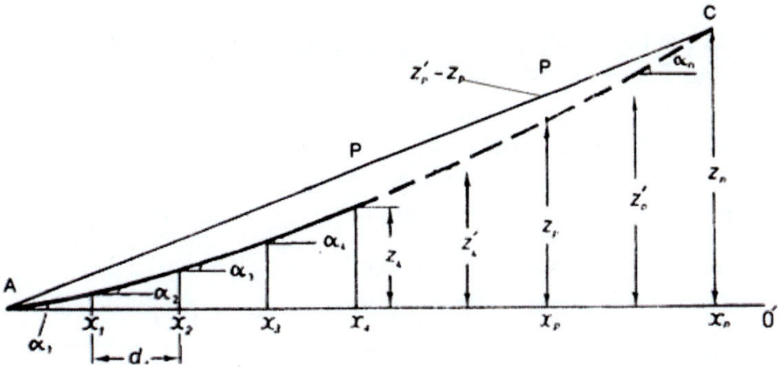

Fig. 8.5 Join of AC

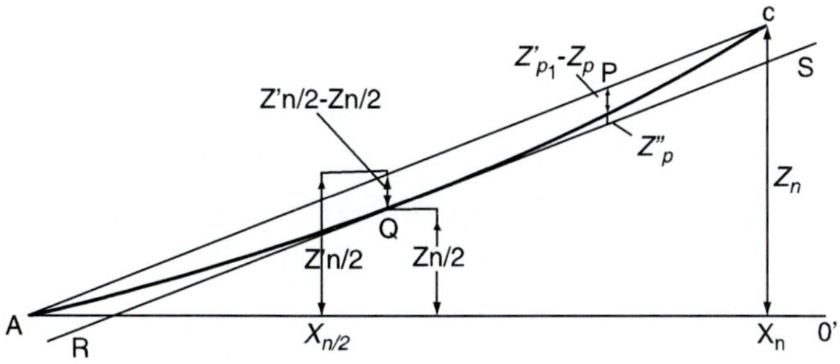

Fig. 8.6 Height of point Q

7. A line RS is drawn though Q and parallel to line AC Fig. 8.6.
8. The heights of the observed points on the curve are deduced with reference to the line RQS. This gives the heights of observed points on the diagonal AC.

For the diagonal BD:

9. Steps from 1 to 3 are repeated and the line BD is drawn.
10. The height of the midpoint of BD is calculated.
11. The ordinate of this point also intersects at Q.
12. A line R′S′ is drawn through Q and parallel to BD.

13. The heights of the observed points on the curve BD with reference to line R′QS′ are deduced. Thus giving the heights of observed points on the diagonal BD.
14. This way, the heights of all the points on the two diagonals are deduced with reference to the plane passing through the two lines RQS and R′QS′.
15. Let this plane be called as *xy* plane.
16. The heights of the points A and C with reference to the plane *xy* will be equal; let it be h. Similarly the heights of the points B and D on the diagonal BD will be equal; say it is equal to a.

For parallel sides AD or BC and AB or DC

17. Steps 1–8 are repeated for observing the heights on the points along the side AB. We know the heights of the points A and B from the *xy* plane; hence height of the points with reference to *xy* plane is determined by adding the proportional height from the *xy* plane.
18. The heights of any other point on line, DC, AD, BC and the two bisecting lines GH and EF are similarly obtained with reference to the same plane *xy*.

The procedure is mathematically given in the following sections.

8.3 Derivation of Formulae for Height

8.3.1 Height of a Point on the Diagonal AC

Let α_p be the auto-collimator reading of the pth point and the spacing between the third foot and the other two feet of the mirror mount is d, and then the height of the pth point, referring to Fig. 8.4 about the arbitrary line AO, is given by

$$z_p = \sum_{r=0}^{r=p} d\alpha_r. \tag{8.1}$$

It should be noted that α_0 is zero by definition. The abscissa x_p along the line AO is pd. Let the last point be $C(x_{2n}, z_{2n})$ for $p = 2n$, giving the ordinate of C as

$$z_{2n} = \sum_{r=0}^{r=2n} d\alpha_r. \tag{8.2}$$

Join AC and then height of the pth point P on the straight line AC

$$z'_p = (p/2n) \sum_{r=0}^{r=2n} d\alpha_r.$$

The difference in ordinates of the pth points on the curve AC and the straight line AC is the difference between z'_p and z_p and, therefore, is given by

$$z'_p - z_p = (p/2n) \sum_{r=0}^{r=2n} d\alpha_r - \sum_{r=0}^{r=p} d\alpha_r. \tag{8.3}$$

Fig. 8.5 may please be referred to for the derivation of (8.3).

Let the middle point on the curve be $Q(x_n, z_n)$ (Fig. 8.6) and we draw the line RS parallel to the line AC (Fig. 8.5). The spacing between the lines AC and RS will be $z'_n - z_n$ and thus the ordinate z''_p of the pth point with respect to RS as axis will be given by

$$z''_p = z'_n - z_n - (z'_p - z_p),$$

$$z''_p = \frac{1}{2} \sum_{r=0}^{r=2n} d\alpha_r - \sum_{r=0}^{r=n} d\alpha_r - \frac{p}{2n} \sum_{r=0}^{r-2n} d\alpha_r + \sum_{r=0}^{r=p} d\alpha_r$$

or

$$z''_p = d \left[\sum_{r=0}^{r=p} \alpha_r - \frac{p}{2n} \sum_{r=0}^{r=2n} \alpha_r + \frac{1}{2} \sum_{r=0}^{r=2n} \alpha_r - \sum_{r=0}^{r=n} \alpha_r \right]. \tag{8.4}$$

From (8.4), putting $p = 0$, we get the height of the end point A

$$z''_A = d \left[\frac{1}{2} \sum_{r=0}^{r=2n} \alpha_r - \sum_{r=0}^{r=n} \alpha_r \right] = b. \tag{8.5}$$

Similarly
putting $p = 2n$, we get the height of the end point C

$$z''_c = d \left[\frac{1}{2} \sum_{r=0}^{r=2n} \alpha_r - \sum_{r=0}^{r=n} \alpha_r \right] = b. \tag{8.5a}$$

Further z''_n for the point $Q = 0$

The heights of the points along the second diagonal BD of Fig. 8.1 are similarly calculated with reference to the line R′QS′. The two lines RQS and R′QS′ define the xy plane and heights of the point z'' with reference to these two lines are the heights of the points on the two diagonals with reference to the plane xy.

8.3.2 Height of a Point on the Diagonal BD

Following the steps as enumerated above, the height of any point on the diagonal BD is given by

$$z''_p = d \left[\sum_{r=0}^{r=p} \alpha'_r - \frac{p}{2n} \sum_{r=0}^{r=2n} \alpha'_r + \frac{1}{2} \sum_{r=0}^{r=2n} \alpha'_r - \sum_{r=0}^{r=n} \alpha'_r \right], \tag{8.6}$$

Where α'_r is the deflection from the mirror.

The value p at the starting point B is zero; therefore a – the height of point B is

$$a = d\left[\frac{1}{2}\sum_{r=0}^{2n}\alpha'_r - \sum_{r=0}^{r=n}\alpha'_r\right] = d\left[\frac{1}{2}\sum_{r=0}^{r=n}\alpha'_r + \frac{1}{2}\sum_{r=n+1}^{r=2n}\alpha'_r - \sum_{r=0}^{r=n}\alpha'_r\right]$$

$$= d\left[\frac{1}{2}\sum_{r=n+1}^{r=2n}\alpha'_r - \frac{1}{2}\sum_{r=0}^{r=n}\alpha'_r\right].$$

At the end point D, $p = 2n$; hence the height of the point D is

$$z''_D = a = d\left[\frac{1}{2}\sum_{r=n+1}^{r=2n}\alpha'_r - \frac{1}{2}\sum_{r=0}^{r=n}\alpha'_r\right]. \tag{8.6a}$$

8.3.3 Height of a Point on the Sides AB

The heights of the points on the two parallel sides AB and CD are calculated with reference to another plane $x'y'$. If z_p is the observed height of a point on line BA, and z'_p be the corresponding height on the line joining the point B to the tip of the height at the point A Fig. 8.6, we get

$$z'_p - z_p = (p/2n_2)\sum_{r=0}^{r=2n_2} d_2\alpha'_r - \sum_{r=0}^{r=p} d_2\alpha_r.$$

Here d_2 is the value of d – perpendicular distance of the front foot from the line joining the other two feet. $2n_2$ are the number of steps taken along the two sides AB or DC.

The height of the curve RQS (Fig. 8.7), given by the measured points above the $x'y'$ plane, is given by the difference between the heights of the line RS above the reference plane $x'y'$ and $\left(z'_p - z_p\right)$.

The height of the point D is a and that of A is b from the reference plane; hence the height of the pth point as counted from D is given by

$$z''' = a + (b-a)(p/2n_2) - (z'_p - z_p)$$

$$= d_2\left[\sum_{r=0}^{r=p}\alpha_r - (p/2n_2)\sum_{r=0}^{r=2n_2}\alpha_r\right] + (p/2n_2)(b-a) + a. \tag{8.7a}$$

If P is counted from the point A, then height of pth point will be given by

$$z''' = d_2\left[\sum_{r=0}^{r=p}\alpha_r - (p/2n_2)\sum_{r=0}^{r=2n_2}\alpha_r\right] + (p/2n_2)(a-b) + b \tag{8.7b}$$

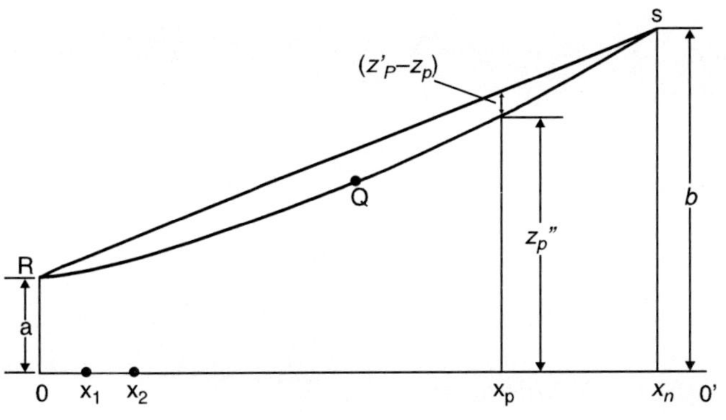

Fig. 8.7 Heights of the points on the line AB

8.3.4 Height of a Point on the Side DC

Following the same steps as for AB and representing α_r by α_r', the height of pth point on DC, if p is counted from D, is given by

$$z''' = d_2\left[\sum_{r=0}^{r=p}\alpha_r' - (p/2n_2)\sum_{r=0}^{r=2n_2}\alpha_r'\right] + (p/2n_2)(b-a) + a. \tag{8.8a}$$

If p is counted from the point C, then the height of pth point is given by

$$z''' = d_2\left[\sum_{r=0}^{r=p}\alpha_r' - (p/2n_2)\sum_{r=0}^{r=2n_2}\alpha_r\right] + (p/2n_2)(a-b) + b. \tag{8.8b}$$

8.3.5 Height of a Point on the Sides AD

Let another mirror with d_3 as the perpendicular distance between the front foot to the line joining the rear two feet and $2n_3$ the number of steps taken along AD. The height of the point D is a and that of A is b from the reference plane; hence the height of the pth point as counted from D from xy plane is given by

$$z''' = d_3\left[\sum_{r=0}^{r=p}\alpha_r - (p/2n_3)\sum_{r=0}^{r=2n_3}\alpha_r\right] + (p/2n_3)(b-a) + a. \tag{8.9a}$$

If p is counted from the point A, then height of pth point is given by

$$z''' = d_3 \left[\sum_{r=0}^{r=p} \alpha_r - (p/2n_3) \sum_{r=0}^{r=2n_3} \alpha_r \right] + (p/2n_3)(a-b) + b. \tag{8.9b}$$

As height of the point A is b from reference plane xy and that of D is a; hence proportional height to be added is

$$(p/2n_3)(a-b) + b.$$

8.3.6 Height of a Point on the Side BC

Similarly the height of the pth point as counted from B is given by

$$z''' = d_3 \left[\sum_{r=0}^{r=p} \alpha'_r - (p/2n_3) \sum_{r=0}^{r=2n_3} \alpha_r \right] + (p/2n_3)(b-a) + a. \tag{8.10a}$$

If the pth point is taken from the point C, then its height is given by

$$z''' = d_3 \left[\sum_{r=0}^{r=p} \alpha_r - (p/2n_3) \sum_{r=0}^{r=2n_3} \alpha_r \right] + (p/2n_3)(a-b) + b. \tag{8.10b}$$

8.3.7 Height of a Point on the Middle Side GH

$$z''' = d_3 \left[\sum_{r=0}^{r=p} \alpha_r - (p/2n_3) \sum_{r=0}^{r=2n_3} \alpha_r \right] + (p/2n_3)(f-c) + c. \tag{8.11}$$

Here c the height of point G is the sum of heights h_1 and h_2. The h_1 is the mean of the heights of points A and B, G being the midpoint of AB, and h_2 is the measured height of G along the line AB. Here h_1 and h_2 are given as

$$h_1 = \frac{a+b}{2},$$

$$h_2 = \frac{1}{2} d_2 \left\{ \sum_{r=0}^{r=n_2} \alpha_r - \sum_{n_2+1}^{r=2n_2} \alpha_r \right\},$$

giving c as

$$c = \frac{1}{2} d_2 \left\{ \sum_{r=0}^{r=n_2} \alpha_r - \sum_{n_2+1}^{r=2n_2} \alpha_r \right\} + \frac{a+b}{2}. \tag{8.12}$$

Similarly f is given as

$$f = \frac{1}{2}d_2\left\{\sum_{r=0}^{r=n_2}\alpha'_r - \sum_{n_2+1}^{r=2n_2}\alpha'_r\right\} + \frac{a+b}{2}. \tag{8.13}$$

8.3.8 Height of a Point on the Middle Side EF

$$z''' = d_2\left[\sum_{r=0}^{r=p}\alpha_r - (p/2n_2)\sum_{r=0}^{r=2n_2}\alpha_r\right] + (p/2n_2)(f' - c') + c'. \tag{8.14}$$

Here c' the height of point E is the sum of heights h'_1 and h'. The h'_1 is the mean of the heights of points A and D, E being the midpoint of AD, and h'_2 is the measured height E along the line AD. Hence h'_1 and h'_2 are given as

$$h'_1 = \frac{a+b}{2}$$

and

$$h'_2 = \frac{1}{2}d_3\left\{\sum_{r=0}^{r=n_3}\alpha_r - \sum_{n_3+1}^{r=2n_3}\alpha_r\right\}.$$

Here c' is given as

$$c' = \frac{1}{2}d_3\left\{\sum_{r=0}^{r=n_3}\alpha_r - \sum_{n_3+1}^{r=2n_3}a_r\right\} + \frac{a+b}{2} \tag{8.15}$$

and f' is given as

$$f' = \frac{1}{2}\,d_3\left\{\sum_{r=0}^{r=n_3}\alpha'_r - \sum_{n_3+1}^{r=2n_3}\alpha'_r\right\} + \frac{a+b}{2}. \tag{8.16}$$

Equations (8.12), (8.13), (8.15) and (8.16) are used to calculate the heights of the end points of GH and EF respectively with reference to the plane xy.

Some of the points may be below the reference plane xy. If desired, the maximum height below the plane xy is added to the heights of all the points. By doing so all the heights are positive and the points are above the reference plane. It is then a very simple matter to judge the flatness of the surface plate under calibration.

8.3.9 Heights of Some Important Points

Referring to Fig. 8.1 and using the symbols used in the section above, heights with respect to plane *xy* of certain important points are as follows:

Point	Height
Q the intersection of the diagonals	0
A and C the ends of the diagonal AC	b
B and D the ends of the diagonal BD	a

G midpoint of AB,
$$c = \frac{1}{2}d_2\left\{\sum_{r=0}^{r=n_2}\alpha_r - \sum_{n_2+1}^{r=2n_2}\alpha_r\right\} + \frac{a+b}{2}, \tag{8.17}$$

H the midpoint of DC
$$f = \frac{1}{2}d_2\left\{\sum_{r=0}^{r=n_2}\alpha'_r - \sum_{n_2+1}^{r=2n_2}\alpha'_r\right\} + \frac{a+b}{2}, \tag{8.18}$$

E midpoint of AD,
$$c' = \frac{1}{2}d_3\left\{\sum_{r=0}^{r=n_3}\alpha_r - \sum_{n_3+1}^{r=2n_3}\alpha_r\right\} + \frac{a+b}{2}, \tag{8.19}$$

F the midpoint of BC
$$f' = \frac{1}{2}d_3\left\{\sum_{r=0}^{r=n_3}\alpha'_r - \sum_{n_3+1}^{r=2n_3}\alpha'_r\right\} + \frac{a+b}{2}. \tag{8.20}$$

Midpoints of EF and GH may not be the same and may not coincide with the point Q. So it may be easily seen that the lines GH and EF do not intersect each other; also none of them may intersect either of the two diagonals.

If the intersection point of the two diagonals at the centre Q above the chosen reference plane is z_1, and if the heights of the lines GH and FE at the centre are z_2 and z_3 respectively, then the mean intersection point can be taken as $(2z_1 + z_2 + z_3)/4 = \bar{z}$. Then correction amount $((\bar{z} - z_1)(1 - 2s_1/w)$ should be added to the points on the diagonals AQC and BQD. Here s_1 is the distance from the centre Q along the diagonal to the corresponding corner and w is the length of the diagonal. The corresponding correction for the paths QGA, QGB, QHD and QHC is $((\bar{z} - z_2)\{1 - 2s_2/(u + v)\}$, where s_2 is the distance of the point from the point Q along either of the paths, and u and v are the lengths of the parallel sides AB or DC and AD or BC, respectively. Similarly the corrections for the paths QFA, QFD, QEB and QEC are $((\bar{z} - z_3)\{1 - 2s_3/(u + v)\}$, where s_3 is the distance of Q along the aforesaid paths.

8.4 Numerical Example

8.4.1 For Diagonals BD or AC

Formulae used for diagonals

$$z''_p = d_1\left[\sum_{r=0}^{r=p}\alpha_r - \frac{p}{2n_1}\sum_{r=0}^{r=2n_1}\alpha_r + \frac{1}{2}\sum_{r=0}^{r=2n_1}\alpha_r - \sum_{r=0}^{r=n_1}\alpha_r\right],$$

$$b = d_1\left[\frac{1}{2}\sum_{r=0}^{r=2n_1}\alpha_r - \sum_{r=0}^{r=n_1}\alpha_r\right] \qquad a = d_1\left[\frac{1}{2}\sum_{r=0}^{r=2n_1}\alpha'_r - \sum_{r=0}^{r=n_1}\alpha'_r\right]$$

Data

$$d_1 = 55mm, n_1 = 12$$

$$(1/2)\sum_0^{2n_1} d_1\alpha_r = 11.275/2 = 5.637, \; 1/2\sum_0^{2n_1} d_1\alpha_r = -9.9$$

$$\sum_0^{n_1} d_1\alpha_r = 4.675 \qquad \sum_0^{n_1} d_1\alpha_r = -11.55$$

Giving $\mathbf{b = 0.96\ \mu m}$ Giving $\mathbf{a = 1.65\ \mu m}$

8.4.1.1 Description of Table 8.1

I column is α the observations
II column is αd
III column is $d\sum_{r=0}^{r=p}\alpha_r$ progressive sum of column II
Last item is $d\sum_{r=0}^{r=2n_1}\alpha_r$ and middle item is $d\sum_{r=0}^{r=n_1}\alpha_r$; these give the values of b and a
IV column is $(p/n_1)d_1\sum_{r=0}^{r=2n_1}\alpha_r$
V column is III – IV + b gives height z of points taken along the diagonal AC

The same five columns are repeated for BD as shown in Table 8.1.

Table 8.1 Observations and calculation of heights of points on the lines BD and AC

	BD					AC				
S. No	I	II	III	IV	V	I	II	III	IV	V
	$\alpha 10^{-4}$ rad	$d_1\alpha$ $d_1 = 55$	$d_1 \Sigma\alpha$	$p/n \sum_0^{2n_1} d_1\alpha_r$	III–IV +0.96	α'	$d_1\alpha'$	$\Sigma d_1\alpha'$	$p/n_1 \sum_0^{2n_1} d_1\alpha'_r$	III–IV +1.65
0	0.0	0.0	0.0	0.0	0.96	0.0	0.0	0.0	0.0	1.65
1	0.0	0.0	0.0	0.940	−0.02	0.0	0.0	0.0	−1.65	3.3
2	0.0	0.0	0.0	1.879	−0.92	−3.0	−1.65	−1.65	−3.3	3.3
3	1.5	0.825	0.825	2.819	−1.03	−5.0	−2.75	−4.4	−4.95	2.2
4	3.0	1.65	2.475	3.758	−0.32	−5.0	−2.75	−7.15	−6.6	1.1
5	2.0	1.1	3.575	4.698	−0.16	−4.0	−2.2	−9.35	−8.25	0.55
6	2.0	1.1	4.675	5.638	0.00	−4.0	−2.2	−11.55	−9.9	0.0
7	2.0	1.1	5.775	6.577	0.16	−2.0	−1.1	−12.65	−11.55	0.55
8	0.5	0.275	6.05	7.517	−0.51	−3.0	−1.65	−14.3	−13.2	0.55
9	1.5	0.825	6.875	8.456	−0.62	−2.0	−1.1	−15.4	−14.85	1.10
10	3.0	1.65	8.525	9.396	0.09	−2.0	−1.1	−16.5	−16.5	1.65
11	2.0	1.1	9.625	10.335	0.25	−1.0	−0.55	−17.05	−18.15	2.75
12	3.0	1.65	11.275	11.275	0.96	−5.0	−2.75	−19.8	−19.8	1.65

8.4.2 For Longer Parallel Sides BA or CD

Formula used for longer parallel sides

$$z_p = d_2\left[\sum_{r=0}^{r=p}\alpha_r - (p/2n_2)\sum_{r=0}^{r=2n_2}\alpha_r\right] + p/2n_2)(b-a) + a \text{ for BA}$$

$$z_p = d_2\left[\sum_{r=0}^{r=p}\alpha_r - (p/2n_2)\sum_{r=0}^{r=2n_2}\alpha_r\right] + p/2n_2)(a-b) + b \text{ for CD}$$

Data

$d_2 = d = 55\,mm$ $n_2 = 10$

$\sum_0^{2n_2} d_2\alpha_r = 2.75$ $\sum_0^{n_2} d_2\alpha_r = 19.25$

$\mathrm{B} = -\sum_0^{2n_2} d_2\alpha_\mathrm{r} + b - a$

$= -2.75 - 0.69 = -3.44$

$\mathrm{B} = -\sum_0^{2n_2} d_2\alpha_\mathrm{r} + a - b$

$= -19.25 + 0.69 = -18.56$

$\sum_0^{n_2} d_2\alpha_r = 0.0$ $\sum_0^{n_2} d_2\alpha_r = 9.35$

$\sum_0^{n_2} d_2\alpha_r - 1/2\sum_0^{2n_2} d_2\alpha_r + 1/2(b+a)$

$= c$

$1.375 - 0.0 + 1.305 = \mathbf{0.07} = c$

$\sum_0^{n_2} d_2\alpha_r - 1/2\sum_0^{2n_2} d_2\alpha_r + 1/2(b+a) = f$

$-9.625 + 9.35 + 1.305 = \mathbf{1.03} = f$

8.4.2.1 Description of Table 8.2

I column is observations
II column is height αd
III column progressive sum of αd

Last item is $d_2 \sum_{r=0}^{r=2n_2} \alpha_r$ and middle item is $d_2 \sum_{r=0}^{r=n_2} \alpha_r$; the knowledge of a and b gives B to be used in column IV and also for the values of c and f.

IV column $(p/2n_2)$B
V Column is III – IV + V + a

Similar 5 steps are taken for CD. From column III, B and f are calculated as before.

8.4.3 For Sides BC or AD

Formula used for heights of shorter parallel sides

$$d_3\left[\sum_{r=0}^{r=p} \alpha_r - (p/2n_3)\sum_{r=0}^{r=2n_3} \alpha_r\right] + (p/2n_3)(b-a) + a \text{ For the side BC,}$$

$$d_3\left[\sum_{r=0}^{r=p} \alpha'_r - (p/2n_3)\sum_{r=0}^{r=2n_3} \alpha'_r\right] + (p/2n_3)(a-b) + b \text{ For the side AD.}$$

Data

$d_3 = 50\,\text{mm}$ $\qquad$ $n_3 = 6$

$\sum_0^{2n_3} d_3\alpha_r = 7.0,\ b-a = 1.65 - 0.96 = 0.69$ $\qquad$ $\sum_0^{2n_3} d_3\alpha_r = 0.5,\ a-b = -0.69$

Table 8.2 Observations and calculation of heights of points on the lines BA and CD

	BA					CD				
S No	Iα	IIαd	IIIΣαd	IVB.p/n	V IV + V + a	Iα	IIαd	IIIΣαd	IVB.p/n	V IV + V + b
0	0.0	0.0	0.0	0.0	1.65	0.0	0.0	0.0	0.0	0.96
1	0.0	0.0	0.0	−0.344	1.31	0.0	0.0	0.0	−1.856	−0.90
2	−2.0	−1.1	−1.1	−0.688	−0.14	3.0	1.65	1.65	−3.712	−1.10
3	−1.0	−0.55	−1.65	−1.032	−1.03	0.0	0.0	1.65	−5.568	−2.96
4	1.0	0.55	−1.1	−1.376	−0.83	6.0	3.3	4.95	−7.424	−1.51
5	2.0	1.1	0.0	−1.720	−0.07	8.0	4.4	9.35	−9.28	1.03
6	0.0	0.0	0.0	−2.064	−0.41	7.0	3.85	13.20	−11.136	3.02
7	0.0	0.0	0.0	−2.408	−0.76	4.0	2.20	15.40	−12.992	3.37
8	1.0	0.55	0.55	−2.752	−0.55	2.0	1.10	16.50	−14.848	2.61
9	2.0	1.1	1.65	−3.096	−0.26	3.0	1.65	18.15	−16.704	2.41
10	2.0	1.1	2.75	−3.44	0.96	2.0	1.1	19.25	−18.56	1.65

Giving $B = -\sum_{0}^{2n_3} d_3\alpha_r + b - a = -6.31$ Giving $B = -\sum_{0}^{2n_3} d_3\alpha_r + a - b$

$= -1.19$

$\sum_{0}^{n_3} d\alpha_r = 2.0$ $\sum_{0}^{n_3} d\alpha_r = -1.5$

$-1/2\sum_{0}^{2n_3} d_3\alpha_r + \sum_{0}^{n_3} d\alpha_r + 1/2(b+a) = c'$ $-1/2\sum_{0}^{2n_3} d_3\alpha_r + \sum_{0}^{n_3} d\alpha_r$

$+1/2\,(b + +) = f'$

$c' = -3.5 + 2.0 + 1.305 = \mathbf{-0.195}$ $f' = -0.25 + (-1.5) + 1.305$

$= \mathbf{-0.445}$

Similar two sets of 5 steps each are followed for BC and AD. Column III in each case will give the corresponding values of B, c' and f' (Table 8.3).

8.4.4 *For Middle Parallel Sides EF and GH*

$$z_p = d_3\left[\sum_{r=0}^{r=p}\alpha_r - (p/2n_3)\sum_{r=0}^{r=2n_3}\alpha_r\right] + (p/2n_3)(f-c) + \text{ for GH}$$

$$z_p = d_2\left[\sum_{r=0}^{r=p}\alpha_r - (p/2n_2)\sum_{r=0}^{r=2n_2}\alpha_r\right] + (p/2n_2)(f'-c') + c' \text{ for EF}$$

Data

$D = 55$ mm, $n_2 = 10$for EF and $d_3 = 55$ mm and $n_3 = 6$ for GH

Table 8.3 Observations and calculation of heights of points on the lines BC and AD

	AD					BC				
SN	α	αd	$\Sigma\alpha d$	B.p/n	IV + V + b	α	αd	$\Sigma\alpha d$	B.p/n	IV + V + a
0	0.0	0.0	0.0	0.0	0.96	0.0	0.0	0.0	0.0	1.65
1	0.0	0.0	0.0	−1.05	−0.09	0.0	0.0	0.0	−0.198	1.45
2	1.0	0.5	0.5	−2.10	−0.64	−5.0	−2.5	−2.5	−0.397	−1.25
3	3.0	1.5	2.0	−3.15	−0.19	2.0	1.0	−1.5	−0.595	−0.44
4	5.0	2.5	4.5	−4.21	−1.25	−2.0	−1.0	−2.5	−0.793	−1.64
5	4.0	2.0	6.5	−5.26	2.20	1.0	0.5	−2.0	−0.992	−1.34
6	1.0	0.5	7.0	−6.31	1.65	5.0	2.5	0.5	−1.19	0.96

Table 8.4 Observations and calculation of heights of points on the lines EF and GH

	EF					GH				
S.N	α	αd	$\Sigma\alpha d$	B.p/n	IV + V + c	α	αd	$\Sigma\alpha d$	B.p/n	IV + V + a
0	0.0	0.0	0.0	0.0	−0.44	0.0	0.0	0.0	0.0	0.07
1	0.0	0.0	0.0	0.734	−0.30	0.0	0.0	0.0	−0.733	−0.8
2	−1.0	−0.55	−0.55	1.48	+0.49	2.0	1.0	1.0	−1.467	−0.54
3	−3.0	−1.65	−2.2	2.22	−0.42	1.0	0.5	1.5	−2.20	−0.77
4	−1.0	−0.55	−2.75	2.96	−0.23	3.0	1.5	3.0	−2.933	0.00
5	−2.0	−1.1	−3.85	3.70	−0.59	3.0	1.5	4.5	−3.667	0.76
6	−0.0	0.0	−3.85	4.44	+0.15	2.0	1.0	5.5	−4.4	1.03
7	−1.0	−0.55	−4.40	5.18	+0.34					
8	−2.0	−1.1	−5.5	5.92	−0.02					
9	−2.0	−1.1	−6.60	6.66	−0.38					
10	−1.0	−.55	−7.1	7.4	−0.19					

$\mathrm{B} = -\sum_0^{2n} d\alpha_r + c - f = +7.1 + (-0.19 + 0.445) = 7.35$ $\mathrm{B} = -\sum_0^{2n} d\alpha_r + f - c = -5.5 + 1.03 + 0.07 = -4.4$. Similar two sets of 5 steps each as given for parallel sides are followed for EF and GH (Table 8.4).

8.5 Fitting a Plane to the Given Data

The most general equation of a plane is given by

$$ax + by + cz = d. \tag{8.21}$$

In the particular case of calibration of a surface plate, it is z which is measured; hence (8.21) can also be expressed as

$$z = ax + by + d, \tag{8.22}$$

where new a, b and d are old a, b, d divided by c. As z is the only measured quantity and all other coordinates are calculated from the geometry of the mean plane xy, the deviation of any point from the mean plane is due to the difference in its z ordinate. If z'_p is the z ordinate of such a point and z_p is the ordinate of the point on the plane then $z'_p - z_p$ is given as

$$z'_p - z_p = z'_p - ax_p - by_p - d.$$

Now if each of these deviations from the mean plane is a random deviation, and if all these random points belong to the same random population, then the probability of all deviations found occurring together is the product of individual probability of happening for any point, which is given by the Gaussian probability distribution. The combined probability P is given by

$$P = \frac{1}{\sigma^n (2\pi)^{n/2}} \prod_{p=1}^{p=n} \exp(-z'_p - z_p)^2/2\sigma^2,$$

$$P = \frac{1}{\sigma^n (2\pi)^{n/2}} \exp\left\{-\sum_{p=1}^{p=n} \frac{(z'_p - z_p)^2}{2\sigma^2}\right\}. \tag{8.23}$$

The expression in (8.23) will be a maximum if the negative power of the exponent is a minimum. The condition, therefore, for the maximum probability is that E, as given in (8.24), is a minimum. E is given by

$$E = \sum_{p=1}^{p=n} (z'_p - z_p)^2. \tag{8.24}$$

To make E a minimum its partial derivatives with respect to variable a, b and d must be zero.

Hence conditions for minimum value E are

$$\begin{aligned} \frac{\delta E}{\delta a} &= 0 \\ \frac{\delta E}{\delta b} &= 0 \\ \frac{\delta E}{\delta d} &= 0. \end{aligned} \tag{8.25}$$

It is seen that in applying the condition that P should be a maximum, known as the "Principle of maximum likelihood", this in turn has led to the requirement that sum of the squares of the deviations should be a minimum. This condition is known as "Principle of least squares" and is often applied directly without using the principle of maximum likelihood.

From (8.25), therefore, we get

$$\begin{aligned} \frac{\delta E}{\delta a} &= -\sum_{p=1}^{p=n} x_p(z'_p - ax_p - by_p - \vec{d}) = 0, \\ \frac{\delta E}{\delta b} &= -\sum_{p=1}^{p=n} y_p(z'_p - ax_p - by_p - d) = 0, \\ \frac{\delta E}{\delta d} &= \sum_{p=1}^{p=n} (z'_p - ax_p - by_p - d) = 0. \end{aligned} \tag{8.26}$$

Rewriting (8.26), we get

$$\sum_{p=1}^{p=n} x_p z'_p = a \sum_{p=1}^{p=n} x_p^2 + b \sum_{p=1}^{p=n} x_p y_p + d \sum_{p=1}^{p=n} x_p, \tag{8.27}$$

$$\sum_{p=1}^{p=n} y_p z'_p = a \sum_{p=1}^{p=n} y_p x_p + b \sum_{p=1}^{p=n} y_p^2 + d \sum_{p=1}^{p=n} y_p, \tag{8.28}$$

$$\sum_{p=1}^{p=n} z'_p = a \sum_{p=1}^{p=n} x_p + b \sum_{p=1}^{p=n} y_p + nd. \tag{8.29}$$

These are called normal equations. There are three equations and three unknown, namely a, b and d, and each unknown can be calculated from these equation by simple calculations.

To eliminate d from (8.27) and (8.29) multiply (8.27) by n and (8.29) by $\sum_{p=1}^{p=n} x_p$ and subtracting we get

$$a \left\{ n \sum_{p=1}^{p=n} x_p^2 - \left(\sum_{p=1}^{p=n} x_p \right)^2 \right\} + b \left\{ n \sum_{p=1}^{p=n} x_p y_p - \sum_{p=1}^{p=n} x_p \sum_{p=1}^{p=n} y_p \right\}$$
$$= n \sum_{p=1}^{p=n} x_p z'_p - \sum_{p=1}^{p=n} x_p \sum_{p=1}^{p=n} z'_p. \tag{8.30}$$

Similarly, eliminating d from (8.28) and (8.29) by multiplying the (8.28) by n and (8.29) by $\sum_{p=1}^{p=n} y_p$ and subtracting we get

$$a \left\{ n \sum_{p=1}^{p=n} x_p y_p - \sum_{p=1}^{p=n} x_p \sum_{p=1}^{p=n} y_p \right\} + b \left\{ \sum_{p=1}^{p=n} y_p^2 - \left(\sum_{p=1}^{p=n} y_p \right)^2 \right\}$$
$$= n \sum_{p=1}^{p=n} y_p z'_p - \sum_{p=1}^{p=n} y_p \sum_{p=1}^{p=n} z'_p. \tag{8.31}$$

Dividing each side of equation (8.30) by n^2, we see that coefficient of a in (8.30) becomes

$$\frac{\sum_{p=1}^{p=n} x_p^2}{n} - \frac{\left(\sum_{p=1}^{p=n} x_p \right)}{n^2}.^2 \tag{8.32}$$

This is nothing but variance of x, so we denote it by V_{xx}.

Similarly coefficient of b in (8.30) becomes

$$\frac{\sum_{p=1}^{p=n} x_p y_p}{n} - \frac{\sum_{p=1}^{p=n} x_p \sum_{p=1}^{p=n} y_p}{n^2} \tag{8.33}$$

This is nothing but the covariance between x and y, so we denote it as Vxy

Following the same principle for (8.31), we write the two equations (8.30) and (8.31) respectively as

$$V_{xx}a + V_{xy}\, b = V_{xz'} \tag{8.34}$$

$$V_{xy}a + V_{yy}b = V_{yz'}. \tag{8.35}$$

Solving (8.35) and (8.36) for a and b, we get

$$a = \frac{V_{xz'}V_{yy} - V_{yz'}V_{xy}}{V_{xx}V_{yy} - V_{xy}^2}, \tag{8.36}$$

$$b = \frac{V_{xx}V_{yz'} - V_{xz'}V_{xy}}{V_{xx}V_{yy} - V_{xy}^2}. \tag{8.37}$$

Substituting the values of a and b in (8.31), we get the value of d giving us

$$d = \frac{\sum_{p=1}^{p=n} z'_p}{n} - \frac{V_{xz'}V_{yy} - V_{yz}V_{xy}}{V_{xx}V_{yy} - V_{xy}^2} \times \frac{\sum_{p=1}^{p=n} x_p}{n} - \frac{V_{yz'}V_{xx} - V_{xy}V_{xz'}}{V_{xx}V_{yy} - V_{xy}^2} \cdot \frac{\sum_{p=1}^{p=n} y_p}{n}. \tag{8.38}$$

The equation of the mean plane is thus

$$\begin{aligned} z = {} & \frac{V_{xz'}V_{yy} - V_{yz'}V_{xy}}{V_{xx}V_{yy} - V_{xy}^2} \cdot x + \frac{V_{xz'}V_{yy} - V_{yz'}V_{xy}}{V_{xx}V_{yy} - V_{xy}^2} \cdot y + \frac{\sum_{p=1}^{p=n} z'_p}{n} \\ & - \frac{V_{xz'}V_{yy} - V_{yz}V_{xy}}{V_{xx}V_{yy} - V_{xy}^2} \cdot \frac{\sum_{p=1}^{p=n} x_p}{n} - \frac{V_{yz'}V_{xx} - V_{xy}V_{xz'}}{V_{xx}V_{yy} - V_{xy}^2} \cdot \frac{\sum_{p=1}^{p=n} y_p}{n}. \end{aligned} \tag{8.39}$$

Now the coordinates of the centroid of all points are

$$\frac{\sum_{p=1}^{p=n} x_p}{n}, \frac{\sum_{p=1}^{p=n} y_p}{n}, \frac{\sum_{p=1}^{p=n} z'_p}{n}. \tag{8.40}$$

If the coordinates of the centroid are substituted for x, y and z in (8.39), it is clearly seen that the equation becomes identically equal to zero; hence it is seen that the plane in (8.39) passes through the centroid.

By shifting the origin to the centroid, we can get the equation of the same plane, which is simpler because of the absence of the constant term; the other coefficients of x and y remain the same.

Usually the overall departure from the mean plane is not much different than those calculated by using the equations derived in Sect. 8.3.1. The mean plane method is, however, useful if the standard deviation of the departure of the surface plate from the mean plane is required.

8.5.1 Standard Deviation from the Mean Plane

The equation of the mean plane, with centroid as origin, may be written as

$$\mathrm{A}X + \mathrm{B}Y - Z = 0, \tag{8.41}$$

$$\text{Let } X_p = x_p - \frac{\sum_{p=1}^{p=n} x_p}{n}, Y_p = y_p - \frac{\sum_{p=1}^{p=n} y_p}{n} Z_p = z_p - \frac{\sum_{p=1}^{p=n} z'_p}{n}. \tag{8.42}$$

The distance of any point (X_p, Y_p, Z_p) from the plane is given by

$$\frac{\mathrm{A}X_p + \mathrm{B}Y_p - Z_p}{(\mathrm{A}^2 + \mathrm{B}^2 + 1)^{1/2}}. \tag{8.43}$$

Thus the standard deviation is given by

$$s = \left\{ \sum_{p=0}^{p=n} \frac{(\mathrm{A}X_p + \mathrm{B}Y_p - Z_p)^2}{(n-2)(\mathrm{A}^2 + \mathrm{B}^2 + 1)} \right\}^{1/2}. \tag{8.44}$$

$n-2$ appears in the denominator, as two parameters A and B have been determined from n points, giving n-2 as effective degrees of freedom.

8.6 Uncertainty in Measurements

8.6.1 Uncertainty in Measured Height of a Point

The instrument used for measurement of angular deviation α is same. Hence Type A standard uncertainty for each α will be the same. Let it be σ_α.

8.6.1.1 Uncertainty in Height of a Point on the Diagonals AC or BD

Notations used: d_1 step distance on AC or BD

d_2 step distance on AB, CD or EF
d_3 step distance on AB, GH or AD
$2n_1$ number of observations taken on AC or BD
$2n_2$ number of observations taken on AB or EF or CDD and
$2n_3$ number of observations taken on AD, GH, BC

The height of pth point on the diagonals AC or BD above the reference plane defined by the lines RQS and R′QS′ from (8.4)

$$z_{CA\ or\ BD} = d_1\left[\sum_{r=0}^{r=p}\alpha_r - \frac{p}{2n_1}\sum_{r=0}^{r=2n_1}\alpha_r + \frac{1}{2}\sum_{r=0}^{r=2n_1}\alpha_r - \sum_{r=0}^{r=n_1}\alpha_r\right]. \tag{8.45}$$

Expressing $\sum_{r=0}^{r=2n_1}\alpha_r$ in 2nd term as the sum of $\sum_{r=0}^{r=p}\alpha_r$ and $\sum_{r=p+1}^{r=2n_1}\alpha_r$, and in 3rd term as the sum $\sum_{r=0}^{r=n_1}\alpha_r$ and $\sum_{r=n_1+1}^{2n_1}\alpha_r$, (8.45) becomes

$$z_{\text{CA or BD}} = d_1\left[\sum_{r=0}^{r=p}\alpha_r - \frac{p}{2n_1}\sum_{r=0}^{r=p}\alpha_r - \frac{p}{2n_1}\sum_{r=P+1}^{r=2n_1}\alpha_r + \frac{1}{2}\sum_{r=n_1+1}^{r=2n_1}\alpha_r - \frac{1}{2}\sum_{r=0}^{r=n_1}\alpha_r\right],$$

$$z_{\text{CA or BD}} = d_1\left[\left(1-\frac{p}{2n_1}\right)\sum_{r=0}^{r=p}\alpha_r - \frac{p}{2n_1}\sum_{r=p+1}^{r=2n_1}\alpha_r + \frac{1}{2}\sum_{r=n_1+1}^{r=2n_1}\alpha_r - \frac{1}{2}\sum_{r=0}^{r=n_1}\alpha_r\right]. \tag{8.46}$$

We know that if z is a linear combination of several Gaussian variables for example

$$z = a_1x_1 + a_2x_2 + a_3x_3 + \cdots + a_nx_n, \tag{8.47}$$

then the variance σ_z^2 of the variable z is given by

$$V_z = a_1^2\sigma_{x1}^2 + a_2^2\sigma_{x2}^2 + a_3^2\sigma_{x3}^2 + \cdots + a_n^2\,\sigma_{xn}^2. \tag{8.48}$$

Measurement of all α_r is subject to uncertainty and thus can be considered as random variables. Further each α_r is obtained by using the same instrument; hence the standard deviation of each α_r will be the same. Let it be σ_α.

Following the law enunciated in (8.48), the (8.46) becomes

$$s^2_{\text{CA or BD}} = \left[p - \frac{p^2}{n_1} + \frac{p^3}{4n_1^2} + \frac{p^2}{4n_1^2}(2n_1 - p) + \frac{1}{4}n_1 + \frac{1}{4}n_1\right]d_1^2\sigma_\alpha^2$$

or

$$s^2_{\text{CA or BD}} = \left[p - \frac{p^2}{2n_1} + \frac{n_1}{2}\right] d_1^2 \alpha_\alpha^2. \tag{8.49}$$

8.6.1.2 Uncertainty in Height of a Point on Lines AB or DC

The height of the pth point on BA, if p is counted from B, is given as

$$z_{\text{AB or DC}} = d_2 \left[\sum_{r=0}^{r=p} \alpha_r - (p/2n_2) \sum_{r=0}^{r=2n_2} \alpha_r \right] + p/2n_2)(b-a) + a.$$

Then by expressing $\sum_{r=0}^{r=2n_1} \alpha_r$ as sum of two summations one from 1 to p and the other from $p+1$ to $2n_1$ and simplifying other terms, the height z_{ABorDC} of the pth point on the line AB or DC is expressed as

$$z_{\text{AB or DC}} = \left[\left(1 - \frac{p}{2n_2}\right) \sum_{r=0}^{r=p} \alpha_r - \frac{p}{2n_2} \sum_{r=p+1}^{r=2n_2} \alpha_r \right] d_2 + a\left(1 - \frac{p}{2n_2}\right) + \frac{p}{2n_2} b. \tag{8.50}$$

Now a and b from (8.5a) and (8.6a) are

$$a = d_1 \left[\frac{1}{2} \sum_{r=n_1+1}^{r=2n_1} \alpha_r - \frac{1}{2} \sum_{r=0}^{r=n_1} \alpha_r \right] \quad \text{and}$$

$$b = d_1 \left[\frac{1}{2} \sum_{r=n_1+1}^{r=2n_1} \alpha'_r - \frac{1}{2} \sum_{r=0}^{r=n_1} \alpha'_r \right]$$

Following (8.48) the variances of a and b

$$\sigma_a^2 = d_1^2 \left[\left(\frac{1}{4}(n_1 + \frac{1}{4} n_1 \right) \right] \sigma_\alpha^2 = d_1^2 \sigma_\alpha^2 n_1/2, \tag{8.51}$$

$$\sigma_b^2 = d_1^2 \left[\left(\frac{1}{4}(n_1 + \frac{1}{4} n_1 \right) \right] \sigma_a^2 = d_1^2 \sigma_\alpha^2 n_1/2. \tag{8.52}$$

Applying the law in (8.48) and substituting the values of variances of a and b in (8.50), we get the variance $\sigma^2_{\text{BA, DC}}$ as

$$\sigma_z^2 = \left[\left\{ \left(1 - \frac{p}{2n_2}\right)^2 p + \frac{p^2}{4n_2^2}(2n_2 - p) \right\} d_2^2 + \left(1 - \frac{p}{2n_2}\right)^2 d_1^2 n_1/2 + \frac{p^2}{4n_2^2}\left(d_1^2 n_1/2\right) \right] \sigma_\alpha^2,$$

$$\sigma^2_{\mathrm{BA,\,DC}} = \left[\left(p - \frac{p^2}{n_2} + \frac{p^3}{4n_2^2} + \frac{p^2}{2n_2} - \frac{p^3}{4n_2^2}\right) d_2^2 + \left(1 - \frac{p}{n_2} + \frac{p^2}{4n_2^2} + \frac{p^2}{4n_2^2}\right) d_1^2 n_1/2 + \right] \sigma_\alpha^2,$$

$$\sigma^2_{\mathrm{BA,\,DC}} = \left[p\left(1 - \frac{p}{2n_2}\right) d_2^2 + \left(1 - \frac{p}{n_2} + \frac{p^2}{2n_2^2}\right) d_1^2 n_1/2\right] \sigma_\alpha^2. \tag{8.53}$$

The expression in (8.53) is also valid for uncertainty in height of the pth point on DC; p is counted from the end D.

8.6.1.3 Uncertainty in Height of a Point on Lines AD and CB

Uncertainty of the pth point on AD or CB, if p is counted from the end A for points on AD or from the point C for points on CB, is obtained by replacing d_2 and n_2 by d_3 and n_3 respectively in (8.53). The expression for uncertainty of the pth point on AD or CB is given by

$$\sigma^2_{\mathrm{AD,\,CB}} = \left[p\left(1 - \frac{p}{2n_3}\right) d_3^2 + \left(1 - \frac{p}{n_3} + \frac{p^2}{2n_3^2}\right) d_1^2 n_1/2\right] \sigma_\alpha^2. \tag{8.54}$$

8.6.1.4 Uncertainty in Height of a Point on Line GH

The expression for the height of pth point on GH from (8.11) is

$$z_{\mathrm{GH}} = d_3 \left[\sum_{r=0}^{r=p} \alpha_r - (p/2n_3) \sum_{r=0}^{r=2n_3} \alpha_r\right] + (p/2n_3)(f - c) + c.$$

The above expression is modified, as in Sect. 8.6.1.1 and may be written as

$$z_{\mathrm{GH}} = d_3 \left[\left(1 - \frac{p}{2n_3}\right) \sum_{r=0}^{r=p} \alpha_r + \frac{p}{2n_3} \sum_{p+1}^{r=2n_3} \alpha_r\right] + c\left(1 - \frac{p}{2n_3}\right) + \frac{p}{2n_3} f. \tag{8.55}$$

Now c from (8.12) is given

$$c = \frac{1}{2} d_2 \left\{\sum_{r=0}^{r=n_2} \alpha_r - \sum_{n_2+1}^{r=2n_2} \alpha_r\right\} + \frac{a+b}{2}$$

and f from (8.13) is given as

$$f = \frac{1}{2}d_2\left\{\sum_{r=0}^{r=n_2}\alpha'_r - \sum_{0}^{r=2n_2}\alpha'_r\right\} + \frac{a+b}{2}.$$

Giving variance of c as

$$\sigma_c^2 = \frac{1}{4}d_2^2(n_2+n_2)\sigma_\alpha^2 + \frac{1}{4}(\sigma_a^2+\sigma_b^2).$$

Substituting the value of σ_a^2 and σ_b^2 from (8.51) and (8.52), we get

$$\sigma_c^2 = \left(\frac{1}{2}d_2^2 n_2 + \frac{1}{4}n_1 d_1^2\right)\sigma_\alpha^2. \tag{8.56}$$

Similarly the variance of f will be given as

$$\sigma_f^2 = \frac{1}{2}\left(d_2^2 n_2 + \frac{1}{2}d_1^2 n_1\right)\sigma_\alpha^2. \tag{8.57}$$

Applying again the law of variances enunciated in (8.48), we get

$$\sigma_{\mathrm{GH}}^2 = p\left(1-\frac{p}{2n_3}\right)d_3^2\sigma_\alpha^2 + \frac{1}{4}\left(1-\frac{p}{n_3}+\frac{p^2}{2n_3^2}\right)\left(2n_2d_2^2+n_1d_1^2\right)\sigma_\alpha^2 \tag{8.58}$$

8.6.1.5 Uncertainty in Height of a Point on Line EF

The uncertainty in height of a point on the line EF is similarly obtained by interchanging n_2 with n_3 and d_2 with d_3. The expression of variance of the height of the pth point on EF is given by

$$\sigma_{\mathrm{EF}}^2 = p\left(1-\frac{p}{2n_2}\right)d_2^2\sigma_\alpha^2 + \frac{1}{4}\left(1-\frac{p}{n_2}+\frac{p^2}{2n_2^2}\right)\left(2n_3d_3^2+n_1d_1^2\right)\sigma_\alpha^2. \tag{8.59}$$

8.6.2 Uncertainty in Difference in Heights of Two Points on the Diagonal

In regard to position of two points, two possibilities exist: one is that that both points are on the same line and the second is that two points lie on different lines.

8.6.3 Points Are on the Same Straight Line

8.6.3.1 The Points Are on the Diagonal AC or BD

Number of points taken on either diagonal is $2n_1$. The height of the pth point on the diagonal AC above the reference plane is given by

$$z_{ACp} = d_1\left[\sum_{r=0}^{r=p}\alpha_r - \frac{p}{2n_1}\sum_{r=0}^{r=2n_1}\alpha_r + \frac{1}{2}\sum_{r=0}^{r=2n_1}\alpha_r - \sum_{r=0}^{r=n_1}\alpha_r\right]. \tag{8.60}$$

Similarly height of any other qth point will be given by

$$z_{ACp} = d_1\left[\sum_{r=0}^{r=q}\alpha_r - \frac{q}{2n_1}\sum_{r=0}^{r=2n1}\alpha_r + \frac{1}{2}\sum_{r=0}^{r=2n1}\alpha_r - \sum_{r=0}^{r=n1}\alpha_r\right]. \tag{8.61}$$

If $q \le p$ then the summation from $r = 0$ to $r = 2n_1$ may be divided into three parts, namely (1) from $r = 0$ to $r = q$, (2) $r = q + 1$ to $r = p$ and (3) $r = p + 1$ to $r = 2n_1$. Thus, (8.60) and (8.61) can be written as

$$z_{ACp} = d_1\left[\sum_{r=0}^{r=q}\alpha_r + \sum_{r=q+1}^{r=p}\alpha_r - \left(\frac{1}{2} - \frac{p}{2n_1}\right)\left\{\sum_{r=0}^{r=q}\alpha_r + \sum_{r=q+1}^{r=p}\alpha_r + \sum_{r=p+1}^{r=2n_1}\alpha_r\right\} - \sum_{r=0}^{r=n_1}\alpha_r\right], \tag{8.62}$$

$$z_{ACq} = d_1\left[\sum_{r=0}^{r=q}\alpha_r + \sum_{r=q+1}^{r=p}\alpha_r - \left(\frac{1}{2} - \frac{q}{2n_1}\right)\left\{\sum_{r=0}^{r=q}\alpha_r + \sum_{r=q+1}^{r=p}\alpha_r + \sum_{r=p+1}^{r=2n_1}\alpha_r\right\} - \sum_{r=0}^{r=n_1}\alpha_r\right]. \tag{8.63}$$

Subtracting (8.63) from (8.62), we get the difference Δ_{AC} as

$$\Delta_{AC} = d_1\left[\left(\frac{q}{2n_1} - \frac{p}{2n_1}\right)\left\{\sum_{r=0}^{r=q}\alpha_r + \sum_{r=p+1}^{r=2n_1}\alpha_r\right\} + \left(1 + \frac{q-p}{2n_1}\right)\sum_{r=q+1}^{r=p}\alpha_r\right]. \tag{8.64}$$

Here also every α_r is subject to uncertainty and thus can be considered as random variables. Further each α_r is obtained by using the same instrument; hence the standard deviation of each α_r will be the same, taking it to be as σ_α.

Now Δ_{AC} is a linear function of α_r; hence the variance σ^2_{AC} of Δ_{AC} as given by (8.48) is given as

$$\sigma_{AC}^2 = \left\{ \frac{(q-p)^2}{4n_1^2}(q+2n_1-p) + \left(1+\frac{(q-p)}{2n_1}\right)^2 (p-q) \right\} d_1^2\sigma_\alpha^2, \quad (8.65)$$

$$\sigma_{AC}^2 = \left\{ 1+\frac{q-p}{2n_1} \right\} (p-q)\sigma_\alpha^2 d_1^2. \quad (8.66)$$

Here we see that variance and therefore the standard uncertainty is not constant, nor is linearly connected with the difference $p - q$. The uncertainty is zero for $p = q$ and for $p = 2n_1$ and q is zero. Further uncertainty is maximum for $p - q$ equal to n_1 and is given as

$$\text{Maximum standard uncertainty} = \left(\sqrt{n_1/2}\right)\sigma_\alpha d_1. \quad (8.67)$$

Same expression is equally true for the other diagonal BD

8.6.3.2 Points Are on the Parallel Side BA

The number of steps taken on the side BA is $2n_2$. The height of the pth point on BA above the datum plane from (8.7a) is given by

$$z_{BAp} = d_2 \left[\sum_{r=0}^{r=p} \alpha_r - (p/2n_2) \sum_{r=0}^{r=2n_2} \alpha_r \right] + (p/2n_2)(b-a)+a.$$

Expressing $\sum_{r=0}^{2n_2} \alpha_r = \sum_{r=0}^{r=p} \alpha_r + \sum_{r=p+1}^{2n_2} \alpha_r$ we get

$$z_{BAp} = \left[\left(1-\frac{p}{2n_2}\right) \sum_{r=0}^{r=p} \alpha_r - \frac{p}{2n_2} \sum_{r=p+1}^{r=2n_2} \alpha_r \right] d_2 + a\left(1-\frac{p}{2n_2}\right) + b\frac{p}{2n_2}. \quad (8.68a)$$

Similarly the height of qth point on BA above the datum is given by

$$z_{BAq} = \left[\left(1-\frac{q}{2n_2}\right) \sum_{r=0}^{r=p} \alpha_r - \frac{q}{2n_2} \sum_{r=p+1}^{r=2n_2} \alpha_r \right] d_2 + a\left(1-\frac{q}{2n_2}\right) + b\frac{q}{2n_2}. \quad (8.68b)$$

Taking $p \geq q$ and q as positive integer, the height difference between the two points is derived by subtracting (8.68b) from (8.68a), and is given by

$$\Delta_{BA} = \left[\frac{q}{2n_2} \sum_{r=q+1}^{r=2n_2} \alpha_r - \frac{p}{2n_2} \sum_{r=p+1}^{r=2n_2} \alpha_r + \left(1-\frac{p}{2n_2}\right) \sum_{r=0}^{r=p} \alpha_r \right.$$

$$-\left(1-\frac{q}{2n_2}\right)\sum_{r=0}^{r=q}\alpha_r\Bigg]d_2+\frac{p-q}{2n_2}(b-a), \tag{8.69}$$

where a and b are

$$a = d_1\frac{1}{2}\left[\sum_{r=n_1+1}^{r=2n_1}\alpha_r - \sum_{r=0}^{r=n_1}\alpha_r\right], \tag{8.70}$$

$$b = d_1\frac{1}{2}\left[\sum_{r=n_1+1}^{r=2n_1}\alpha'_r - \sum_{r=0}^{r=n_1}\alpha'_r\right]. \tag{8.71}$$

Substituting the values of a and b in (8.69) and collecting terms of same summations, we get

$$\Delta_{\mathrm{BA}} = \left[\left(\frac{q-p}{2n_2}\right)\sum_{r=0}^{r=q}\alpha_r + \left(1-\frac{p-q}{2n_2}\right)\sum_{r=q+1}^{r=p}\alpha_r + \left(\frac{q-p}{2n_2}\right)\sum_{r=p+1}^{r=2n_2}\alpha_r\right]d_2$$
$$+\left(\frac{p-q}{2n_2}\right)\frac{d_1}{2}\left[\sum_{r=n_1+1}^{r=2n_1}\alpha'_r - \sum_{r=n_1+1}^{r=2n_1}\alpha_r + \sum_{r=0_1}^{r=n_1}\alpha_r - \sum_{r=0}^{r=n_1}\alpha'_r\right]. \tag{8.72}$$

Here also Δ_{BA} is a linear function of α_r so the variance $(\sigma_{BA})^2$ by (8.48) is given by

$$\sigma^2_{\mathrm{BA}} = \left[\frac{(p-q)^2}{4n_2^2}(2n_2-p+q)+\left(1-\frac{p-q}{2n_2}\right)^2(p-q)\right]\sigma^2_\alpha d_2^2$$
$$+\frac{(p-q)^2}{4n_1^2}\frac{d_1^2}{4}(4n_1)\sigma^2_\alpha. \tag{8.73}$$

Equation (8.73) on simplification becomes

$$\sigma^2_{\mathrm{BA}} = \left[(p-q)\left(1-\frac{p-q}{2n_2}\right)d_2^2+\frac{(p-q)^2n_1d_1^2}{4n_2^2}\right]. \tag{8.74}$$

It may be noted that the first term has the same form as that for a diagonal, and the second term represents the contribution to the uncertainty from the heights of the two ends of diagonals BD and AC, which arises from the uncertainties of a and b.

In this case also by putting $p-q$ equal to h and differentiating (8.74) with respect to h and equating the differential coefficient equal to zero we see that that the uncertainty is a maximum at h and is given by

$$h = \frac{2n_2}{(2 - \mathrm{n}_1 d_1^2 / n_2 d_2^2)} \tag{8.75}$$

and maximum value of standard uncertainty σ_{BA} is given by

$$\sigma_{\mathrm{BA}} = \frac{\sqrt{(2n_2)} d\sigma_{\mathrm{a}}}{\sqrt{\{2(2 - n_1 d_1^2 / n_2 d_2^2)\}}}. \tag{8.76}$$

8.6.3.3 The Points Are on the Parallel Side CD

Equations (8.74), (8.75) and (8.76) hold good for any points on CD, where p and q are numbered from the end C

8.6.3.4 The Points Are on the Parallel Sides BC or AD

Equations similar to (8.74), (8.75) and (8.76) will hold good for any two points on sides BC or AD, except that $2n_2$ is replaced by $2n_3$. Here $2n_3$ is the number of observations taken on sides BC or AD.

8.6.3.5 The Points Are on the Central Side GH

The height of the pth point on the GH line above the datum plane is given by

$$z_{\mathrm{GH}p} = \left[\left(1 - \frac{p}{2n_3}\right)\sum_{r=0}^{r=p}\alpha_r - \frac{p}{2n_3}\sum_{r=p+1}^{2n_3}\alpha_r\right] d_3 + c\left(1 - \frac{p}{2n_3}\right) + \frac{fp}{2n_3}. \tag{8.77}$$

Similarly the height of qth point on the line GH above the datum plane is given by

$$z_{\mathrm{GH}q} = \left[\left(1 - \frac{q}{2n_3}\right)\sum_{r=0}^{r=q}\alpha_r - \frac{q}{2n_3}\sum_{r=q+1}^{2n_3}\alpha_r\right] d_3 + c\left(1 - \frac{q}{2n_3}\right) + \frac{qf}{2n_3}. \tag{8.78}$$

Here $2n_3 \geq p \geq q \geq 0$.

The difference of height Δ_{GH} between these two points is obtained by subtracting (8.78) from (8.77), giving us

$$\Delta_{\text{GH}} = \left[\left(\frac{q-p}{2n_3}\sum_{r=0}^{r=q}\alpha_r\right) + \left(1-\frac{p-q}{2n_3}\right)\sum_{r=q+1}^{r=p}\alpha_r - \left(\frac{p-q}{2n_3}\right)\sum_{r=p+1}^{r=2n_3}\alpha_r\right]d_3 + \left(\frac{p-q}{2n_3}\right)(f-c). \tag{8.79}$$

Here c is the height of the middle point of BA above the datum plane and is given by

$$c = \left[\frac{1}{2}\sum_{r=0}^{n_2}\alpha_r - \frac{1}{2}\sum_{n_1+1}^{2n_2}\alpha_r\right]d_2 + \frac{a+b}{2}. \tag{8.80}$$

By putting $p = n_2$ in (8.68) and substituting the values of a and b from (8.70) and (8.71), c is given as

$$c = \left[\frac{1}{2}\sum_{r=0}^{r=n_2}\alpha_r - \frac{1}{2}\sum_{n_2+1}^{2n_2}\alpha_r\right]d_2 + \frac{1}{4}\left[\sum_{n_1+1}^{r=2n_1}\alpha_r - \sum_{r=0}^{n_1}\alpha_r\right]d_1 + \frac{1}{4}\left[\sum_{n_1+1}^{r=2n_1}\alpha'_r - \sum_{r=0}^{n_1}\alpha'_r\right]d_1, \tag{8.81}$$

and f is the height of the midpoint of CD from the datum plane and is given as

$$f = \left[\frac{1}{2}\sum_{r=0}^{n_2}\alpha'_r - \frac{1}{2}\sum_{n_1+1}^{2n_2}\alpha'_r\right]d_2 + \frac{a+b}{2}. \tag{8.82}$$

From (8.81) and (8.82), we get

$$f - c = \left[\frac{1}{2}\sum_{r=0}^{r=n_2}\alpha'_r - \frac{1}{2}\sum_{n_1+1}^{2n_2}\alpha'_r - \frac{1}{2}\sum_{r=0}^{r=n_2}\alpha_r + \frac{1}{2}\sum_{n_1+1}^{2n_2}\alpha_r\right]d_2. \tag{8.83}$$

Substituting the values of $f - c$ from (8.83) in (8.79), we get

$$\Delta_{\text{GH}} = \left[\left(\frac{q-p}{2n_3}\sum_{r=0}^{r=q}\alpha_r\right) + \left(1-\frac{p-q}{2n_3}\right)\sum_{r=q+1}^{r=p}\alpha_r - \left(\frac{p-q}{2n_3}\right)\sum_{r=p+1}^{r=2n_3}\alpha_r\right]d_3 + \frac{(p-q)}{2n_3}\left[\frac{1}{2}\sum_{r=0}^{r=n_2}\alpha'_r - \frac{1}{2}\sum_{n_1+1}^{2n_2}\alpha'_r - \frac{1}{2}\sum_{r=0}^{r=n_2}\alpha_r + \frac{1}{2}\sum_{n_1+1}^{2n_2}\alpha_r\right]d_2.$$

Applying (8.48) as before, we get the variance $(\sigma_{GH})^2$ as

$$\sigma_{GH}^2 = \left[(p-q)\left\{1 - \frac{p-q}{2n_3}\right\} d_3^2 + \left\{ \frac{(p-q)^2 +_2}{4n_3^2} d_2^2 \right\} \right] \sigma_a^2. \tag{8.84}$$

Variance is a maximum at

$$h = (p-q) = \frac{2n_3}{2 - n_2 d_2^2 / n_3 d_3^2}, \tag{8.85}$$

giving maximum standard uncertainty σ_{GH} as

$$(\sigma_{GH})_{max} = \frac{d_3\sqrt{n_3}\sigma_a}{\sqrt{(2 - d_2^2 n_2 / n_3 d_3^2)}}. \tag{8.86}$$

8.6.3.6 The Points are on the Central Side EF

Following similar steps as above, the variance in difference in heights of the points along the EF is given by

$$\sigma_{EF}^2 = \left[(p-q)\left\{1 - \frac{p-q}{2n_2}\right\} d_2^2 + \left\{ \frac{(p-q)^2}{4n_2^2} n_3 d_3^2 \right\} \right] \sigma_\alpha^2. \tag{8.87}$$

Equation (8.87) is obtained by interchanging n_3 with n_2 and d_3 with d_2 in (8.84).

The variance is maximum when $p - q = h$ is given by

$$h = (p-q) = \frac{2n_2}{2 - n_3 d_3^2 / n_2 d_2^2} \tag{8.88}$$

and maximum standard uncertainty is

$$(\sigma_{EF})_{max} = \frac{d_2\sqrt{n_2}\sigma_a}{\sqrt{(2 - n_3 d_3^2 / n_2 d_2^2)}}. \tag{8.89}$$

8.6.4 *Points on Two Different Lines*

8.6.4.1 One Point on the Diagonal BD and the Other on Longer Sides BA or CD

If the two points are on two different lines, say, one on BD and another on BA, then variance $\sigma_{BA,BD}^2$ in difference in heights is found out by taking following steps:

Find the height of the pth point on BD; p is being numbered from the end B
Find the height of the qth point on BA; q is also numbered from the end B
Find the difference in height of the two points; the two cases arrive (1) $p \geq n_1$ and (2) $p \leq n_1$
Find the variance; by applying the law for the variances enunciated in (8.48) we get

For $p \geq n_1$

$$\sigma^2_{\mathrm{BA,BD}} = \left[q(1 - q/n_2)\, d_2^2 + \frac{q^2 n_1}{8n_2^2} d_1^2\right]\sigma_\alpha^2 + \left[\left(1 - \frac{p}{2n_1}\right)\left(p - \frac{qn_1}{n_2}\right) + \frac{q^2 n_1}{8n_2^2}\right] d_1^2 \sigma_\alpha^2. \tag{8.90}$$

For $p \leq n_1$

$$\sigma^2_{\mathrm{BA,BD}} = \left[q(1 - q/2n_2) d_2^2 + \frac{q^2 n_1}{8n_2^2} d_1^2\right]\sigma_\alpha^2 + \left[p\left(1 - \frac{q}{2n_2} - \frac{p}{2n_1}\right) + \frac{n_1 q^2}{8n_2^2}\right] d_1^2 \sigma_\alpha^2. \tag{8.91}$$

Equations (8.90) and (8.91) are also applicable for the points lying on CD and CA; p, q are numbered from C.

Variances for the points on BD and BC or AC and AD are obtained by replacing n_2 by n_3 and d_2 by d_3 in (8.90) and (8.91). In the case of points on BD and BC, p and q are numbered from B. In the case AC and AD, p and q are numbered from A.

8.6.4.2 Points on the Diagonal and Any Other Non-diagonal Lines GH or EF

1. One point lies on BD and another on GH.
   Following the three steps enunciated in Sect. 8.6.4.1, the variance $\sigma^2_{\mathrm{GH,BD}}$ of the difference in heights of the two points is given by
   For $p \geq n_1$

$$\sigma^2_{\mathrm{GH,BD}} = \left[q\left(1 - \frac{q}{2n_3}\right) d_3^2 + \frac{n_2}{2}\left\{1 + \frac{q}{n_3}\left(\frac{q}{2n_3} - 1\right)\right\} d_2^2 + \left\{\frac{3p}{2} - \frac{p^2}{2n_1} - \frac{3n_1}{4}\right\} d_1^2\right]\sigma_\alpha^2. \tag{8.92}$$

   For $p \leq n_1$

$$\sigma^2_{\mathrm{GH,BD}} = \left[q\left(1 - \frac{q}{2n_3}\right) d_3^2 + \frac{n_2}{2}\left\{1 + \frac{q}{n_3}\left(\frac{q}{2n_3} - 1\right)\right\} d_2^2 + \left\{\frac{p}{2} - \frac{p^2}{2n_1} - \frac{n_1}{4}\right\} d_1^2\right]\sigma_\alpha^2. \tag{8.93}$$

   Equations (8.92) and (8.93) will also hold good for two points: one on diagonal AC and the other on GH.
2. One point is on any diagonal and another on line EF

By interchanging n_3 with n_2, and d_2 with d_3, we get the variance of difference in heights of two points, out of which one lies on any diagonal and the other on EF, giving us

For $p \geq n_1$

$$\sigma^2_{\mathrm{EF,BD}} = \left[q\left(1 - \frac{q}{2n_2}\right) d_2^2 + \frac{n_3}{2}\left\{1 + \frac{q}{n_2}\left(\frac{q}{2n_2} - 1\right)\right\} d_3^2 + \left\{\frac{3p}{2} - \frac{p^2}{2n_1} - \frac{3n_1}{4}\right\} d_1^2 \right] \sigma^2_{\alpha}.$$

For $p \leq n_1$

$$\sigma^2_{\mathrm{EF,BD}} = \left[q\left(1 - \frac{q}{2n_2}\right) d_2^2 + \frac{n_3}{2}\left\{1 + \frac{q}{n_2}\left(\frac{q}{2n_2} - 1\right)\right\} d_3^2 + \left\{\frac{p}{2} - \frac{p^2}{2n_1} - \frac{n_1}{4}\right\} d_1^2 \right] \sigma^2_{\alpha}.$$

8.6.4.3 Points Are on the Central Lines EF and GH

Height of pth point on GH from (8.11) is

$$z_{\mathrm{GH}p} = d_3 \left[\sum_{r=0}^{r=p} \alpha_r - (p/2n_3) \sum_{r=0}^{r=2n_3} \alpha_r \right] + (p/2n_3)(f - c) + c \tag{8.94}$$

and the height of qth point on EF

$$z_{\mathrm{EF}q} = d_2 \left[\sum_{r=0}^{r=q} \alpha_r - (q/2n_2) \sum_{r=0}^{r=2n_2} \alpha_r \right] + (q/2n_2)(f' - c') + c'. \tag{8.95}$$

So the difference in the height of two points is

$$\begin{aligned} \Delta_{\mathrm{GH,EF}} = {} & d_3 \left[\sum_{r=0}^{r=p} \alpha_r - \frac{p}{2n_3} \sum_{r=0}^{2n_3} \alpha_r \right] - d_2 \left[\sum_{r=0}^{p} \alpha_r - \frac{q}{2n_2} \sum_{r=0}^{r=2n_2} \alpha_r \right] \\ & + \frac{p}{2n_3}(f - c) - \frac{q}{2n_2}(f' - c') + (c - c'). \end{aligned} \tag{8.96}$$

Equation (8.96) may be written as

$$\begin{aligned} \Delta_{\mathrm{GH,EF}} = {} & d_3 \left[\left(1 - \frac{p}{2n_3}\right) \sum_{r=0}^{r=p} \alpha_r - \frac{p}{2n_3} \sum_{r=p+1}^{2n_3} \alpha_r \right] \\ & + d_2 \left[\left(1 - \frac{q}{2n_2}\right) \sum_{r=0}^{r=p} \alpha_r - \frac{q}{2n_2} \sum_{r=p+1}^{2n_2} \alpha_r \right] \\ & + \frac{p}{2n_3}(f - c) - \frac{q}{2n_2}(f' - c') + (c - c'). \end{aligned} \tag{8.97}$$

The variance of left-hand side will be the sum of individual variances of five terms on the right-hand side of the equation (8.97). By applying the law of variances given in (8.48), we get variances of each term

So Variance V_1 of first term is

$$\begin{aligned} V_1 &= d_3^2\left[\left(1-\frac{p}{2n_3}\right)^2 p\sigma_a^2+\frac{p^2}{4n_3^2}(2n_3-p)\sigma_a^2\right] \\ &= d_3^2\sigma_a^2\left[p-\frac{p^2}{n_3}+\frac{p^3}{4n_3^2}+\frac{p^2}{2n_3}-\frac{p^3}{4n_3}\right] \\ &= d_3^2\sigma_\alpha^2\left[p\left(1-\frac{p}{2n_3}\right)\right]. \end{aligned} \quad (8.98)$$

Similarly the variance V_2 of second term will be

$$V_2 = d_2^2\sigma_\alpha^2\left[q\left(1-\frac{q}{2n_2}\right)\right]. \quad (8.99)$$

Now third term may be written as

$$\mathrm{Term}_3 = \frac{p}{2n_3}(f-c) = \frac{p}{2n_3}\left[\frac{1}{2}\sum_{r=0}^{r=n_2}\alpha_r'-\frac{1}{2}\sum_{n_1+1}^{2n_2}\alpha_r'-\frac{1}{2}\sum_{r=0}^{r=n_2}\alpha_r+\frac{1}{2}\sum_{n_1+1}^{2n_2}\alpha_r\right]d_2,$$

giving variance V_3 of this term as

$$\begin{aligned} V_3 &= \frac{p^2}{16n_3^2}d_2^2\sigma_\alpha^2(n_2+n_2+n_2+n_2) \\ &= \frac{p^2n_2+{}_2^2\sigma_\alpha^2}{4n_3^2}. \end{aligned} \quad (8.100)$$

Interchanging n_2 with n_3 and d_2 with d_3, we get the variance V_4 of the fourth term as

$$V_4 = \frac{q^2n_3d_3^2\sigma_\alpha^2}{4n_2^2}. \quad (8.101)$$

Now the fifth term can be written in the summation form as

$$c-c' = \frac{1}{2}\left[\sum_{r=0}^{r=n_2}\alpha_r - \sum_{r=n_2+1}^{2n_2}\alpha_r-\right]d_2 - \frac{1}{2}\left[\sum_{r=0}^{r=n_3}\alpha_r - \sum_{r=n_3+1}^{2n_3}\alpha_r-\right]d_3, \quad (8.102)$$

giving the variance V_5 of $c - c'$ as

$$V_5 = \frac{1}{4}(n_2 + n_2) +_2^2 \sigma_\alpha^2 + \frac{1}{4}(n_3 + n_3)\, d_3^2\sigma_\alpha^2$$
$$= \frac{1}{2}(n_2 d_2^2 + n_3 d_3^2)\sigma_\alpha^2 \tag{8.103}$$

Combining (8.98), (8.99), (8.100), (8.101) and (8.103), we get the variance $\sigma^2_{\mathrm{GH,EF}}$ of difference in heights of two points, one on GH and another EF as

$$\sigma^2_{\mathrm{GH,EF}} = d_3^2\sigma_\alpha^2\left[p\left(1 - \frac{p}{2n_3}\right)\right] + d_2^2\sigma_\alpha^2\left[q\left(1 - \frac{q}{2n_2}\right)\right]$$
$$+ \frac{p^2 n_2 d_3^2\sigma_\alpha^2}{4n_3^2} + \frac{q^2 n_3 d_2^2\sigma_\alpha^2}{4n_2^2} + \frac{1}{2}(n_2 d_2^2 + n_3 d_3^2)\sigma_\alpha^2.$$

$$\sigma^2_{\mathrm{GH,EF}} = \left[\mathrm{p}\left(1 - \frac{\mathrm{p}}{2\mathrm{n}_3}\right)\mathrm{d}_3^2 + \mathrm{q}\left(1 - \frac{\mathrm{q}}{\mathrm{n}_2}\right)\mathrm{d}_2^2 + \frac{\mathrm{p}^2\mathrm{n}_2\mathrm{d}_3^2}{4\mathrm{n}_3^2} + \frac{\mathrm{q}^2\mathrm{n}_3\mathrm{d}_2^2}{4\mathrm{n}_2^2} + \frac{1}{2}\left(\mathrm{n}_2\mathrm{d}_2^2 + n_3\mathrm{d}_3^2\right)\right]\sigma_\alpha^2 \tag{8.104}$$

8.7 Type A Evaluation of Uncertainty

The standard uncertainty of measurement of angular changes, symbolized as σ_α, is calculated by repeated observations. We may have repeated readings either at one point a larger number of times or may be a few repetitions but at different points on the surface. For a laboratory it may be easier to take three repetitions at all points of the surface plate, calculate the variance σ_α^2 and record it. Repeating this exercise on several surface plates will give the pooled variance, square root of which can be used as standard uncertainty in angular measurements. Uncertainty in height measurement will then be d times the uncertainty in angular measurement. For example d_1 in our case is 55 mm and d_2 is 50 mm.

If the standard deviation in angular measurement σ_α is 10^{-5} radians, then uncertainty through Type A evaluation is $= 10^{-5} \times 50\,\mathrm{mm} = 0.5\,\mu\mathrm{m}$.

8.8 Type B Evaluation of Standard Uncertainty

8.8.1 Uncertainty Due to the Finite Digital Readout of the Measuring Instrument

Normally the measuring instruments have least count of $10\,\mu\mathrm{rad}(10\,\mu\,\mathrm{m/m})$. It means the real observation may lie anywhere within the semi-range of $5\,\mu\mathrm{rad}$.

Observations follow rectangular distribution with semi-range of 5 μrad. So standard uncertainty U_d due to digital readout is given as

$$U_d = 5 \times 10^{-6} \cdot 50\,\text{mm}/\sqrt{3} = 0.14\,\mu\text{m}.$$

8.8.2 Uncertainty Due to Certified Accuracy of the Measuring Instrument

Uncertainty given in the certificate is 10 μrad at 95% confidence level coverage factor 1.96 standard uncertainty in height due to this source U_c as

$$U_c = 10 \times 10^{-6} \times 50\,\text{mm}/1.96 = 0.26\,\mu\text{m}.$$

8.8.3 Uncertainty Due to Unstability of the Instrument

Certified error due to unstability of the measuring instrument is given as 10 μrad, hence uncertainty component due to this cause U_s as

$$U_s = 10 \times 10^{-6} \times 50\text{mm}/\sqrt{3} = 0.29\,\mu\text{m}.$$

8.8.4 The Error in Non-Uniformity of Temperature

If the surface plate is not maintained at uniform temperature, especially in the vertical direction, the thermal expansion of the material of the plates the upper surface will rise or fall depending upon the difference in temperature. W is the thickness, t is temperature difference and α is the thermal coefficient then change in height is given by

$$W \cdot t \cdot \alpha$$

For example if $W = 15\,\text{cm}$, t be 0.2 °C and $\alpha = 11.5 \times 0^{-6}/°\text{C}$.

Uncertainty component U_t due to this is given

$$0.2 \times 0.15 \times 11.5 \times 10^{-6} = 0.345\,\mu\text{m}.$$

Assuming rise or fall of a point on the surface follows the rectangular distribution with semi-range of 0.345 standard uncertainty in height U_t due to this cause is given by

$$U_t = 0.345/\sqrt{3} = 0.199\,\mu\text{m} = 0.20\,\mu\text{m}.$$

Hence combined standard uncertainty is

$$u_c = \sqrt{0.5^2 + 0.14^2 + 0.26^2 + 0.29^2 + 0.20^2} = 0.68\,\mu\text{m}.$$

8.8.5 *Effective Degrees of Freedom*

Effective degrees of freedom ν_{eff} are given as

$$\nu_{\text{eff}} = \frac{0.68^4}{0.5^2/9 + 0.14^4/\infty + 0.26^4/\infty + 0.29^4/\infty + 0.19^4/\infty} = 30.8.$$

8.8.6 *Extended Uncertainty*

At the confidence level of 95%, Student's t factor for 31 degrees of freedom is 2.01; hence extended uncertainty u_e is given as

$$u_e = 2.01 \times 0.68\,\mu\text{m} = 1.36\,\mu\text{m}.$$

References

1. C.F. Dietrich, *Uncertainty, Calibration and Probability* (Adam Hilger, New York, 1991)

Chapter 9
Uncertainty in Mass Measurement

9.1 Balances

There are essentially two types of balances. One type of balances is pure comparators, whose measurable properties are its sensitivity or the mass value of the smallest scale interval and repeatability. These mass comparators give only the difference of mass values between the two weights. Equal-arm two-pan freely oscillating balances also fall into this category. For such mass comparators, one has to calculate the sensitivity or the mass value of the smallest graduation and repeatability.

The other type of balances is direct weighing instruments that indicate the mass of a body placed on its pan. Besides the value of its smallest scale interval and repeatability, it has many other measurable properties such as display scale, built-in weights and linearity. Most of the modern day balances are damped type; their ability to return to zero of the display scale on removal of the load is also an important measurable property.

9.2 Choice of Standard Weights

The OIML [1] has recommended that the maximum permissible error (mpe) tolerable in the standard weight should not be larger than the one third of the mpe allowed for the under-test weighing instrument. According to OIML Recommendation R-76 [2], all non-automatic weighing instruments have been categorized into four classes. Class I weighing instruments are of the highest class and have best resolution, and cover weighing instruments have verification interval of 1 mg or more. Weighing instruments having verification interval of less than 1 mg are not covered by R-76. Extending the rule of [1] for balances having smallest scale interval of less than 1 mg, the weights used for calibration of such balances should, therefore, have the mpe of not more than one third of the mass value of the smallest scale interval of the weighing instrument under test. In case of very high precision

S.V. Gupta, *Measurement Uncertainties*, DOI 10.1007/978-3-642-20989-5_9,

weighing instruments and if this condition is difficult to obtain then it is ensured that the uncertainty in mass value of the weight is less than the mass value of the smallest scale interval of the weighing instrument under test. In this case measured mass value of the weight is used.

9.3 Calibration of Balance

The following tests, depending upon the type of the balance, are to be conducted:

1. Repeatability
2. Sensitivity or calibration of smallest scale interval
3. Return of the indicator to zero of the balance scale
4. Calibration of smallest built-in weight
5. Calibration of built-in weights or weight check at full, $^3/_4$, $^1/_2$ and $^1/_4$ scale capacity
6. Linearity check
7. Check for off-centre placement of weight (Corner test)
8. Discrimination test

9.3.1 Repeatability

Repeatability consists of 10 (ten) weighings with the same mass without any intermediary zero adjustments. For direct reading balance, calculate the differences between the direct readings of the scale with and without the test weight placed on the balance pan. In case of freely oscillating balances, the rest points with weight on the pan are determined. Calculate the standard deviation of these rest points. The test is carried out at half and full capacity of the balance. Customary the magnitude of the calculated standard deviation should be less than one third of the mass value of the smallest scale interval. However in case of high precision balances, manufacturer's specifications are followed.

9.3.1.1 Repeatability for Direct Reading Balances

Observation sheets for repeatability at half load for a 1 kg balance at half load are given in Table 9.1

Note: The balance is initially so adjusted that the reading at the zero load is at a few divisions so that negative shift due to variation of zero is taken care of. It may be noted that for repeatability test, mass value of the load used need not be known. Only requirement is that its mass value does not change during the test.

To get mass indicated by the balance of the load, mean of the two consecutive indications at no load is subtracted from the indication with load.

Table 9.1 Observations and calculation sheet

S. No	Indication with no load y mg	Indication with load g	Difference g	$x - \bar{x} = d$ 10^{-2} mg	d^2 10^{-4} mg	y^2
1	0.5	500.0011	500.00055	−12	144	0.25
2	0.6	500.0012	500.00065	−2	4	0.36
3	0.5	500.0012	500.00070	5	25	0.25
4	0.5	500.0013	500.00085	22	484	0.25
5	0.4	500.0011	500.00065	−2	4	0.16
6	0.5	500.0012	500.00065	−2	4	0.25
7	0.6	500.0012	500.00070	5	25	0.36
8	0.4	500.0011	500.00065	−2	4	0.16
9	0.5	500.0012	500.00065	−2	4	0.25
10	0.6	500.0012	500.00065	−2	4	0.36
	0.5					0.25
Sum	5.6			0	702	2.90

From Table 9.1, we get

Mean value of the load 500.00067 g

and

Standard deviation $(702/9)^{1/2} \times 10^{-2} = 0.088$ mg.

Last column of Table 9.1 gives the squares of scale readings y at no load. Using this sum we can calculate the standard deviation at zero load as follows:

$$\text{Standard deviation of zero shift} = \left[\{2.90 - (5.6)^2/11\}/10\right]^{1/2}$$
$$= [\{2.90 - 2.849\}/10]^{1/2} = 0.07 \text{ mg.}$$

The uncertainty from type A evaluation for repeatability is 0.088 mg and for return of indication to zero is 0.07 mg.

Type B uncertainty in repeatability or return of zero will be due to the uncertainty in the mass value of the smallest scale interval. That is standard uncertainty of the mass standard by which the smallest scale interval has been calibrated and this will be uncertainty by type B evaluation in repeatability test.

Type B uncertainty has arisen because we have observed the scale readings in terms of its graduations and assumed that one smallest scale interval is 0.1 mg.

9.3.2 Calibration of Smallest Scale Interval

A calibrated standard weight of at least 10–100 times the value of the smallest interval is chosen. Five sets of observations are taken by placing the weight and

Table 9.2 Smallest scale interval

S. No	Indications in terms of smallest scale intervals		Difference	Value of smallest scale interval mg
	No load	Load 10 mg		
1	5	105	100	0.1
2	4	104	100	0.1
3	5	106	101	0.099
4	5	105	100	0.1
5	5	105	100	0.1
			Mean value 0.1 mg	

removing it and noting the scale indication. Set of observations taken are shown in Table 9.2.

The uncertainty in mass standard of 10 mg has been given as ±0.010 mg. Contribution per scale interval, therefore, is 0.0001 mg. Standard uncertainty through type B evaluation method in assigning the mass value to the smallest scale interval is 0.0001 mg.

For mass comparators showing only the difference in mass of two weights, type A uncertainty is 0.088 mg and type B uncertainty is 0.0001 mg.

For balances having built-in weights, these weights are calibrated against standard weights and components of uncertainty are calculated by type A as well as type B evaluation methods.

For mass comparators we calculate the shift in rest points (scale reading) for loading and unloading of a small load and report the mass value of the smallest scale interval.

9.3.3 *Calibration of Built-in Weight*

9.3.3.1 Measurement Model

The mathematical model for calibration of weights on conventional mass basis:

$$C = S - (R - Z),$$

where C is the calculated correction, S is the value of the standard mass from its certificate, R is the mean of two readings for the same verified point and Z is themean of the two zero readings.

9.3.3.2 Measurement Data

In order to minimize errors introduced by standard weights, it would be preferable, if possible, to use a single weight rather than a summation.

For our example we have

Mass nominal value (g)	Mass value (g)	Source
500	500.00013	Weight calibration certificate

Observations	
Reading at no load	0.0000
First reading at the load	500.0001
Second reading at the load	500.0002
Reading at no load	0.0002

Calculations:

Mean of observation of standard mass
$R = 500.00015\,\text{g}$
Mean of zero readings
$Z = 0.0001\,\text{g}$

Correction:

$$C = S - (R - Z) = 500.00013 - (500.00015 - 0.0001)\text{g} = 0.00008\,\text{g} = 0.08\,\text{mg},$$

rounded to one tenth of the smallest graduation, i.e. 0.01 mg.

9.3.4 Uncertainty in Balance Calibration

The major sources of uncertainty in calibration of built-in weight are as follows:

a. Uncertainty of the standard mass, u_S

Uncertainty of the standard mass found from the calibration certificate is 0.05 mg. If the coverage factor is taken as 2, then standard uncertainty of the standard is

$$u_S = \frac{0.05}{2} = 0.025\,\text{mg}. \tag{9.1}$$

Note: In this situation we have assumed a normal distribution with coverage factor 2. In case of using more than one standard weight, the standard uncertainty of the combination is the sum of absolute value of the standard uncertainty of each weight of the combination.

b. Drift of the standard mass since last calibration, u_D

Customary the variation of mass in standard weights, over the calibration interval, does not exceed 10% of the maximum permissible error. For our 500 g standard weight (maximum permissible error 2.5 mg), the drift value is 0.25 mg. Drift value follows a rectangular distribution, giving u_D as

$$u_D = \frac{0.25}{\sqrt{3}} = 0.144\,\text{mg}. \tag{9.2}$$

c. Uncertainty due to the value of the smallest scale interval of the balance U_{div}

The value of smallest graduation in the above example is 0.1 mg. A rounding error of $0.1/2 = 0.05$ mg is assumed. The true value follows rectangular distribution; therefore, standard uncertainty due to this cause is

$$u_{div} = \frac{0.05}{\sqrt{3}} = 0.0289\,\text{mg}. \tag{9.3}$$

d. Uncertainty due to repeatability, u_{repeat}

The balance repeatability has been verified and the calculated standard deviation is 0.088 mg. The repeatability follows a normal distribution:

$$u_{repeat} = 0.088\,\text{mg}. \tag{9.4}$$

Uncertainty budget for 500 g built-in weight is given in Table 9.3.

9.3.5 Linearity Check

It is carried out by continuously loading the balance and taking observation at each step till the maximum capacity of the balance. The observations are repeated while unloading in the reverse order. Sometimes we may see that observations at identical loads while the loading and unloading are not the same. This may be due to lack of repeatability or hysteresis. A graph between the load (x-axis) and indication by the

Table 9.3 Uncertainty budget for 500 g built-in weight

Uncertainty component	Meas. unit	Distr.	U	n	Divisor	$u(x)$	c	$c \times u(x)$	$c \times u(x)^2$ 10^{-4}
u_S	mg	Norm.	0.05	30	2	0.025	1	0.025	6.25
u_D	mg	Rect.	0.25	8	1.7321	0.144	1	0.144	20.73
u_{div}	mg	Rect.	0.05	1000	1.7321	0.029	1	0.029	8.41
u_{repeat}	mg	Rect.	0.088	9	1	0.088	1	0.088	77.44
Sum									112.83
Combined standard uncertainty (RSS)									0.106 mg
Coverage factor for 95% confidence level									2
Expanded uncertainty (mg)									0.212

n is the degree of freedom, c is the coefficient factor and $u(x)$ is the uncertainty component balance (y-axis) is drawn. A straight line from zero load to the maximum load is drawn. The maximum difference between the ordinates on the straight line and the indication at any load is the linearity error. The correction due to this error is applied while using the balance. Standard uncertainty of each load (standard weights) is one component of type B uncertainty, which should be taken into account while calculating the uncertainty of mass determined by the balance. This uncertainty may differ from one load to another.

9.3.6 Check for Off-Centre Placement of Weight/Corner Test

For balances of high sensitivity and small capacity (mass comparators), make concentric circles on a paper and three radii subtending 120° with each other, cut the paper of the size of the pan of the balance under test and place it on the pan. Uncertainty in mass value of the standard weight should be less than one third of the value of the smallest scale division. Take observations at the centre and extreme of one radius, and repeat the process along all the three radii. Maximum difference between central and extreme position is the error due to eccentricity. Weight should be of half the capacity of the balance, rounded off to the nearest single weight available in the set. Avoid using more than one weight for corner test. In normal use the weight should be placed in the centre of the pan.

9.3.7 Discrimination Test

The test is to ensure the correct change in digital indication. The test is carried out at three loads, namely minimum, half and full capacity loads. The balance is loaded and 10 extra weights each of one tenth of the mass value of the verification interval are placed. Let the indication be I. The additional weights are removed one by one till the indication is unambiguously reduced by one scale interval (digit); i.e. indication now is I–d. Place gently a mass of 1.4 times the scale interval; the indication should unambiguously increase by one scale interval.

This test is an acceptance test. If a balance fails in this test, it is to be rejected.

Here we may note that basic properties of the balance are its repeatability and sensitivity so uncertainty of the balance is the combination of these uncertainties; other sources of uncertainty depend upon load and the standards used and other applicable properties. Hence it is important that standard weights used in calibration of balance should be of appropriate accuracy and stability. For analogue balances, uncertainty due to the smallest graduation need not be considered separately.

However, other sources of error are variation of environmental conditions, which cause variable buoyancy correction, arm length of the balance and formation of air convectional currents.

9.4 Uncertainty in Calibration of Weights

When the mass M_u of a weight under calibration is in equilibrium with that of the standard weight of known mass M_s, then equilibrium equation is

$$M_u = M_s + (V_u - V_s)\sigma + (I_u - I_s). \tag{9.5}$$

Here V_u and V_s are respectively the volume of the under calibration and standard weights, and I_u and I_s are the indication of the balance in terms of same mass units as M_s.

The second term of the right-hand side of (9.5) is known as buoyancy correction.

Applying the law of variances to (9.5) and taking square of uncertainty equal to the variance, we get

$$u_u^2 = u_s^2 + (V_u - V_s)^2 u_\sigma^2 + (u_{V_u}^2 + u_{V_s}^2)^2 \sigma^2 + u_R^2. \tag{9.6}$$

R is taken as difference of two indications.

In the calibration of a stainless steel weight against the national prototype of platinum iridium, the following data are given:

$u_R = 2.5\,\mu$g by type A evaluation with 9 degrees of freedom; Uncertainty in standard is $10.5\,\mu$g at 95% confidence level; $V_s = 46.39355\,\text{cm}^3$; $V_u = 124.5364\,\text{cm}^3$; $u_{V\text{u}} = 0.0011$; $u_{V\text{s}} = 0.00004\,\text{cm}^3$; $\sigma = 1.1983\,\text{mg/cm}^3$ and $u_\sigma = 0.0001\,\text{mg/cm}^3$

As uncertainty is $10.5\,\mu$g at 95% confidence level, so standard uncertainty $u_s = 10.5/1.96 = 5.4\,\mu\text{g}$

Substituting the data in (9.6), we get

$$\begin{aligned} u_u^2 &= (5.4) \times 10^{-12} + (124.5364 - 46.38355)^2 \times (0.0001 \times 10^{-3})^2 \\ &\quad + \{(0.0011)^2 + (0.00004)^2\} \times (1.1983 \times 10^{-3})^2 + (2.5)^2 \times 10^{-12} \\ &= (29.16 + 61.06 + 0.17 + 6.25) \times 10^{-12} = 96.64 \times 10^{-12}\,\text{g}^2, \end{aligned}$$

giving

$$u_u = 9.83 \times 10^{-6}\text{g}.$$

However, most of the weights are calibrated and verified to lie in certain class of mpe (maximum permissible errors) and with the required uncertainty. Hence in the calibration of weights, environmental parameters and other measurements are carried out to meet the requirements of the mpe class. By thumb rule uncertainty should not be more than one third of the mpe. Hence the corrections applicable and the uncertainty requirement are also fixed keeping in view the mpe class of the weight under test.

9.5 Measurement Requirements for Various Corrections

9.5.1 Buoyancy Correction

Equation (9.5) may also be expressed in terms of the density of two weights as

$$M_u(1 - \sigma/\rho_u) = M_s(1 - \sigma/\rho_s) + (I_u - I_s).$$

$$M_u = M_s(1 - \sigma/\rho_s)/(1 - \sigma/\rho_u) + (I_u - I_s) = M_s + M_s(\sigma/\rho_u - \sigma/\rho_s) + (I_u - I_s).$$

The buoyancy correction per unit mass therefore is

$$\sigma\,(1/\rho_u - 1/\rho_s). \tag{9.7}$$

Normally the weights, in daily use and of denomination greater than a gram, are made of materials of density from 8,000 to $8,400\,\text{kg/m}^3$.

Buoyancy correction per unit mass is

$$c = \left(\frac{1}{8000} - \frac{1}{8400}\right) \times 1.2 \cong 7.14 \text{ parts per million}, \tag{9.8}$$

say one part per hundred thousand.

So for weights, having mpe equal to or more than 1 part per 10,000, buoyancy correction is not necessary. This means the buoyancy correction is not necessary for weights of OIML Classes M1 and lower.

Fractional weights up to 100 mg are made of similar materials whose density does not vary much. Still smaller weights are normally made of aluminium. Hence the buoyancy correction is not applied in fractional weights. However, if the fractional weights are made of platinum, then buoyancy correction is to be applied.

Quite often weights, from 1 g and up, are calibrated by a calibration laboratory with an uncertainty of one part per million. For applying buoyancy corrections, density of air should be calculated by using 2007 CIPM equation [3] from the measured environmental conditions. As the differences in volumes are small, uncertainty requirement of measuring pressure, temperature and relative humidity is not very stringent. For example, measurement requirement is within a mm of Hg (100 Pa) for pressure, within 0.1 °C for temperature and within 10% for relative humidity.

9.5.2 MPE and Correction

If the correction applicable is less than or equal to one tenth of the maximum permissible error (mpe), then it may not be necessary to apply it. This rule may be applied for calibration of weights.

9.5.3 Calibration of Weights Against Pt–Ir Standard

In a special case, when national prototype (in Pt–Ir) is used for calibrating transfer standards of mass which are of density ranging from 8,400 to $8,000\,\text{kg/m}^3$ not only buoyancy correction is to be applied but air composition and environmental parameters need to be known accurately.

Let buoyancy correction is denoted by B then

$$B = (V_s - V_u)\sigma,$$

$$\frac{\Delta B}{B} = \frac{\Delta(V_s - V_u)}{V_s - V_u} + \frac{\Delta\sigma}{\sigma}. \tag{9.9}$$

In case a weight of density $8,000\,\text{kg/m}^3$ is calibrated against a platinum iridium standard of mass of density $21,557\,\text{kg/m}^3$, the value of B is given roughly

$$B = (125 - 46)1.2\,\text{mg} = 94.8\,\text{mg}. \tag{9.10}$$

Hence from (9.10) for 1 μg accuracy in buoyancy correction

$$\frac{\Delta B}{B} = \frac{1}{95,000} \approx 10^{-5}. \tag{9.11}$$

To attain such a relative uncertainty, we must know the volume of the two weights and also the air density with an uncertainty of 1 in 10^{-5}.

Roughly air density σ is given as

$$\sigma = \frac{PM_a}{ZRT}\left[1 - x_v\left(1 - \frac{M_v}{M_a}\right)\right], \tag{9.12}$$

where M_a and M_v are the molar mass of dry air and water vapours, respectively, x_v is the mole fraction of water vapours and Z is the compressibility factor, giving

$$\frac{\Delta\sigma}{\sigma} = \frac{\Delta P}{P} + \frac{\Delta M_a}{M_a} + \frac{\Delta c}{c} - \frac{\Delta Z}{Z} - \frac{\Delta R}{R} - \frac{\Delta T}{T}. \tag{9.13}$$

Here c is factor $\left[1 - x_v\left(1 - \frac{M_v}{M_a}\right)\right]$, a correction due to humidity

9.5.4 Components of Relative Uncertainty for Air Density

9.5.4.1 Due to Pressure Alone

For weighing purpose the pressure value of interest is around 100,000 Pa (atmospheric pressure).

Hence

$$\frac{\delta\sigma}{\sigma} = \frac{\Delta P}{100{,}000} = 10^{-5}\delta p\,\mathrm{Pa}^{-1}. \tag{9.14}$$

Here δp is measured in pascal (Pa).

9.5.4.2 Due to Temperature Alone

Temperature in a laboratory for weighing is around 27 °C = 300 K.

Hence

$$\frac{\delta\sigma}{\sigma} = \frac{\delta T}{T} = \frac{\delta T}{300} = 3.3 \times 10^{-3}\,\delta T\,\mathrm{K}^{-1}, \tag{9.15}$$

where δT is the temperature interval in degree Celsius (°C) or in K.

9.5.4.3 Due to Humidity Alone

Now the c, in terms of molar masses of water and air and the molar fraction x_v, may be expressed as

$$c = \left[1 - x_v\left(1 - \frac{M_v}{M_a}\right)\right] = \left[1 - x_v\left(1 - \frac{18.0}{28.97}\right)\right] = [1 - 0.378\,x_v]. \tag{9.16}$$

But x_v the molar volume of water vapours in terms of relative humidity h is related to

$$h = \frac{x_v}{x_{sv}}. \tag{9.17}$$

But x_v in terms of the saturation vapour pressure at the temperature and pressure of air is given as [4]

$$x_v = h \times f(p,t) \times p_{sv}/p. \tag{9.18}$$

1. If $t = 27$°C or 300 K

then

$p_{sv}(t) = 3{,}566$ Pa at $p = 100{,}000$ Pa, $f(p,\ t) = 1.0042$ [4].
Differentiating (9.18), we get

$$\delta x_v = \delta h \times 1.0042 \times 35{,}666/100{,}000,$$

giving

$$\delta x_v = 0.0358 \times \delta h. \tag{9.19}$$

From (9.16)

$$c = 1 - 0.3780 \times 0.0358\,h \tag{9.20}$$

and

$$\delta c = -0.378 \times 0.0358\,\delta h = -0.0135 \times \delta h,$$

$$\frac{\delta c}{c} = -13.5 \times 10^{-3} \delta h. \tag{9.21}$$

2. If $t = 20°\mathrm{C}$

then

$$p_{\mathrm{sv}}(t) = 2{,}338\ \mathrm{Pa\ at}\ p = 100{,}000, \quad f(p, t) = 1.0040, \tag{9.22}$$

giving $x_{\mathrm{v}} = 0.02347$.

Hence

$$\delta c = -0.378 \times 0.02356\,\delta h = 0.0089\,\delta h,$$

$$\frac{\delta c}{c} = -9 \times 10^{-3} \delta h. \tag{9.23}$$

9.5.4.4 Due to Change in CO_2 Alone

The value of molar mass of dry air [3] is 28.96546×10^{-3} kg per mole if the mol fraction of CO_2 present is 0.0004. In general for other mole fractions of CO_2 the expression for molar mass of air is

$$M_{\mathrm{a}} = 28.96546 + 12.011\,(x_{\mathrm{CO_2}} - 0.0004). \tag{9.24}$$

$$\frac{\delta \sigma}{\sigma} = \frac{\delta M_{\mathrm{a}}}{M_{\mathrm{a}}} = \frac{12.011}{28.96546} \delta x_{\mathrm{CO_2}} = 0.41 \delta x_{\mathrm{CO_2}}. \tag{9.25}$$

9.5.4.5 Due to Change in Molar Mass of Air

Molar mass of air is 28.96546 g [3] with relative uncertainty of a few parts per million. Change in molar mass of dry air due to changed Argon composition is 6.6×10^{-5} [5], which has been accounted for in 2007 CIPM equation for moist air density [3].

9.5.4.6 Due to Change in Compressibility Factor and R

Compressibility factor Z [6] can be calculated with an uncertainty of better than one in a million. From latest CODATA [7], the value of $R(8.314472)$ is known with a relative uncertainty of 1.7×10^{-6}.

9.6 Measurement Requirements for p, T, and h

9.6.1 For 10 μg Uncertainty in 1 kg

For calibration of 1 kg stainless steel weight or a weight of near about same density against the platinum–iridium standard with an uncertainty of 10 μg, the density or volume of the weights and density of air inside the balance chamber must be known within an uncertainty of 1 in 10^4; i.e.

$$\frac{\Delta\sigma}{\sigma} = 10^{-4}.$$

Using (9.14), (9.15, 9.21 or 9.25), uncertainty requirements are

$$\begin{aligned}\delta p &= 10\,\text{Pa},\\ \delta T &= 1/33 = 0.03\,\text{K},\\ \delta h &= 0.1 \text{ i.e. with in few percent of relative humidity.}\end{aligned}$$

9.6.2 For 1 μg Uncertainty in 1 kg

For calibration of a stainless steel weight or a weight of near about similar density against the platinum–iridium standard with an uncertainty of 1 μg, the density or volume of the weights and density of air inside the balance chamber must be known within an uncertainty of 1 in 10^5.

$$\frac{\Delta\sigma}{\sigma} = 10^{-5}.$$

Uncertainty requirements of parameters of air density are

$$\begin{aligned}\delta p &= 1\,\text{Pa},\\ \delta T &= 1/330 = 0.003\,\text{K},\\ \delta h &= 0.01 \text{ i.e. with in 1\% of relative humidity.}\end{aligned}$$

9.6.3 Measurement Requirement for CO_2

We know from (9.25) that

$$\delta x_{co_2} = 2.44\frac{\delta\sigma}{\sigma}.$$

9.6.3.1 For 10 μg Uncertainty in 1 kg

For 10 μg uncertainty, i.e. $\frac{\delta\sigma}{\sigma} = 10^{-4}$ tolerable uncertainty in molar fraction of CO_2, is $2.44 \times 10^{-4} = 0.00024$. Molar fraction in normal air is not expected to vary by more than 0.00024; hence molar fraction of carbon dioxide needs to be monitored, but correction may not be necessary.

From air density table [8], we know that

Air density at 20°C at 760 mm of Hg and relative humidity 0.5 with CO_2 molar fraction 0.0005 is given as 1.199 266 mg/cm^3

and

Air density at 20°C at 760 mm of Hg and relative humidity 0.5 with CO_2 molar fraction 0.0003 is 1.199 167 mg/cm^3.

So calibrating a stainless steel 1 kg standard against the national prototype of Pt–Ir will have difference in buoyancy corrections of 10 μg, which is about the same as of uncertainty. However normally, the change in mole fraction of CO_2 present in air is not expected by more than 0.0001.

9.6.3.2 For 1 μg Uncertainty in 1 kg

For 1 μg uncertainty, required relative uncertainty in air density is $\frac{\delta\sigma}{\sigma} = 10^{-5}$. Hence tolerable uncertainty in molar fraction of CO_2 is only 0.000024; hence molar fraction needs to be measured and necessary correction needs to be applied if molar fraction of CO_2 is different from 0.0004 by an amount of 0.000024 or more.

9.7 Uncertainty in Calibration of Weights Under Legal Metrology

Weights covered under legal Metrology follow OIML Recommendation 111 [9]. Reference, Secondary and Working Standards of Legal Metrology are respectively of class E2, F1, and F2, respectively. Commercial weights are of class M1 for bullion trade and M3 for coarser work.

9.7.1 Reference Standard Weights (Class E2)

The reference standard weights are in class E2 maximum permissible errors and calibrated by the National Physical Laboratory India against the standards of known mass and uncertainty u_s, which is about one third of 1 mg. Uncertainty due to repetitions by type A evaluation is estimated. Uncertainty component due to given uncertainty in NPL standard weights is applied.

The buoyancy correction is applied and uncertainty in the knowledge of buoyancy correction is another component of uncertainty. This includes uncertainty in base density (Molar composition) of air, measurement of various environmental parameters to calculate air density, volume of weights or density of NPL standard and reference standard weights.

9.7.2 Secondary Standard Weights (Class F1)

The secondary standard weights are calibrated against reference standard weights. The material of reference and secondary standard weights is same at present, so no buoyancy correction needs to be applied. Only uncertainty to be applied is due to (1) reference standard weight as given by the NPL in its certificate and (2) repetitions. Normally mass value of each weight is not given. It is assured that mass values of each weight lies within he maximum permissible of class F1, namely E_{F1}.

9.7.3 Working Standard Weights (Class F2)

Working standards are verified against Secondary standards. Standard Uncertainty in secondary standards is taken as $\pm E_{\mathrm{F1}}/\sqrt{3}$. Repetition uncertainty is 2 times the standard deviation (SD) of at least ten repeated values. Total standard uncertainty u_{sec} is given by

$$u_{\mathrm{sec}} = \sqrt{E_{\mathrm{F1}}^2/3 + (2\mathrm{SD})^2}.$$

In case of digital balances an additional uncertainty due to balance which is taken as $d/\sqrt{6}$ is also considered.

9.7.4 Commercial Weights (Class M1)

9.7.4.1 Class M1 and M2 Weights

Class M1 and M2 weights are verified against working standard weights; no buoyancy correction is applied. Only uncertainty components due to weights and balance used are to be calculated. Uncertainty due to working standard weight is taken as $E_{\mathrm{F2}}/\sqrt{3}$ and due to digital balance is taken as $d/\sqrt{6}$. Then the standard uncertainty u_{w} is given as

$$u_{\mathrm{w}} = \sqrt{E_{\mathrm{F2}}^2/3 + d^2/6}.$$

9.7.4.2 Class M3 Weights

For class M3 weights such as cast iron weights, only uncertainty is due to balance used; if it is two-pan balance then uncertainty may be taken as zero; for digital balance it is $d/\sqrt{6}$.

Note: Here d is the mass value of the last digit of the balance used. Hence uncertainty I M3 weights is given

$$u_\mathrm{w} = \sqrt{E_{\mathrm{M1}}^2/3 + d^2/6}.$$

References

1. OIML Recommendation R-47, Standard Weights for testing high Capacity Weighing Instruments (1979)
2. OIML Recommendation R-76, Non-automatic weighing instruments (1992)
3. A. Picard, R.S Davis, M. Glaser, K. Fujii, Revised formula for the density of moist air (CIPM-2007). Metrologia **45**, 149–155 (2008)
4. P. Giacomo, Equation for the determination of the density of moist air. Metrologia **18**, 33–40 (1981)
5. S.Y. Park, J.S. Kim, J.B. Lee, M.B. Esler, R.S. Davis, R.I. Wielgosz, A re-determination of the argon content of air buoyancy corrections in mass standard comparisons. Metrologia **41**, 387–395 (2004)
6. R.W. Hyland, A correlation for the second interaction virial coefficients and enhancement factors for moist air. J. Res. Nat. Stand. **79A**, 551–560 (1975)
7. P.J. Mohr, B.N. Taylor, D.B. Newell, CODATA values of physical constants. Rev. Mod. Phys. **80**, 633–730 (2008)
8. S.V. Gupta, A treatise on mass metrology. under publication
9. OIML, Weights of classes E1, E2, F1, F2,M1, M2, M3", OIML-R-111 (1994)

Chapter 10
Uncertainty in Volumetric Measurement

10.1 Introduction

There are three methods of calibrating volumetric measures [1]:

1. Gravimetric method used for volumetric glassware; range is 1 mm^3 to 50 dm^3.
2. Volumetric comparison for larger vessels 100 dm^3 to 10 m^3.
3. Strapping (dimensional measurement) method for any volumetric measure more than 10 m^3.

In gravimetric method, the mass of pure medium (water or mercury) contained or delivered under specified conditions is measured. Corrections are applied due to density of the medium at different temperatures, buoyancy correction and decrease in the volume/capacity of the measure to a given standard temperature [1]. The medium used is mercury for volumetric measures smaller than 1 cm^3; for measures of higher capacity the double distilled water is used as transfer medium.

In volumetric comparison, water is transferred from the standard measure of known volume to the measure under test. In case of larger measures, multiple transfer method is used. In either case, the standard measure should be calibrated for volume delivered. For multiple filling, the standard measure should be an integral sub-multiple of the measure under test and there should be another smaller measure of graduated type, which is used to estimate the volume required to completely fill the measure under test. In case of larger measures of delivery type, the standard measure should be of content type. The water is transferred from larger measure to the standard measure either one time or several times as the case may be. For delivery measures calibrated by multiple delivery method, a small graduated measure content type is used if the delivery measure under test has a capacity larger than the integral multiples of standard content measure and delivery type is used for the measure under test having capacity less than the integral multiple of standard content measure.

For larger vessels, such as storage tanks, ships and barges, dimensions are measured and geometry is applied to calculate the capacity. Due allowance is taken

S.V. Gupta, *Measurement Uncertainties*, DOI 10.1007/978-3-642-20989-5_10,

of various fixtures inside the tank and deadwood volume. Quite often, volume of liquid contained per unit height is measured and the calibration table containing volume versus height in the tank is prepared.

10.2 Uncertainty Using Gravimetric Method

10.2.1 Type A Evaluation of Standard Uncertainty

Standard uncertainty through Type A: The standard deviation of the repeated results taken under the similar conditions gives the standard uncertainty. At least three values of computed results of the volume should be taken. The variance is estimated and is pooled with the previous variance to obtain the pooled variance, if available. After some time with pool of say 100 variances, the pooled variance so obtained may also be used to assess the quality of the results. If the deviations of the two results are less than the pooled variance then the mean of two calculated results will be the volume of the measure and standard uncertainty will be equal to the square root of pooled variance.

This method is also applied for all measured parameters such as that of temperature, pressure and relative humidity or any other factor which affects the volume/capacity measurement.

10.2.2 Type B Evaluation of Standard Uncertainty

The mass m of the medium required to fill up to the certain mark or delivered from the specified mark, as the case may be, is given as

$$m(1-\sigma/D) = 1{,}000 \times V_s\,[1+\alpha(t-t_s)]\,(\rho_t-\sigma). \tag{10.1}$$

Here

σ is the density of air at the time of measurement, kg/m^3; for the time being it is considered to be constant
ρ_t is the density of the medium at temperature of measurement, kg/m^3
D is density of the standard weights used in weighing the water, kg/m^3
α is thermal coefficient of the material of the measure under test per °C
t is the temperature of water °C
t_s is the standard temperature °C
V_s is the volume of the measure under test m^3

Validity: The above relationship is also valid for

(a) When mass is measured in g, and density in g/dm^3 then capacity is in dm^3
(b) When mass is measured in mg, and density in mg/cm^3 then capacity is in cm^3

Sources of errors are:

1. Measurement of mass of the medium (water or mercury)
2. Temperature of the medium (water or mercury)
3. Temperature gradient within the medium (water or mercury)
4. Uncertainty in measurement of mass due to balance alone
5. Uncertainty in the mass value of the standard weights used
6. Density of air, which is affected by ambient temperature, pressure and relative humidity
7. Density of water
8. Density of weights
9. Coefficient of thermal expansion of the material of the measure
10. Evaporation of water vapours
11. Spillage of water during transfer

Uncertainty components in the measurement of temperature of water are calculated by using both Type A and Type B evaluation methods.

For density of air, pressure, temperature and relative humidity are measured, but uncertainty only through Type B evaluation method is calculated.

Density of water is taken from water tables or density versus temperature relation [2], so only Type B uncertainty is calculated. The overall standard uncertainty in the knowledge of density of water is less than one part per million [2].

Density of standard weights is taken from the certificate; hence only Type B evaluation method is used.

Coefficient of thermal expansion of the material of the measure is also taken from the literature, so only Type B evaluation is used.

10.2.3 Mathematical Model for Volume (Gravimetric Method)

From (10.1) above, by doing some mathematical operations, V is expressed in terms of other parameters as

$$V = m(1 - \sigma/D)\{1 - \alpha(t - t_s)\}(\rho + \sigma) \tag{10.2}$$

or

$$\log V = \log m + \log(1 - \sigma/D) + \log\{1 - \alpha(t - t_s)\} + \log(\rho + \sigma).$$

Differentiating partially V with respect to each parameter, we get

$$\frac{1}{V}\frac{\delta V}{\delta m} = \frac{1}{m} \Rightarrow \frac{\delta V}{V} = \frac{\delta m}{m}. \tag{10.3}$$

$$\frac{1}{V}\frac{\delta V}{\delta \sigma} = \left[\frac{-1/D}{1 - \sigma/D} + \frac{1}{\rho + \sigma}\right] \Rightarrow \frac{\delta V}{V} = \left[\frac{1}{\rho + \sigma} - \frac{1}{D - \sigma}\right]\delta\sigma. \tag{10.4}$$

Neglecting σ from the denominator as σ is very small in comparison to D as well as to ρ, we get

$$\frac{\delta V}{V} = \left[\frac{1}{\rho} - \frac{1}{D}\right]\delta\sigma, \tag{10.5}$$

$$\frac{1}{V}\frac{\delta V}{\delta D} = \frac{-\sigma/D^2}{1-\sigma/D} \Rightarrow \frac{\delta V}{V} = \frac{-\sigma}{D(D-\sigma)}\delta D \approx \frac{-\sigma}{D^2}\delta D, \tag{10.6}$$

$$\frac{1}{V}\frac{\delta V}{\delta \alpha} = \frac{-(t-t_s)}{1-\alpha(t-t_s)} \Rightarrow \frac{\delta V}{V} = \frac{-(t-t_s)}{1-\alpha(t-t_s)}\delta\alpha \approx -(t-t_s)\delta\alpha, \tag{10.7}$$

$$\frac{1}{V}\frac{\delta V}{\delta t} = \frac{-\alpha}{1-\alpha(t-t_s)} \Rightarrow \frac{\delta V}{V} = \frac{-\alpha\delta t}{1-\alpha(t-t_s)} \approx -\alpha\delta t, \tag{10.8}$$

$$\frac{1}{V}\frac{\delta V}{\delta \rho} = \frac{1}{\rho-\sigma} \Rightarrow \frac{\delta V}{V} = \frac{\delta\rho}{\rho-\sigma} \approx \frac{\delta\rho}{\rho}. \tag{10.9}$$

10.2.3.1 Relative Uncertainty Due to Change in Temperature Alone

One may notice that if there is a temperature change then there is a change in density is water as well as in air.

The change in density of water [2] at 25°C for $\delta t = 0.5$°C is 0.1295 kg/m^3 and that of air change is only 0.002143 kg/m^3.

$$\begin{aligned}\frac{\delta V}{V} &= -\frac{\alpha\delta t}{1} + \delta\sigma\left[\frac{1}{\rho} - \frac{1}{D}\right] + \frac{\delta\rho}{\rho-\sigma}\\ &= -3\times10^{-5}\times0.5 + 0.002143\,(1.003 - 0.125)\,10^{-3} + 0.1300\times10^{-3}\\ &= -0.15\times10^{-4} + 0.19\times10^{-5} + 1.3\times10^{-4} = 1.17\times10^{-4}.\end{aligned} \tag{10.10}$$

The uncertainty due to temperature alone is quite significant; hence corrections to be applied are given in steps of 0.1°C.

Second term in (10.10) is due to the change in air density when temperature changes by 0.5°C. The correction due to variation in air density is to be considered for larger capacity measures.

10.2.3.2 Relative Uncertainty Due to Change in Density of Weights Alone

The density of weights varies from 8,400 to 8,000 kg/m^3

$$\delta D = 400\,\text{kg/m}^3$$

then from (10.6)

$$\frac{\delta V}{V} = \frac{1.2}{D^2}\delta D = 1.2 \times 400/(8{,}200)^2 = 7.14 \times 10^{-6}. \quad (10.11)$$

10.2.3.3 Relative Uncertainty Due to Change in Coefficient of Expansion Alone

$$\frac{\delta V}{V} = (t - t_s)\delta\alpha = 10 \times 3 \times 10^{-6} = 3 \times 10^{-5}. \quad (10.12)$$

In literature normally coefficient of linear expansion $30 \times 10^{-6}/°C$ is given with an uncertainty of 10% and difference in temperature is around 10°C.

10.2.3.4 Uncertainty Due to Setting the Level at the Graduation Mark (Meniscus Setting)

If the setting error is δh, and the diameter of the volumetric measure is d, then error in volume

$$\delta V = d^2 \times \delta h/4. \quad (10.13)$$

10.3 Examples of a Few Measures

Volumetric measures are graduated with a single mark or have a graduated scale. The glassware such as pipettes and flasks are one-mark volumetric measure. For simplicity, let us start with a one-mark bulb delivery pipette of capacity 10 cm^3. Most important attributes of such a pipette are capacity and delivery time.

10.3.1 Uncertainty in Calibration of Capacity of a One-Mark Pipette

The pipette under test is filled up to the graduation mark with water, which with the proper precaution is delivered in a clean pre-weighed beaker and weighed again. Apparent mass of water is determined and the proper correction is added to it to give the volume of water delivered at the standard temperature 27°C. This whole process is repeated three times and standard deviation of volume of water delivered is obtained. The results are shown in Table 10.1:

Table 10.1 Observations and Calculation Sheet

Mass of water	Temperature °C	Correction g	Volume cm^3	Deviation cm^3	Deviation^2 10^{-10} cm^6
9.966 04	22.3	0.034 35	10.000 39	0.000 12	144
9.966 26	22.3	0.034 35	10.000 61	0.000 10	100
9.965 98	22.4	0.034 56	10.000 54	0.000 13	9
		Mean	10.000 51	Sum	253

Standard deviation 0.000 11 cm^3

$$\text{Type A Standard Uncertainty} = 0.000\,11\,\text{cm}^3 = 0.11\,\text{mm}^3. \tag{10.14}$$

Degree of freedom of measurement is only 2.

10.3.1.1 Type B Evaluation of Uncertainty

Due to Balance

Balance used in measurement of mass was of 20 g capacity with a digital readout of 0.01 mg. It has a repeatability of 0.05 mg assumed to be calculated with fairly large number of degree of freedom.

A digital indicator implies that true value may lie anywhere with equal probability within the value of the digital indication. Hence true value follows a rectangular distribution of semi-range equal to half the value of the least digital interval.

In this case value of least digital interval is 0.01 mg; hence its semi-range is 0.05 mg. The standard deviation; i.e. the standard uncertainty of a rectangular distribution with semi range 0.005 mg, is given as

$$0.005/\sqrt{3}\,\text{mg}.$$

So combined standard uncertainty of the balance u_B alone is given as

$$u_B = 0.05(1 + 1/10\sqrt{3}) = 0.05\,\text{mg}.$$
$$\text{This corresponds to the volume of } 0.050\,\text{mm}^3. \tag{10.15}$$

As repeatability is taken for sufficiently large number of weighing and uncertainty due to digital readout has been calculated from rectangular distribution, its degree of freedom is infinity.

Due to Weights

Standard uncertainty in weights of the balance as reported in the certificate is 0.01 mg. Hence standard uncertainty u_w in terms of volume is given as

$$u_w = 0.01\,\text{mm}^3. \tag{10.16}$$

Here also degree of freedom is infinity.

Due to Thermometer

Temperature has been measured with a calibrated thermometer in 0.1°C with an uncertainty of 0.01°C, the uncertainty due to the thermometer reading $0.05/\sqrt{3}$°C, standard uncertainty in temperature $\{(0.01)^2 + (0.05)^2/3\}^{1/2} = 0.030$°C and contribution to volume 0.351 mm^3.

Uncertainty u_t due to thermometer with 0.1°C graduations is given as

$$u_t = 0.0060\,\text{mm}^3. \tag{10.17}$$

Here also degree of freedom is infinity.

Due to Thermal Coefficient

Relative uncertainty as given above $\Rightarrow \frac{\delta V}{V} = \frac{-(t-t_s)}{1-\alpha(t-t_s)}\delta\alpha.$

Here α is taken as 25×10^{-6}. Normally, in literature, we find values of linear coefficient within 10% uncertainty; i.e. the semi-range of the α the volume expansion coefficient is $\pm 3 \times 10^{-6}$. There are several possibilities

1. The stated value of the α follows normal distribution and range has been given with a coverage factor of 1, 2 or 3. In that case to find the standard deviation of α, one may divide by 1, 2, or 3 respectively as the case may be.
2. The stated value of α is supposed to follow rectangular distribution. In this case we will divide the semi-range by $\sqrt{3}$ to obtain the standard deviation.
3. Similarly we may suppose that the stated value of the α follows triangular or trapezium distributions. In those cases the semi-range is to be divided by $\sqrt{6}$ or by some other factor $\sqrt{\beta/6}$) to obtain the corresponding standard deviation.

There is no specific consensus, at the International level, as how to treat the data from the literature. However many metrologists tend to assume rectangular distribution in such cases.

In this particular case we are assuming that if range is in terms of standard deviation then relative uncertainty

$$\frac{\delta V}{V} = 4.6 \times 3 \times 10^{-6} = 13.8 \times 10^{-6}.$$

Giving uncertainty u_α due to uncertainty in α as

$$u_a = 1.38 \times 10^{-5}\,\text{cm}^3 = 0.0138\,\text{mm}^3. \tag{10.18}$$

Here also degree of freedom is infinity.

Due to Density of Water

Density of water is known with a relative standard uncertainty of 0.4×10^{-6} [2], so uncertainty u_ρ in the $10\,\text{cm}^3$ pipette is given as

$$u_\rho = 0.004\,\text{mm}^3. \tag{10.19}$$

Here also degree of freedom is infinity.

Total Uncertainty

Hence total standard uncertainty for $10\,\text{cm}^3$ pipette

$$\begin{aligned}
&= \sqrt{u_r^2 + u_b^2 + u_w^2 + u_t^2 + u_\alpha^2 + u_\rho^2} \\
&= \sqrt{(0.11)^2 + (0.05)^2 + (0.01)^2 + (0.0138)^2 + (0.06)^2 + (0.004)^3} \\
&= \sqrt{0.0121 + 0.0025 + 0.0001 + 0.00019 + 0.0036 + 0.0000} \\
&= 0.136\,\text{mm}^3.
\end{aligned} \tag{10.20}$$

$$\text{Relative standard uncertainty} = 0.139 \times 10^{-4}. \tag{10.21}$$

If we wish to give an expanded uncertainty to certain level of confidence, the uncertainty is to be multiplied by Student's t factor, which depends upon the effective degree of freedom and the confidence level of the total measurement.

$$\text{Effective degree of freedom} = \nu_{\text{eff}} = \frac{u^4(V)}{\sum_{i=1}^{i=n} \frac{u_i^4(y)}{\nu_i}}$$

$$= \frac{3.8 \times 10^{-4}}{(1.4641 \times 10^{-4}/2) + 1.16 \times 10^{-5}/\infty + 10^{-8}/\infty + 4 \times 10^{-8}/\infty) - 1.310^{-5}/\infty}$$

$$= 5.2 \cong 5. \tag{A}$$

For 95% confidence level, the Student's t factor for 5 degrees of freedom is 2.5706.

Expanded uncertainty at 95% Confidence level $= 2.57 \times 0.136\,\text{mm}^3 = 0.350\,\text{mm}^3$.
Statement of capacity of the pipette will be as follows:
Volume of water delivered at 27°C $= 10.00051\,\text{cm}^3 \pm 0.350\,\text{mm}^3$ at $k = 2.57$.

Any subsequent user will take the stated value and the value will follow the normal distribution having a standard deviation of $0.136\,\text{mm}^3$; i.e. standard uncertainty is $0.136\,\text{mm}^3$.

10.3.2 *Calibration of a Burette*

A burette, besides other test, is also calibrated for its (delivery) capacity at least at four points of its scale. For example four points for a 50 cm^3 burette may be from 0 to 50, 0 to 40, 0 to 30 and 0 to 20 cm^3. In this case, to make even three repetitions is too time consuming. Hence only two observations are taken and the method of pooled variance is used to find standard uncertainty by Type A evaluation method. The method is elaborated in Table 10.2.

Degree of freedom here is 4, as one degree of freedom has been consumed in calculating the mean of each of four sets of observations.

Standard deviation, therefore, is $10^{-3}\sqrt{581/4} = 0.012\,\text{cm}^3$, giving uncertainty u_r in repeatability as

$$u_r = 0.012\,\text{cm}^3. \tag{10.22}$$

Components of other uncertainty through Type B evaluation will remain similar as discussed above. Hence other components of uncertainty are evaluated from (10.4) to (10.10). Total standard uncertainty in the burette is then the root of the sum of the squares of all component uncertainties.

10.3.2.1 Evaporation Loss

Evaporation losses are not easily assessable, because of a wide variety of reasons of evaporation. However, in a recent paper uncertainty due to evaporation in a 100 cm^3 measure was estimated as 0.14 mm^3. So in the case of 10 cm^3 pipette it may be roughly estimated as 0.014 mm^3.

Table 10.2 Observations and Calculation Sheet

S No	From to	Mass of water g	Temp. °C	Correction g	Volume cm^3	Mean cm^3	Deviations cm^3	Square $10^{-6}\,cm^6$
1	0 to 50	49.8146	25.6	0.2083	50.023		−0.11	121
2	0 to 50	49.8338	25.6	0.2083	50.046		0.12	144
						50.034		
3	0 to 40	39.8687	25.6	0.1666	40.035		−0.10	100
4	0 to 40	39.8884	25.6	0.1666	40.055		0.10	100
						40.045		
5	0 to 30	29.9142	25.6	0.1250	30.039		−0.003	9
6	0 to 30	29.9203	25.6	0.1250	30.045		0.003	9
						30.042		
7	0 to 20	19.9235	25.6	0.0833	20.007		−0.007	49
8	0 to 20	19.9379	25.6	0.0833	20.021		0.007	49
						20.014	sum	581

10.4 Uncertainty Using Volumetric Comparison

When large number of vessels/measures of very high capacity is required to be calibrated, and uncertainty requirements are not too stringent, the volumetric method is used. In this method, the capacity of the under-test measure is compared with that of the standard of known capacity.

The volumetric method is applicable when two measures are different types; namely one is content type and another is delivery type.

While calibrating each capacity measure is kept in such a way that graduation marks are in the horizontal plane. In case of non-graduated measures, the axis of the delivery measure is kept vertical if it is overflow type. As the marks on either measure are normal to their respective axis, so care should be taken that the content measure is kept on a horizontal ground and the delivery measure is kept in vertical position.

10.4.1 Multiple and One-to-One Transfer Methods

If the capacity of the measure under test and that of the standard is equal then one-to-one transfer or direct comparison method is used.

If content measure is of larger capacity than standard measure, then as stated above, a standard of delivery type, whose capacity is a sub-multiple of the capacity of the measure under test, is used and multiple filling is carried out.

10.4.2 Corrections Applicable in Volumetric Method

Corrections are applied due to (1) coefficients of thermal expansion of materials of the two measures, (2) different reference temperatures for which the measures have been calibrated and (3) different temperatures of the two measures.

10.4.3 Reference Temperatures

There are two cases (1) reference temperature is same for the two measures and (2) the reference temperature for each measure is different.

10.4.3.1 Reference Temperatures Are Equal

Let V, α, ρ and t_r respectively stand for volume, coefficient of expansion, density of water and reference temperature and subscripts s and u are used respectively for standard and under-test measures.

The capacity of the measure under test is nominally equal to n times the capacity of the standard measure. If V_s and V_u are capacities of the standard and under-test measure, then

$$V_u = nV_s + v. \tag{10.23}$$

Here v is the difference in capacity between the nominal and measured value at the time of calibration. This includes the correction due to the actual value of the capacity of the standard measure.

Assuming that there is no loss of liquid during transfer, the mass of the liquid in the under-test measure will be equal to the mass of the liquid transferred from the standard measure. If ρ_{st}, ρ_{ut} is the density of the transfer liquid and t_s and t_u are the temperatures of liquid in standard and under-test measure, respectively, then

$$(nV_{st} + v)\rho_{st} = V_{ut} \times \rho_{ut}. \tag{10.24}$$

If V_{sr} and V_{ur} are their respective capacities at reference temperature, then

$$(nV_{sr} + v)[1 + \alpha_s(t_s - t_r)]\rho_{st} = V_{ur}[1 + \alpha_u(t_u - t_r)]\rho_{ut},$$

giving

$$V_{ur} = (n \times V_{sr} + v)[1 + \alpha_s(t_s - t_r)]\rho_{st}/[1 + \alpha_u(t_u - t_r)]\rho_{ut}. \tag{10.25}$$

Taking logarithm of both sides, we get

$$\log V_{ur} = \log(n \times V_{sr} + v) + \log[1 + \alpha_s(t_s - t_r)] + \log\rho_{st} - \log[1 + \alpha_u(t_u - t_r)] - \log\rho_{ut}. \tag{10.26}$$

Differentiating partially, step by step, with respect to each variable, we get

$$\frac{1}{V_{ur}}\frac{\delta V_{ur}}{\delta V_{sr}} = \frac{1}{V_{sr}} \Rightarrow \delta V_{ur} = V_{ur}\frac{\delta V_{sr}}{V_{sr}}. \tag{10.27}$$

For the purpose of estimating uncertainty v is neglected in the denominator of (10.27), as v is very small in comparison of V_s.

$$\frac{1}{V_{ur}}\frac{\delta V_{ur}}{\delta\alpha_s} = \frac{(t_s - t_r)}{[1 + \alpha_s(t_s - t_r]} \Rightarrow \delta V_{ur} = V_{ur}\frac{\delta\alpha_s(t_s - t_r)}{[1 + \alpha_s(t_s - t_r]}, \tag{10.28}$$

$$\frac{1}{V_{ur}}\frac{\delta V_{ur}}{\delta t_s} = \frac{\alpha_s}{[1 + \alpha_s(t_s - t_r]} \Rightarrow \delta V_{ur} = V_{ur}\frac{\alpha_s\delta t_s}{[1 + \alpha_s(t_s - t_r]}, \tag{10.29}$$

$$\frac{1}{V_{ur}}\frac{\delta V_{ur}}{\delta\rho_{st}} = \frac{1}{\rho_{st}} \Rightarrow \delta V_{ur} = V_{ur}\frac{\delta\rho_{st}}{\rho_{st}}, \tag{10.30}$$

$$\frac{1}{V_{ur}}\frac{\delta V_{ur}}{\delta \alpha_u} = -\frac{(t_u - t_r)}{[1 + \alpha_u(t_u - t_r]} \Rightarrow \delta V_{ur} = -V_{ur}\frac{\delta\alpha_u(t_u - t_r)}{[1 + \alpha_u(t_u - t_r]}, \quad (10.31)$$

$$\frac{1}{V_{ur}}\frac{\delta V_{ur}}{\delta t_u} = -\frac{\alpha_u}{[1 + \alpha_u(t_u - t_r]} \Rightarrow \delta V_{ur} = -V_{ur}\frac{\delta\alpha_u(t_u - t_r)}{[1 + \alpha_u(t_u - t_r)]}, \quad (10.32)$$

$$\frac{1}{V_{ur}}\frac{\delta V_{ur}}{\delta \rho_{ut}} = -\frac{1}{\rho_{ut}} \Rightarrow \delta V_{ur} = -V_{ur}\frac{\delta\rho_{ut}}{\rho_{ut}}. \quad (10.33)$$

Normally measures of such sizes are calibrated to see if the capacity of the measure under test is within the prescribed tolerance. Once the capacity of the measure is within the tolerance limits no further measurements are taken. So uncertainty of capacity measurement also is derived by Type B evaluation method. As the tolerance is $\pm a$, the probability of the capacity of the measure under test lying anywhere within the $V - a$ and $V + a$ is equal. Hence the capacity value of the measure follows a rectangular distribution with semi-range a. Hence standard uncertainty due to tolerance alone is $a/\sqrt{3}$. In some cases, only positive tolerance is allowed; i.e. the capacity of the measure under test shall not be less than the nominal capacity. If positive tolerance allowed is b, then the standard uncertainty of capacity measurement will be $b/\sqrt{3}$.

Normally such measures are periodically verified and that too by the same laboratory. Though the laboratory does not report the actual value of the capacity of the measure, it keeps the data of the measured capacity. Another method, therefore, is to compare the present measured capacity of the measure with the earlier measured values. A sustained record of the capacity values of the measures may serve to indicate the quality of the present measurement and the maintenance of the measure.

10.4.3.2 Numerical Example

Let us consider the calibration of a 100 dm^3 measure with the help of 25 dm^3 automatic pipette. Relative standard uncertainty of the automatic pipette is 5×10^{-5}. Other data are as follows:

$\alpha_s = 54 \times 10^{-6}, \delta\alpha_s = 5 \times 10^{-5}, \delta\rho_s = 1 \times 10^{-6}$ (from the use of tables), the measure under test is of stainless steel with $\alpha_u = 33 \times 10^{-3}$ and standard uncertainty $\delta\alpha_u = 3 \times 10^{-5}$ (taken from the literature).

Temperatures are $t_s = 30°\text{C},\ t_u = 32°\text{C},\ t_r = 20°\text{C}$, giving

$$t_s - t_r = 10°\text{C}$$

and

$$t_u - t_r = 12°\text{C}.$$

Thermometers used are graduated in 0.1°C and having digital readout and have been calibrated with a repeatability of 0.01°C (from the certificate).

Uncertainty due to digital readout is $0.05/\sqrt{3}\,^{\circ}\mathrm{C}$. Giving total standard uncertainty δu_t as

$$u_t = \sqrt{0.01^2 + 0.03^2} = 0.030.$$

Relative uncertainty components are:

Source	Value	Refer to
Due to standard measure	5×10^{-5}	(10.27)
Due to coefficient of expansion of standard	$5 \times 10^{-6} \times 10 = 5 \times 10^{-5}$	(10.28)
Due to thermometer with standard	$54 \times 10^{-6} \times 0.030 = 1.62 \times 10^{-6}$	(10.29)
Due to density of water taken from tables	1×10^{-6}	(10.30)

Density of water depends upon temperature so if there is an uncertainty in thermometer it will reflect back in density also; moreover density of water comes two times in the expression, so its contribution to uncertainty will be twice, giving

$$= 2 \times 0.03 \times 2.11 \times 10^{-4} = 1.27 \times 10^{-5}.$$

Due to coefficient of expansion of under-test $3 \times 10^{-6} \times 12 = 3.6 \times 10^{-5}$. (10.34)

Due to thermometer with under-test measure $33 \times 10^{-6} \times 0.031 = 1.023 \times 10^{-6}$. (10.35)

Due to density of water in under-test 1×10^{-6}. (10.36)

Total relative standard uncertainty

$$= 10^{-5}\sqrt{25 + 25 + 0.026 + 0.01 + 1.61 + 12.96 + 0.01 + 0.01} = 8.03 \times 10^{-5}$$
$$\cong 8 \times 10^{-5}. \quad (10.37)$$

For extended uncertainty, first effective degrees of freedom are calculated by using equation (A) and then determining the Student's t factor from the Student's tables for the desired confidence level. The product of relative standard uncertainty with Student's t factor will give the extended uncertainty.

10.5 Uncertainty in Calibration of Storage Tanks

We have discussed uncertainty in calibration of volumetric measures using gravimetric method. Gravimetric method, because of the use of distilled water, is pretty costly; hence it is used for capacity measure with maximum capacity of

50 dm^3. For larger capacity measures, volumetric comparison method is used. We have described the uncertainty calculations in the calibration of capacity measures calibrated by volumetric comparison method. The volumetric comparison method is employed for measures having a capacity of few thousand dm^3. Still larger capacity measures, going to few thousand cubic metres, are calibrated by dimensional method. The measures of such a high capacity are storage tanks of different shapes and orientations. Measures of still higher capacity are barges and ships.

10.5.1 Storage Tanks

The storage tanks are

- Vertical cylindrical storage tanks with fixed roof
- Vertical cylindrical storage tanks with floating roof
- Horizontal cylindrical storage tanks
- Spheres and spheroids

The bottom of such storage tank may be flat, conical, truncated, hemispherical, elliptical or domed shape.

Before we proceed further, it will be prudent to discuss some important terms.

10.5.2 Some Important Terms

1. Tank strapping: This is a term used for the overall procedures of measurement to determine dimensions of the storage tank. It includes the following measurements:
   - Depth: Shell height, oil height, ring height, equalizer line height and gauging height
   - Thickness of tank walls
   - Circumferences at specified heights
2. Deadwood: Deadwood is any object within the tank, including a floating roof, which displaces liquid and thus reduces the capacity of the tank, including any permanent appurtenances on the outside of the tank, such as cleanouts boxes or manholes, which increase the capacity of the tank
3. Gauge table (calibration table): Table consisting of volume versus gauge height from the datum plate (datum point)

10.5.3 Maximum Permissible Error in Storage Tanks

The maximum permissible errors, recommended for storage tanks by OIML (the International Organization of Legal Metrology) through OIML-R71 [3], are as follows:

$$\pm\,0.2\%\text{ for vertical tanks}$$
$$\pm\,0.3\%\text{ for horizontal tanks and}$$
$$\pm\,0.5\%\text{ for spherical or spheroid tanks}$$

10.5.4 Maximum Permissible Errors in Tape Measures

Before proceeding further let us consider the maximum permissible errors in the tape measures and the circumference measurement

According to Legal Metrology (General) Rule 2011 [4], the measuring tapes are classified as Class I, Class II and Class III. The class-wise maximum permissible errors are as follows:

$$\begin{aligned}&\text{Class I} \pm (0.1 + 0.1\,L)\,\text{mm}\\&\text{Class II} \pm (0.3 + 0.2\,L)\,\text{mm}\\&\text{Class III} \pm (0.6 + 0.4\,L)\,\text{mm}\end{aligned} \tag{10.38}$$

where L is the length in metres rounded off to the next higher whole number. For example if the mean length of C_i is 110.345 m then value of L is 111.

10.5.5 Maximum Permissible Errors in Circumference Measurement

When observations are repeated, mean should lie within specified maximum permissible errors (MPE). In India, by Legal Metrology Act 2010 [4], the MPE for circumference measurements are as follows:

Measuredlength	MPE
Upto 30 m	±2 mm
Over 30and up to 50 m	±4 mm
Over 50and up to 70 m	±6 mm
Over 70but up to 90 m	±8 mm
Over 90 m	±10 mm

(10.39)

10.6 Principle of Preparing Gauge Table (Calibration Table)

1. The intervals of dip at which the tables are made should not be too great; otherwise there will be inaccuracies in interpolating the value of volume at a particular dip not listed in the table. Normally 5 cm interval is sufficient,

along with a proportional table. The table is calculated on the basis of average difference for the chosen interval. Interval of the proportional table should be in mm. Such tables are able to give volumes in dm^3 (litres) for any given or measured depth. However for tanks with lap joints, the proportional table is based on the average difference for each course separately. Levels affected by bottom irregularities and deadwood are not included in calculating the average difference in volume per unit depth used in for preparing the proportional table. This table is not applicable for interpolations of these levels.

2. The tables may be set out more fully if greater speed in calculation is desired. But it should be remembered that the table set out in one page is quicker in use than the one occupying several pages.
3. It should be kept in mind that no liquid measurement requires better relative accuracy of one part in ten thousand. Commercial table never requires a fraction of litre; any table, which is able to calculate within one litre, is more than sufficient.

Keeping these points in view, 5 cm interval with difference table has been found to be adequate.

Broadly speaking, there are two methods of calibration of tanks, namely

1. Dimensional measurements
2. Volumetric

However, more often than not, the combination of both the methods is used for calibrating a storage tank.

Dimensional measurements are carried out either by external strapping or internal strapping.

10.6.1 External Strapping

1. Dimensions by internal strapping
2. Optical reference line
3. Optical triangulation
4. Electro-optical method.

We may measure external diameter at prescribed positions of a course by encircling the tank with a linear tape measure, and repeat the same procedure for each course of the tank. The method is known as measuring dimensions by external strapping. To overcome the objects that are permanent part of the tank, step-over is used and necessary corrections are applied to give the outer circumference. The internal circumference is obtained by measuring the thickness of the shell of the tank and applying necessary correction.

If C_0 is the outer circumference, obtained after applying corrections

- For different coefficients of linear expansion of the tank and the tape measure used

- For different reference temperatures of the tape measure and at which tank is to calibrated and
- For step-over correction if any

then C_i the internal circumference of the tank is given by

$$C_i = C_o - 2\pi \times t. \tag{10.40}$$

Cross section of the tank is then given by

$$A = C_i^2/4\pi \times s.$$

This gives
Capacity of the cylindrical tank per metre $= C_i^2/4\pi$ in m^3,
Or

$$= 1{,}000 \times C_i^2/4\pi \text{ dm}^3.$$

Similarly capacity of cylindrical tank per centimetre $= 10\,C_i^2/4\pi$ dm^3.
Here height is taken as 1 cm. Substituting the value of π above, we get

$$\text{Capacity per cm} = 0.795778\,C_i^2 \text{ dm}^3/\text{cm}. \tag{10.41}$$

Here it should be remembered that C is still in metres.

From above we see that source of measurement uncertainty is the inner circumference. The uncertainty in t, the thickness of the shell, is ignored because of its small value in relation to circumference. Normally the value of thickness is 0.01% of circumference. Steel tape measures are used to measure the circumference.

In the following paragraphs, we calculate the uncertainty of the square of the measured circumference, i.e. of area of cross section A.

10.6.1.1 Type A Evaluation of Uncertainty

Repeating the measurement of the circumference of the tank at the same position is normally not feasible. However circumference, for each course, is measured at three places. These three measurements are used to calculate dispersions from their respective means. Though the mean circumference of different courses may be different, but dispersions from their respective means due to non-repeatability of the measurements may be pooled together for estimating the standard deviation. This gives the uncertainty evaluated by Type A method.

Let the tank under calibration have eight courses and internal circumferences be calculated at three positions of each course. The dispersions and their squares are calculated for each mean and pooled together to calculate the standard uncertainty for non-repeatability of measurements. The mean is the value of the internal circumferences at that place. A numerical example to elucidate the method is given in Table 10.3.

Table 10.3 Calculation sheet for dispersion in circumference measurement

Course No	C_i in m	Mean C_i in m	Deviation in mm	Square of deviations
8 Top	110.0238		−0.3	0.09
8 Middle	110.0188	110.0241	−5.3	28.09
8 Bottom	110.0298		5.7	32.49
7 Top	110.0688		−1.6	2.56
7 Middle	110.0688	110.0704	1.6	2.56
7 Bottom	110.0738		3.4	11.56
6 Top	110.0690		2.3	5.29
6 Middle	110.0650	110.0667	−1.7	2.89
6 Bottom	110.0660		−0.7	0.49
5 Top	110.0491		10.7	114.49
5 Middle	110.0641	110.0598	−4.3	18.49
5 Bottom	110.0661		−6.3	39.69
4 Top	109.9831		−4.0	16.0
4 Middle	109.9861	109.9871	−1.0	1.0
4 Bottom	109.9921		5.0	25.0
3 Top	110.0503		−2.0	4.0
3 Middle	110.0553	110.0523	3.0	9.0
3 Bottom	110.0513		−1	1.0
2 Top	109.9407		−0.7	0.49
2 Middle	109.9487	109.9414	7.3	53.29
2 Bottom	109.9347		−6.7	44.49
1 Top	110.0371		4.3	18.49
1 Middle	110.0371	110.0328	4.3	18.75
1 Bottom	110.0241		−8.7	75.69
Sum of squares of dispersions				525.89
Degrees of freedom = 8(3–2) = 16				

Standard deviation of internal circumference measurements $\sqrt{525.26/16} = 5.73\,\text{mm}$.

Standard uncertainty from Type A evaluation of circumference measurements $U_{\text{ci}} = 5.73\,\text{mm}$

But area of cross section A is given by

$$A = \frac{C_{\text{i}}^2}{4\pi}.$$

Type A relative standard uncertainty of A is

$$U_A = 2U_{\text{ci}}/C_{\text{i}} = 2 \times 5.73/110 \times 10^3 = 0.1 \times 10^{-3}. \tag{10.42}$$

10.6.1.2 Type B Evaluation of Uncertainty

Expressing area of cross section A in terms of outer circumference and shell thickness t, we get $A = \frac{C_i^2}{4\pi}$.

Having known the permissible errors in circumference measurements, we should find out the appropriate tape measure, which should be used in calibration of such tanks.

Let us assume that circumference C_o is measured with standard tape measure of Class II [4] having permissible error as

$$\text{MPE} = \pm(0.3 + 0.2L)\,\text{mm}.$$

Where L is length in metres rounded off to the next higher whole number. For example if the mean length of C_o is 110.345 m then L is 111. Total permissible error in C_o is

$$0.3 + 22.2 = \pm 22.5\,\text{mm}.$$

The error of ±22.5 mm in 110 m circumference is more than the permissible error prescribed for circumference measurements – Sect. 10.5.5. Hence we should use only Class I tape measure. Maximum permissible error in Class I tape measure used for aforesaid circumference measurement will be

$$(0.1 + 11.1)\,\text{mm} = 11.2\,\text{mm}.$$

This implies that the true value of the C_o may lie anywhere, with equal probability, in the range of 110.345 m + 11.2 mm and 110.345 m − 11.2 mm. That is true value will follow a rectangular distribution with semi-range equal to 11.2 mm. Standard deviation S.D. of such a distribution is given

$$\text{S.D.} = \text{semi-range}/\sqrt{3} = 11.2/1.7321 = 6.5\,\text{mm}.$$

Standard uncertainty U_{Co}, therefore, will be 6.5 mm, which will be same for C_i – the internal circumference.

Now let us calculate uncertainty in cross-sectional area A by Type B evaluation method. The area A is given by

$$A = C_i^2/4\pi$$

$$\log A = 2\log C_i - \log 4\pi.$$

Differentiating, we get

$$\frac{\delta A}{A} = \frac{2\delta C_i}{C_i},$$

giving relative standard uncertainty in A from Type B evaluation as

$$U_{\mathrm{B}} = 2 \times 6.5 \times 10^{-3}/110 = 0.11 \times 10^{-3}. \tag{10.43}$$

Using (10.42) and (10.43), the combined relative standard uncertainty U is given by

$$= \sqrt{U_A^2 + U_{\mathrm{B}}^2} = 10^{-3} \times \sqrt{0.1^2 + 0.11^2} = 0.148 \times 10^{-3}. \tag{10.44}$$

10.6.2 Internal Strapping

If internal strapping is used for measuring the diameter of the tank, then the tape measure, due to its own weight, even after applying a mandatory tension of 50 ± 5 N, will not remain horizontal. Hence a correction due to sag is applied.

10.6.2.1 Correction Due to Sag

Assuming that the tape will take a shape of a catenary

The correction Z [5] due to the sag is given as

$$Z = W^2 S^3/24T^2 \text{ in m}, \tag{10.45}$$

where

S is span of the tape in m
T is tension applied in kg force
W is the mass of the tape in kg/m
Putting together tape-related variable as K, K is given as

$$K = W^2/24T^2. \tag{10.46}$$

For a tape of 10 mm wide and 0.25 mm thick, made of steel having a density of 7,850 kg/m^3, the values of K for different values of tension applied to it are

T	K
4.4 kg	8.29×10^{-5}perm^2
4.5 kg	7.92×10^{-5}perm^2
4.6 kg	7.58×10^{-5}perm^2

For a length of 40 m the sag at 4.5 kg tension will be 5.0688 mm. This is the correction in diameter measurement. This correction is to be subtracted from the observed reading.

However, no correction in the measurement of outer circumference due to sagging is required as the tape, in this case, is in contact of the tank surface and its horizontality is monitored.

The evaluation of uncertainty in calibration will be calculated as enunciated above. One may easily notice that uncertainty in area of cross section *A* will be twice as that in diameter measurement.

10.7 Tank Deformation

When vertical tank is full then hydrostatic pressure on the lower courses will be more than on the upper one. The hydrostatic pressure will increase the tank diameter, thereby reducing the height. The reduction in height of the courses will cause lowering of the upper part of the shell. Referring to Fig. 10.1, the relative

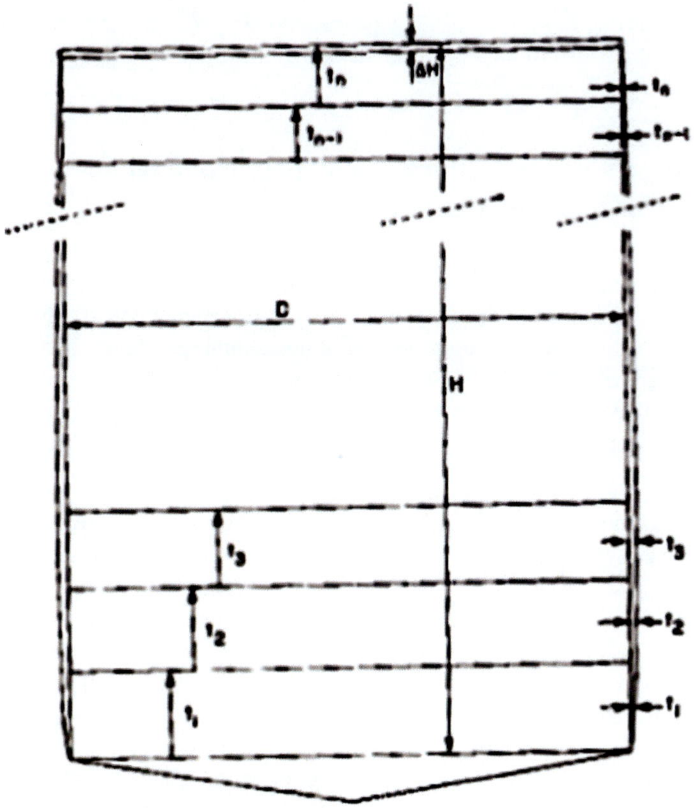

Fig. 10.1 Deformation in Vertical Storage Tank

reduction of tank height is calculated by the formula given below with the following notations [5, 6].

ρ – density of the liquid expressed in kg/m^3,
D – diameter of the tank in m,
E – modulus of elasticity in N/m^2,
μ is the Poisson's ratio,
h_n is height and t_n is the height of the nth course (ring) counted from bottom.
H is height of the tank in m

Then $\Delta H/H$ relative reduction in height is expressed as

$$\begin{aligned}\Delta H/H = (D\rho g/4\mu E)[H/t_1 &+ \{(H-h_1)^2/H\}(1/t_2 - 1/t_1)\\ &+ \{(H-h_1-h_2)^2/H\}(1/t_3 - 1/t_2)\\ &+ \cdots\{H-(h_1+h_2+\cdots+h_{n-1})\}^2/H\}(1/t_n - 1/t_{n-1})]. \end{aligned} \quad (10.47)$$

The corrections due to sagging of the tape measure and deformation of the tank are quite small; hence uncertainty components due to these causes will be negligible.

10.8 Uncertainty in Calibration of a Hydrometer by Comparison Method

In comparison method, two hydrometers of same range and surface tension are taken. Scale of one of the hydrometer is already calibrated by hydrostatic method. In general, observations are not repeated in routine calibrations; hence all uncertainty components are calculated by Type B evaluation.

Let e be value of smallest graduation of hydrometer under test (HUT)

Sources of uncertainty

1. Error in setting the HUT at the desired graduated line, semi-range of which may be taken as $1/4e$. This means the actual setting of the hydrometer may be anywhere in this range with equal probability. The standard uncertainty U_t (standard deviation) $e/4\sqrt{3}$
2. Observation error in standard hydrometer, semi-range of which may be taken as half the value of the smallest graduation of the Std. hydrometer. Normally smallest graduation of Std is $1/2$ of HUT. Therefore, the actual setting of the standard hydrometer may be anywhere in the semi-range $\pm 1/4$ of e with equal probability. The standard uncertainty U_s (standard deviation) $e/4\sqrt{3}$
3. Certificate correction $1/4$ of the smallest graduation of Std. Hydrometer. This means the actual setting of the hydrometer may be anywhere in this range $\pm 1/8$ of e with equal probability. The standard uncertainty U_{s1}(standard deviation) $e/8\sqrt{3}$

4. Combined standard uncertainty U is given as

$$U_c = \sqrt{U_t^2 + U_s^2 + U_{s1}^2} = \sqrt{\frac{(0.25e)^2}{3} + \frac{(0.25e)^3}{3} + \frac{(0.125e)^2}{3}} = 0.216e \tag{10.48}$$

As all the standard deviations are calculated from a continuous rectangular distribution, degrees of freedom are infinite in each case. Hence effective degrees of freedom are infinity. So extended uncertainty at 95.4% confidence level is

$$2 \times 0.216e = 0.432e. \tag{10.49}$$

Uncertainty given by NPL is $0.5e$.

Uncertainty due to corrections

$$1. \text{Temperature correction} = \gamma\rho(t_s - t_u). \tag{10.50}$$

Maximum value of $t_s - t_u$ is 5°C, ρ is 2.5 and γ is $30 \times 10^{-6}/°C$. Uncertainty is due to variation in the actual value of γ, which normally is not greater than 10%, i.e. 3×10^{-6}, giving the uncertainty in correction as

$$3 \times 10^{-6} \times 2.5 \times 5 = 37.5 \times 10^{-6} = 0.000037\text{g/cm}^3. \tag{10.51}$$

Smallest graduation in hydrometer is $0.02/50 = 0.0004\ \text{g/cm}^3$

So uncertainty due to this cause is negligible.

$$2. \text{ Surface tension correction} = \frac{(\rho\pi D(t - t_s)}{Mg} \text{for cgs unit.} \tag{10.52}$$

$$\text{For SI units the surface tension correction} = \frac{10^3(\rho\pi D(t - t_s)}{Mg}. \tag{10.53}$$

The correction is small; hence uncertainty may be taken as negligible for routine work.

1. Meniscus correction:

If the hydrometer under test (HUT) is for opaque liquids, for example lactometers, the observations are taken where top of the meniscus meets the stem of the hydrometer. If such a hydrometer is tested in the transparent liquid then a correction known as meniscus correction (MC) is given as follows:

$$\text{MC} = \frac{T \times R}{\pi\rho Lg}\left[(1 + 2gD^2\rho)^{1/2} - 1\right] \text{in cgs units.} \tag{10.54}$$

$$\text{MC} = 10^3\frac{T \times R}{\pi\rho Lg}\left[(1 + 2gD^2\rho)^{1/2} - 1\right] \text{in SI units.} \tag{10.55}$$

Meniscus correction for a surface tension $T = 75\,\text{mN/m}$ varies from one to two times the value of the smallest scale interval. Uncertainty in this correction is not expected by more than 5%, and hence may be taken as negligible.

In equations from (10.53) to (10.55) notations are as follows:

γ = the coefficient of linear expansion
ρ = the nominal value of the density of the hydrometer at the middle of its scale
t_s, t_u are the reference temperatures for the standard and under-test hydrometers
R is the range of the scale
T is the surface tension
L is the length of scale
D is the diameter of the stem and
g is the acceleration of gravity

References

1. S.V. Gupta, *Comprehensive Volume and Capacity Measurement* (New Age International, New Delhi, 2006)
2. S.V. Gupta, *Practical Density Measurement and Hydrometry* (Institute of Physics, London, 2002)
3. OIML R-71, Fixed storage tanks-general requirements (2008)
4. S.V. Gupta, *A Treatise on Legal Metrology*, Schedule VI, Part VI (Commercial Law, New Delhi, 2011)
5. ISO 7507–6, Calibration of Vertical storage Tanks (monitoring) (1997)
6. OIML R 85, Automatic Level Gauges for Measuring the level of liquid in Fixed Storage Tanks (1998)

Chapter 11
Uncertainty in Calibration of Some More Physical Instruments

11.1 Uncertainty in Calibration of Slip Gauges

Slip gauges are available in various grades namely grade 00, grade 0, grade 1 and grade 2. There are national and international standard specifications, which, besides many other things, specify parallelism between end faces and their flatness. These specifications also specify that the grade 00 and grade 0 slip gauges are to be calibrated by interferometer. If there is more than one slip gauge, then the cleaned gauges are wrung on the base platen of the interferometer and are allowed to attain temperature inside the interferometer. A temperature of 20 ± 0.5°C is maintained inside the interferometer. Interferometer fringes are formed between the light reflected from the platen and upper end of the slip gauge. Fringe fractions for the different monochromatic light sources of known wavelength are determined. The difference between the observed and the known fractions for the given nominal length is matched for coincidence on the slide rule to give the estimated length of the gauge at the ambient temperature.

Corrections are applied due to:

1. Difference in temperature of the gauge from the reference temperature of 20°C.
2. Difference between the conditions of ambient air and standard air.

For first correction, (i) the value of the coefficient of expansion of the material of slip gauge and (ii) its possible variation are required. To apply the correction, (2) exact ambient temperature is required. Hence, uncertainty will arise due to assumed coefficient of expansion and the resolution and calibration uncertainty of the thermometer. The second correction involves refractive index of air, which depends upon the composition, temperature, pressure, humidity of air. Thus, uncertainty will arise from each measurement. The other sources of uncertainty are:

S.V. Gupta, *Measurement Uncertainties*, DOI 10.1007/978-3-642-20989-5_11,

11.1.1 Fringe Fraction

Fringe patterns are observed for three different wavelengths. For one wavelength, fringe fraction is observed then wave length is changed and fringe fraction is read, wavelength is again changed and fringe fraction is read. This process is repeated three times so that three fringe fractions are available for each wavelength. Dispersions are calculated for each of three means. This way we will get nine dispersions with 6 degrees of freedom $3(3-1) = 6$; one freedom is lost in obtaining the mean fraction for each wavelength. Standard deviation is equal to the square root of the squares of dispersions divided by 6. For the purpose of example only let the three laser radiations chosen have wavelength are $\lambda_1 = 0.4669\,\mu\text{m}$, $\lambda_2 = 0.5435\,\mu\text{m}$, $\lambda_3 = 0.6120\,\mu\text{m}$. The arbitrary data and calculations are shown as an example in the Table 11.1.

Standard deviation $= 10^{-2}\sqrt{(0+25+25+25+0+25+0+25+25)/6} = 5/100$
Standard deviation of mean for $\lambda_1 = 5/100\sqrt{3} = 0.0289 \Rightarrow 0.0289\times 0.4669 = 0.0135\,\mu\text{m}$
Standard deviation of mean for $\lambda_2 = 5/100\sqrt{3} = 0.0289 \Rightarrow 0.0289\times 0.5435 = 0.0157\,\mu\text{m}$
Standard deviation of mean for $\lambda_3 = 5/100\sqrt{3} = 0.0289 \Rightarrow 0.0289\times 0.6120 = 0.0177\,\mu\text{m}$
Uncertainty due to the observing fraction of the fringe (Type A evaluation)

$$\sqrt{(0.0135)^2 + (0.0157)^2 + (0.0177)^2/3} = 0.0157\,\mu\text{m}$$

Table 11.1 Observations of fringe fraction and dispersions

S No	Fringe fraction for	Dispersion	$(\text{Dispersion})^2$
	λ_1		
1	30/100	0	0
2	35/100	5/100	$25/10^4$
3	25/100	5/100	$25/10^4$
Mean	30/100		
	λ_2		
1	55/100	5/100	$25/10^4$
2	60/100	0	0
3	65/100	5/100	$25/10^4$
Mean	60/100		
	λ_3		
1	75/100	0	0
2	80/100	5/100	$25/10^4$
3	70/100	5/100	$25/10^4$
Mean	75/100		

Let it be called u_A giving

$$u_A = \pm 0.0157\,\mu\text{m}. \tag{11.1}$$

11.1.2 Actual and Ambient Temperatures of the Slip Gauge

The measured temperature of air is taken as the temperature of the gauge. The difference between measured temperature and actual temperature of the gauge may be ±0.02°C. This affects in transferring gauge length to the reference temperature of 20°C. Taking coefficient of expansion as $11.5 \times 10^{-6}/°C$, the uncertainty due to this cause alone u_{T1} is given as

$$u_{T1} = \pm 11.5 \times 10^{-6} \times 0.02\,L = \pm 0.23\,L\,\mu\text{m. Here } L \text{ is metres.} \tag{11.2}$$

The uncertainty in temperature measurement will be due to two causes, namely (1) the uncertainty in calibration of thermometer and (2) due to its finite resolution. The standard uncertainty in calibration may be taken as 0.01°C. The contribution due to uncertainty in calibration of the thermometer therefore will be

$$u_{T2} = \pm 11.5 \times 10^{-6} \times 0.01\,L = \pm 0.115\,L\,\mu\text{m}. \tag{11.3}$$

The uncertainty u_{T3} due to resolution of thermometer with 0.01°C graduation is obtained by assuming that actual value of the temperature follows a rectangular distribution with 0.01°C as its semi-range, hence u_{T3} is given

$$u_{T3} = \pm 0.115\text{L}/\sqrt{3}\,\mu\text{m} = \pm 0.066L\,\mu\text{m}. \tag{11.4}$$

11.1.3 Coefficient of Linear Expansion

Further Coefficient of expansion may vary by 10% of the assumed value viz. $\pm 1 \times 10^{-6}/°C$. Hence, semi-range R_{T1} of error due to this cause only is given as

$R_{T1} = \pm 1 \times 10^{-6}\,\Delta\theta \cdot L$. Normally, $\Delta\theta$ is not more than 0.5°C, hence the semi-range R_{T1} is given by

$$R_{T1} = \pm 0.5\,L\,\mu\text{m}.$$

Here, the actual value coefficient of expansion may lie with equal probability, therefore follows a rectangular distribution, hence uncertainty u_{T5} is given by

$$u_{T5} = \pm 0.5\,\text{L}/\sqrt{3} = \pm 0.289\,L\,\mu\text{m}. \tag{11.5}$$

11.1.4 Refractivity of Air

Sources of errors and uncertainty are

(1) Pressure: calibration and reading
(2) Temperature: calibration and reading
(3) Vapour pressure of H_2O
(4) Variation in composition of CO_2

It has been given [1–3] that semi-ranges of error due to change in refractivity of air:

For pressure measurement within ±0.1 mm of Hg is 0.03 *L* μm
For temperature measurement within ±0.01°C is 0.009 *L* μm
For humidity measurement, through wet and dry bulb thermometers with 0.1°C graduations, is 0.006 *L* μm.

The uncertainty contribution due to each will again be due to calibration uncertainty and resolution of each of the instruments [3].

11.1.4.1 Pressure of Air

Due to Calibration: Taking standard uncertainty as 0.1 mm of Hg, the uncertainty u_{P1} is

$$u_{P1} = \pm 0.03\, L\, \mu\text{m}. \tag{11.6}$$

Due to resolution: Taking the semi-range of rectangular distribution as 0.1 mm of Hg the uncertainty u_{P2} is given as

$$u_{p2} = \pm 0.03\, L/\sqrt{3}\, \mu\text{m} = \pm 0.0173\, L\, \mu\text{m}. \tag{11.7}$$

11.1.4.2 Temperature of Air

Due to calibration: Taking standard uncertainty as 0.01°C, the uncertainty u_{T6} is given as

$$u_{T6} = 0.009\, L\, \mu\text{m}.$$

Due to resolution: Taking the semi-range of rectangular distribution as 0.01°C, u_{T7} is

$$u_{T7} = \pm 0.009\, L/\sqrt{3} = \pm 0.052\, L\, \mu\text{m}. \tag{11.8}$$

11.1.4.3 Humidity of Air

Due to calibration: Taking standard uncertainty as 0.1°C in wet bulb thermometer, u_{H1} is given as

$$u_{H1} = \pm 0.006\, L\, \mu\text{m}. \tag{11.9}$$

Due to resolution: Taking resolution of the wet bulb thermometer as 0.1°C, u_{H2} is given as

$$u_{H2} = \pm 0.006\, L/\sqrt{3} = \pm 0.0035\, L\ \mu\text{m}. \tag{11.10}$$

11.1.5 Phase Change Due to Reflections from the Platen and End Face of the Gauge

There is reflection from the platen and from the exposed end of the gauge. If the two surfaces are of same finish and material, the error due phase change will be zero. If not then phase change correction affects the optical paths hence the value of length of the gauge block. If the platen is of steel and gauge is of tungsten carbide or otherwise then error in length may lie any where in between ±0.0008 μm [1], hence uncertainty u_{ph} is given as

$$u_{ph} = \pm 0.008/\sqrt{3} = \pm 0.0046\ \mu\text{m}. \tag{11.11}$$

11.1.6 Wringing of the Gauge with the Platen

Normally length of a gauge includes the finite thickness of wringing film. This film is usually 0.012 μm. No correction is applied for the thickness of this film. However, the error in wringing [1] is within ±0.005 μm. Hence, uncertainty component u_w is given as

$$u_w = \pm 0.005/\sqrt{3} = \pm 0.0029\ \mu\text{m}. \tag{11.12}$$

11.1.7 Interferometer Parameter

Some errors creep in due to geometry of the interferometer such as:

1. *Obliquity correction*: For example NPL Hilger gauge interferometer is based on Fizeau principle. In this the viewing aperture is off axis of the incident beam and is rectangular in shape of length L and height h. Distance between the centres of the illuminating and viewing apertures is S. The viewing aperture is situated in the focal plane of the collimating lens of focal length f. The obliquity correction is given by

$$C_o = \left[\frac{S^2}{8L^2} + \frac{L^2 + h^2}{24f^2}\right] \mu\text{m}.$$

In a typical case obliquity correction is

$$C_o = 0.000025\, L\ \mu\text{m}. \tag{11.13}$$

2. *Finite source size*: The typical error due to finite source size is

$$C_s = \pm 0.0000125\,\mu\text{m}. \tag{11.14}$$

3. *Optics of interferometer used*: The wave-front correction in a typical NPL Hilger interferometer is

$$C_f = \pm 0.0025\,\mu\text{m}. \tag{11.15}$$

Normally these corrections are not exactly known for each and every interferometer, so these may be taken as some short of tolerances within which the error may lie. Assuming that actual error due to any of these causes lie any where within the semi-ranges stated in (11.13)–(11.15) with equal probability i.e. the error due to each follows a rectangular distribution with semi-range indicated in (11.13)–(11.15), the uncertainty due to these causes are:

$$u_{C_o} = \pm 0.000025\,L/\sqrt{3} = \pm 0.000014\,L\,\mu\text{m}. \tag{11.16}$$

$$u_{C_s} = \pm 0.0000125\,L/\sqrt{3} = \pm 0.000007\,L\,\mu\text{m}. \tag{11.17}$$

$$u_{C_f} = \pm 0.0025\,L/\sqrt{3} = \pm 0.00144\,L\,\mu\text{m}. \tag{11.18}$$

11.1.8 Parallelism of End Faces and Their Flatness

By definition the length of the gauge is the distance between the centres of its end faces. However fringe fraction may not be estimated exactly at the centre of the fringe pattern, so an uncertainty may crop in due to non flatness and parallelism of the end faces of the gauge. For grade 00 slip gauges, tolerance on flatness as well as non-parallelism of the end faces is 0.05 μm. If the fringe fraction is not estimated at the centre but say it is off centre by tenth of the height of the fringe pattern. The semi-range of uncertainty due each of these causes may be ±0.005 μm. The actual non-flatness or non-parallelism will follow rectangular distribution i.e. uncertainty u_p due to non parallelism is given as

$$u_p = \pm 0.005/\sqrt{3} = \pm 0.0029\,\mu\text{m}. \tag{11.19}$$

Uncertainty due to non-flatness u_f is given as

$$u_f = \pm 0.005/\sqrt{3} = \pm 0.0029\,\mu\text{m}. \tag{11.20}$$

After replacing L in every equation wherever it appears by the nominal value of the gauge, combined uncertainty for the gauge block is the square root of the sum of squares of the uncertainty components enumerated in equations (1 to 12 and 16 to 20) and listed in Table 11.2.

Table 11.2 All uncertainties are in micrometres (μm)

Length	(1)	(2)	(3)	(4)	(5)	(6)	(7)	(8)	(9)
L m	0.0157 L	0.23 L	0.115 L	0.066 L	0.289 L	0.03 L	0.0173 L	0.052 L	0.006 L
25 mm	0.0157	0.0058	0.0029	0.0016	0.0072	0.0008	0.0004	0.0013	0.0002
	(10)	(11)	(12)	(16)	(17)	(18)	(19)	(20)	
L m	0.0035 L	0.0046 L	0.0029 L	$14 \times 10^{-6} L$	$7 \times 10^{-6} L$	0.00144 L	0.0029 L	0.0029 L	
25 mm	0.0001	0.0001	0.0029	0.0000	0.0000	0.0000	0.0029	0.0029	

$$u_c^2 = (0.0157)^2 + (0.0058)^2 + (0.0029)^2 + (0.0016)^2 + (0.0072)^2 + (0.0008)^2 + (0.0004)^2 + (0.0013)^2 + (0.0002)^2 + (0.0001)^2 + (0.0001)^2 + (0.0029)^2 + 0.0 + 0.0 + 0.0 + (0.0029)^2 + (0.0029)^2 = 37072 \times 10^{-8} (\mu\text{m})^2.$$

$$u_c = 0.0192\,\mu\text{m}. \tag{11.21}$$

The uncertainty derived in (11.1) has been evaluated by Type A method and has six degrees of freedom. All other uncertainty components have been evaluated by Type B evaluation and have infinite degree of freedom. Effective degree of freedom ν_{eff} is given as

$$\nu_{\text{eff}} = \frac{(0.0192)^4}{(0.0157)^4/6} = 2.2 \times 6 \approx 13. \tag{11.22}$$

Coverage factor at 95.45% confidence level is = 2.21, hence Expanded uncertainty at 95.45% confidence level u_e becomes

$$u_e = 2.21 \times 0.0192 = 0.0424\,\mu\text{m}. \tag{11.23}$$

11.2 Uncertainty in Calibration of a Micrometer Against a Standard Slip Gauge

A micrometer is calibrated at several points of its scale with grade "0" slip gauge. For the purpose of illustrating as to how the uncertainty in its calibration is calculated, 25 mm point of its scale is taken as an example. In the following paragraphs, quite a few examples have been taken from the NABL document No 141 [8].

11.2.1 Particulars of Standard Gauge and Micrometer Under-Test

Length of the slip gauge at 20°C = 20.00010 ± 0.00008 mm as given in the certificate

Ambient temperature is 23°C
Smallest graduation of mercury in glass thermometer is 1°C
Smallest scale interval of micrometer under test is 0.0001 mm

11.2.2 Mathematical Model

$$L_{\text{UT}}\left[1 + \alpha_{\text{UT}}(t_1 - t_s)\right] + O + C = L_{\text{ST}}\left[1 + \alpha_{\text{ST}}(t_2 - t_s)\right]. \tag{11.24}$$

Here, L_{UT} and L_{ST} are lengths of under test and standard at reference temperature of t_s. The t_1 and t_2 are respectively the temperature of the micrometer and slip gauge. α_{UT} and α_{ST} are the coefficients of linear expansion of the slip gauge and micrometer respectively. O is mean value of the difference in observations from its nominal value of the opening of the jaw of the micrometer and its scale and C is the correction at scale point under test.

$$C = L_{ST}\left[1 + \alpha_{ST}(t_2 - t_s)\right] - L_{UT}\left[1 + \alpha_{UT}(t_1 - t_s)\right] - O. \qquad (11.25)$$

11.2.3 *Sources of Uncertainty and Values of Uncertainty Components*

11.2.3.1 Coefficients of Expansion

α_{ST} and α_{UT}, these are assumed to be equal, which may not be true. The value of each α_{ST} and α_{uT} is taken as $11.5 \times 10^{-6}\,\mathrm{K}^{-1}$, which may differ by 10%; α_{ST} may differ from α_{UT} by say 20%.

For $L_{ST} = L_{UT} = 25$ mm and $t - t_s = 3$°C, semi-range of uncertainty components are:

$$\begin{aligned} \pm 25 \times 11.5 \times 10^{-6} \times 3 \times 10/100 &= \pm 86.25 \times 10^{-6}\,\mathrm{mm} \\ &= \pm 0.086\,\mu\mathrm{m} \text{ for variation in } \alpha_{UT}. \\ \pm 25 \times 11.5 \times 10^{-6} \times 3 \times 20/100 &= \pm 172.5 \times 10^{-6}\,\mathrm{mm} \\ &= \pm 0.172\,\mu\mathrm{m} \text{ for difference in } \alpha_{UT} \text{ and } \alpha_{ST}. \end{aligned}$$

Assuming that actual value of α_{ST} or α_{UT} or the difference between them follow rectangular distribution with infinite degree of freedom, standard uncertainty due to each source respectively is:

$$U_1 = \pm 0.086/\sqrt{3} = \pm 0.050\,\mu\mathrm{m}, \qquad (11.26)$$

$$U_2 = \pm 0.172/\sqrt{3} = \pm 0.099\,\mu\mathrm{m}. \qquad (11.27)$$

11.2.3.2 Temperature Measurement

Temperatures t_1 and t_2 may differ by 1°C also the error in temperature measurement may be say 1°C. Hence, uncertainty components are

$$\begin{aligned} \pm 25 \times 11.5 \times 10^{-6} \times 1 &= \pm 287 \times 10^{-6}\,\mathrm{mm} \\ &= \pm 0.287\,\mu\mathrm{m} \text{ for difference in temperatures,} \end{aligned}$$

$$\pm 25 \times 11.5 \times 10^{-6} \times 1 = \pm 287 \times 10^{-6}\,\text{mm}$$
$$= \pm 0.287\,\mu\text{m for temperature measurement.}$$

Assuming that actual values of t_1 or t_2 follow rectangular distribution with infinite degrees of freedom, standard uncertainty due to each source is

$$U_3 = \pm 0.287/\sqrt{3} = \pm 0.165\,\mu\text{m}, \quad (11.28)$$

$$U_4 = \pm 0.287/\sqrt{3} = \pm 0.165\,\mu\text{m}. \quad (11.29)$$

11.2.3.3 Micrometer Under-Test

Faces of micrometer may not be flat. Tolerance in lack of flatness is normally of 0.5 μm, the actual lack of flatness will follow rectangular distribution with infinite degrees of freedom, giving standard uncertainty as

$$U_5 = \pm 0.5/\sqrt{3} = \pm 0.288\,\mu\text{m}. \quad (11.30)$$

Faces of micrometer may not be parallel. Tolerance for lack of parallelism is say 0.5 μm, the actual lack of parallelism will follow rectangular distribution with infinite degrees of freedom, giving standard uncertainty as

$$U_6 = \pm 0.5/\sqrt{3} = \pm 0.288\,\mu\text{m}. \quad (11.31)$$

11.2.3.4 Slip Gauge

Uncertainty in the value of standard slip gauge is 0.08 μm. Slip gauge was calibrated by measurement against another standard. Hence, its measured value will follow a normal distribution with infinite degrees of freedom. Let us assume stated uncertainty is at confidence level of 95.45% with coverage factor of two. The standard uncertainty, therefore, is given as

$$U_7 = \pm 0.08/\sqrt{2} = \pm 0.040\,\mu\text{m}. \quad (11.32)$$

11.2.3.5 Measured Value of *O*

Uncertainty in measured value of O from the set of observations is the standard deviation of the mean of observations. The details of observation and calculations are given in Table 11.3.

Standard deviation of the mean $= 0.547/\sqrt{5} = 0.245\,\mu\text{m}$.

Table 11.3 Observations and calculation sheet

S. No.	Observations x_i μm	$\bar{x} - x_i$ μm	$(\bar{x} - x_i)^2$ μm^2
1	1	0.4	0.16
2	0.0	0.6	0.36
3	1	0.4	0.16
4	0.0	0.6	0.36
5	1	0.4	0.16
Mean $= 3/5 = 0.6\,\mu$m		Standard deviation $= (1.20/4)^{1/2} = 0.547\,\mu$m	

Hence, standard uncertainty u_A from Type A method is the standard deviation of the mean and is given as

$$u_A = \pm 0.245\,\mu\text{m}. \tag{11.33}$$

11.2.4 Combined Standard Uncertainty

$$u_c = \sqrt{(0.050)^2 + (0.099)^2 + (0.165)^2 + (0.165)^2 + (0.288)^2 + (0.288)^2 + (0.040)^2 + (0.245)^2}$$
$$= \sqrt{0.294264} = 0.542\,\mu\text{m}. \tag{11.34}$$

11.2.5 Effective Degree of Freedom

$$\nu_{\text{eff}} = \frac{(u_c)^4}{\sum u_i^4/4} = \frac{(0.543)^4}{(0.245)^4/4 + 0} = 24.1 \times 4 = 96 \approx \infty. \tag{11.35}$$

11.2.6 Expanded Uncertainty

Coverage factor at 95.45% confidence level is 2, hence expanded uncertainty at 95.45% confidence level is given by

$$u_e = 0.543 \times 2 = 1.083\,\mu\text{m}. \tag{11.36}$$

11.3 Uncertainty in Pressure Measuring Instruments

11.3.1 Primary Standard of Pressure

Definition: When the force F is applied on the area A, then the pressure P generated is

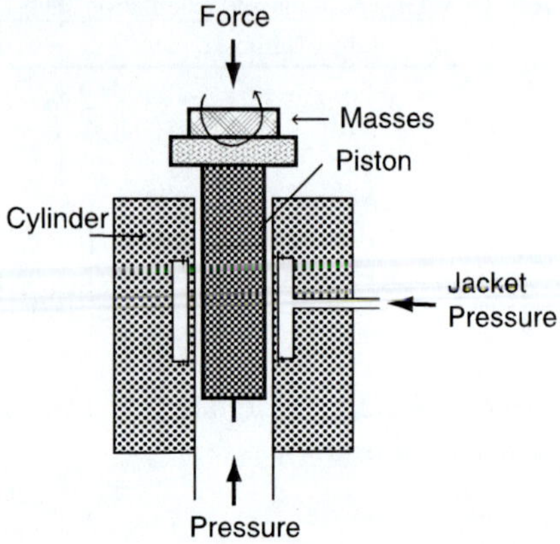

Fig. 11.1 Primary piston gauge

$$P = \frac{F}{A}. \tag{11.37}$$

Primary Standard: The Controlled clearance piston gauge is normally used as primary pressure standard. A sketch of the primary standard piston gauge is shown in Fig. 11.1. The area of the piston gauge is A and the central load applied on the piston is F.

However, there are numerous factors and environmental quantities that affect the equality of the equation (11.37). Some of them affect the force acting on the piston gauge, and some on its effective area A, while there are others, which contribute directly to the pressure. Taking into the consideration of all the effects, simple equation (11.37) transforms to:

$$P = \frac{\left(\sum_{i=1}^{i=n} m_i\right) g_{\text{local}}(1 - \sigma/\rho) + T \times C + T_{\text{w}}}{A_0(1 + bP)\left\{1 + (\alpha_{\text{p}} + \alpha_{\text{c}})(t - t_{\text{s}}\right\}\left\{1 + d(P_{\text{z}} - P_{\text{j}}\right\}}. \tag{11.38}$$

Here

$\left(\sum_{i=1}^{i=n} m_i\right) g_{\text{local}}$ is the sum of forces applied by the weights m_1 to m_{n}

σ is density of air
ρ is density of weights
T is surface tension of the fluid
C is the circumference of the piston where it emerges out of the fluid
T_{w} is the tare weight of the piston weight carrier (PWC)
A_{o} is area of the piston
t and t_{s} are the actual and reference temperatures, respectively
α_{p} and α_{c} are the coefficients of linear expansion of piston and cylinder, respectively
d is the pressure distortion coefficient of the jacketed cylinder

P_j changes the clearance between cylinder and piston, therefore $d \times P_j$ is the amount of change in clearance

If P_j increases, the clearance decreases, and thus there will be a pressure P_z at which the clearance becomes zero or in other words $d \times P_z$ is the total clearance at the measured pressure P_M. In practice, both P_z and P_j are obtained by the fall rate of the piston [4]. A simple method to obtain these terms is to determine the $V^{1/3}$ as a function of P_j. Here, V is the rate of fall of piston.

Taking logarithms of both sides of (11.38), we get

$$\log P = \log\left[\sum_{i=1}^{i=n} M_i g_{\text{local}}\left(1-\frac{\sigma}{\rho}\right) + T \times C + \Gamma_w\right] - \log(1+bP) - \log A_o$$
$$- \log\left[1 + (\alpha_p + \alpha_c)(t - t_s)\right] - \log\left[1 + d(P_z - P_j)\right]. \tag{11.39}$$

Partial differentiation of (11.39) gives

$$\left(\frac{\delta P}{P}\right)_{M_i} = \frac{\sum \delta M_i g_{\text{local}}(1-\sigma/\rho)}{\left[\sum_{i=1}^{i=n} M_i g_{\text{local}}\left(1-\frac{\sigma}{\rho}\right) + T \times C + \Gamma_w\right]} \approx \sum \delta M_i / M_i, \tag{11.40}$$

$$\left(\frac{\delta P}{p}\right)_{g_{\text{local}}} = \frac{\sum M_i \delta g_{\text{local}}(1-\sigma/\rho)}{\left[\sum_{i=1}^{i=n} M_i g_{\text{local}}\left(1-\frac{\sigma}{\rho}\right) + T \times C + \Gamma_w\right]} = \frac{\delta g_{\text{local}}}{g_{\text{local}}}, \tag{11.41}$$

$$\left(\frac{\delta P}{P}\right)_{\sigma} = \frac{-\sum M_i g_{\text{local}} \delta\sigma/\rho}{\left[\sum_{i=1}^{i=n} M_i g_{\text{local}}\left(1-\frac{\sigma}{\rho}\right) + T \times C + \Gamma_w\right]} \approx \frac{-\delta\sigma}{\rho(1-\sigma/\rho)} \approx -\frac{\sigma}{\rho}\frac{\delta\sigma}{\sigma}, \tag{11.42}$$

$$\left(\frac{\delta P}{P}\right)_{\rho} = \frac{\sum M_i g_{\text{local}} \sigma \delta\rho/\rho^2}{\left[\sum_{i=1}^{i=n} M_i g_{\text{local}}\left(1-\frac{\sigma}{\rho}\right) + T \times C + \Gamma_w\right]} \approx \frac{\sigma\delta\rho}{\rho^2(1-\sigma/\rho)}$$
$$\approx (\sigma/\rho)(\delta\rho/\rho). \tag{11.43}$$

The terms $T \times C$ and T_w are small in comparison with applied loads, so these have been neglected in deriving the different uncertainty components given in (11.40)–(11.43)

$$\left(\frac{\delta P}{P}\right)_b = -\frac{P\delta b}{1+bP} \approx P\delta b, \tag{11.44}$$

$$\left(\frac{\delta P}{P}\right)_{A_o} = -\frac{\delta A_o}{A_o}, \tag{11.45}$$

$$\left(\frac{\delta P}{P}\right)_{\alpha_p} = \frac{(t-t_s)\delta\alpha_p}{[1+(\alpha_p+\alpha_c)(t-t_s)]} \approx (t-t_s)\delta\alpha_p, \tag{11.46}$$

$$\left(\frac{\delta P}{P}\right)_{\alpha_c} = -\frac{(t-t_s)\delta\alpha_c}{[1+(\alpha_p+\alpha_c)(t-t_s)]} \approx (t-t_s)\delta\alpha_c, \tag{11.47}$$

$$\left(\frac{\delta P}{P}\right)_{(t-t_s)} = -\frac{(\alpha_p+\alpha_c)\delta(t-t_s)}{[1+(\alpha_p+\alpha_c)(t-t_s]} \approx (\alpha_p+\alpha_c)\delta(t-t_s), \tag{11.48}$$

$$\left(\frac{\delta P}{P}\right)_{d} = -\frac{(P_z-P_j)\delta d}{1+d(P_z-P_j)} \approx (P_z-P_j)\delta d, \tag{11.49}$$

$$\left(\frac{\delta P}{P}\right)_{P_z} = -\frac{d\delta P_z}{1+d(P_z-P_j)} \approx d\delta P_z, \tag{11.50}$$

$$\left(\frac{\delta P}{P}\right)_{P_j} = \frac{d\delta P_j}{1+d(P_z-P_j)} \approx d\delta P_j, \tag{11.51}$$

$$\begin{aligned} U_c\left(\frac{\delta P}{P}\right) &= \Big[(dU_{P_j})^2 + (dU_{P_z})^2 + \{(P_z-P_j)U_d\}^2 + \{(\alpha_p+\alpha_c)U_t\}^2 \\ &\quad + \{Ug_{\text{local}}/g_{\text{local}})^2 + \Big\{\sum_{i=1}^{i=n}(U_{M_i}/M_i)^2 + \{(\sigma/\rho)U_\sigma/\rho\}^2 + (U\sigma/\rho)^2 \\ &\quad + (PU_b)^2 + (U_{A_o}/A_o)^2 + \{(t-t_s)U_{\alpha_p}\}^2 + \{(t-t_s)U\alpha_c\}^2\Big\}\Big]^{1/2}, \end{aligned}$$

$$\begin{aligned} U_c\left(\frac{\delta P}{P}\right) &= \Big[(dU_{P_j})^2 + (dU_{P_z})^2 + \{(P_z-P_j)U_d\}^2 + \{(\alpha_p+\alpha_c)U_t\}^2 \\ &\quad + \{Ug_{\text{local}}/g_{\text{local}})^2 + \sum_{i=1}^{i=n}U_{c_i}/M_i)^2 + (PU_b)^2 + (U_{A_o}/A_o)^2 \\ &\quad + \{(t-t_s)U_{\alpha_p}\}^2 + \{(t-t_s)U\alpha_c\}^2\Big]^{1/2}. \end{aligned} \tag{11.52}$$

Here, U_{ci} is the combined standard uncertainty due to all the weights used.

From (11.38), we see that pressure P is not a linear function of various affecting parameters. So the Law of propagation of uncertainties used in (11.52) is not quite exact.

For the purposes of Type A uncertainty, fall rate of the piston is measured as a function of P_j to determine the optimum operating jacket pressure, which gives a close clearance between the piston and the cylinder and stall the jacket pressure. To determine the operating jacket pressure, the cube root of the fall rate is plotted against the jacket pressure P_j at different loads P_M. Repeating the measurement with different gaseous media [5], the value of P_z is determined by least square fitting of P_z versus P_M. The standard deviation of such an exercise is taken as the standard uncertainty calculated by Type A evaluation method. The Type A uncertainty as

Table 11.4 Type B uncertainty components

S No	Uncertainty due to	Nominal value	Uncertainty component	Distribution	Uncertainty ppm
1	M	52.0429 kg	$U(M)$	Normal	1.1
2	g_{local}	$9.7912413\,\text{ms}^{-2}$	$U(g_{\text{local}})$	Rectangular	1.2
3	A_{o}	$1.000642 \times 10^{-4}\,\text{m}^2$	$U(A_{\text{o}})$	Normal	2.5
4	α_{c}	$1.2 \times 10^{-5}/\text{K}$	$U(\alpha_{\text{c}})$	Rectangular	1.3
5	α_{p}	$4.42 \times 10^{-5}/\text{K}$	$U(\alpha_{\text{p}})$	Rectangular	1.3
6	$t - t_{\text{s}}$	296 K	$U(t - t_{\text{s}})$	Normal	0.5
7	b	$-5.963 \times 10^{-13}/\text{Pa}$	$U(b)$	Rectangular	0.2
8	P	5.09×10^6 Pa	$U(P)$	Normal	0.2
9	P_{z}	2.53×10^7 Pa	$U(P_{\text{z}})$	Normal	0.1
10	d	$9.81 \times 10^{-12}/\text{Pa}$	$U(d)$	Normal	10.5
11	P_{j}	5.75×10^6 Pa	$U(P_{\text{j}})$	Normal	0.5

reported by Dr A. K. Bandyopadhyay [6] for the piston gauge used as primary standard at NPL India, is 3 ppm, which in absolute terms is equal to 16 Pa.

The uncertainty components are evaluated by Type B evaluation using (11.39) to (11.51). The combined uncertainty is determined by taking the square root of sum of the squares of the uncertainties as enumerated in (11.39)–(11.51).

Different uncertainty components by Type B evaluation, for the piston gauge used as primary standard at NPL India, are given in Table 11.4.

The square root of the sum of squares of the component uncertainties given in last column gives the combined relative uncertainty. In this case, it comes out as 11×10^{-6}. In absolute terms, the uncertainty is $5.09 \times 10^6 \times 11 \times 10^{-6}\,\text{Pa} = 56\,\text{Pa}$.

Combined standard uncertainty

$$U_{\text{c}} = \sqrt{56^2 + 16^2} = 58\,\text{Pa}. \tag{11.53}$$

If the chosen level of confidence is 95%, then the value k – the coverage factor is 2 giving an extended combined uncertainty as 116 Pa at 5.09×10^6 Pa.

11.3.2 Transfer Standards

In addition to the primary standard of pressure, several piston gauges are used for calibrating the gauges of other calibrating laboratories. Nomenclature of such gauges may be Secondary standards [6]. However, there is a problem of adopting this name because in the field of Legal Metrology, the words Secondary standards are used for standards maintained at the state level. The author feels that name transfer standard or simply NPL standards may be used for the standards maintained at NPL and used for calibrating the instruments of other laboratories.

Pressure (P_{Transfer}) generated by the transfer gauge is given as:

$$P_{\text{Trasnfer}} = \frac{g_{\text{NPL}}\left(\sum_{i=1} M_i\right)(1 - \sigma/\rho) + T \times C + \Gamma_{\text{w}}}{A_{\text{o}}(1 + \lambda P)\left\{1 + (\alpha_{\text{c}} + \alpha_{\text{p}})(t - t_{\text{s}})\right.}. \tag{11.54}$$

Transfer standards of pressure gauges are calibrated against the primary standard piston gauge by cross-float method. The basic principle is to connect the primary and Transfer gauge either directly or through a differential pressure cell. Initially, both the gauges are loaded simultaneously to a constant pressure and allowed to attain the equilibrium pressure. The gauges are isolated from the rest of the system. Equilibrium condition is said to be reached if there is no fluid flow through the common pressure line. The equilibrium is attained by adjusting the fractional masses on the primary gauge so that its original fall rate is obtained. In that case,

$$P_{\text{Transfer}} = P_{\text{pri}}.$$

With the same load on the Transfer gauge, several equilibrium conditions are obtained by adjustment of mass on the primary standard gauge. The standard deviation of the load on the primary gauge gives uncertainty for Transfer gauge by type A evaluation method. The relative uncertainty by Type A method obtained at NPL is 4×10^{-6}.

Alternately, we may express effective area of the piston of the transfer gauge as

$$A_{\text{o}}(1 + \lambda) = \frac{g_{\text{local}}\left(\sum M_i(1 - \sigma/\rho) + T \times C + \Gamma_{\text{w}}\right)}{P_{\text{pri}}\left[1 + (\alpha_{\text{p}} + \alpha_{\text{c}})(T - t_{\text{s}})\right]}. \tag{11.55}$$

Different loads are applied to the transfer gauge and equilibrium is obtained by adjusting the loads on the primary gauge satisfying the condition (11.55). Thus, replacing P_{Transfer} by P_{pri} in (11.55), the best estimate of A_{o} and $A_{\text{o}}\lambda$ is obtained by least square method. The average of the squares of residual errors gives the variance evaluated by Type A method. Its square root is another component of Type A uncertainty, say it value is 26 ppm.

Type B uncertainty is obtained exactly in the same way as has been obtained for primary gauge. In addition, the uncertainty in the P_{pri} is also taken into consideration.

The uncertainty components are given below in Table 11.5

Table 11.5 Type B uncertainty components for transfer standard gauge

S No	Uncertainty source	Nominal value	Distribution of value	Uncertainty component	Uncertainty ppm
1	M	40 kg	Normal	$U(M)$	1.02
2	g_{local}	$9.7912413\ \text{ms}^{-2}$	Rectangular	$U(g_{\text{local}})$	1.02
3	α_{c}	$1.2 \times 10^{-5}/\text{K}$	Rectangular	$U(\alpha_{\text{c}})$	0.3
4	α_{p}	$4.42 \times 10^{-5}/\text{K}$	Rectangular	$U(\alpha_{\text{p}})$	0.1
5	$t - t_{\text{s}}$	296 K	Normal	$U(t - t_{\text{s}})$	0.16
6	P_{pri}	4×10^6 Pa	Normal	$U(P)$	11
7	λ	3×10^{-12} Pa	Rectangular	$U(\lambda)$	1.5

The square root of the sum of squares of the component uncertainties given in the last column of the Table 11.5 gives $U_{\mathrm{TypeB}} = 11.2 \times 10^{-6}$. Combining with type A uncertainty, we get

$$U_c = 10^{-6}\sqrt{11.2^2 + 4^2 + 26^2} = 29 \times 10^{-6}. \tag{11.56}$$

11.3.3 Dead Weight Pressure Gauge Tester

Dead Weight Pressure Gauge tester (DWT) is essentially a piston gauge, output of which is fed to a dial pressure gauge under test. Calibration of a DWT means to determine the effective area of the piston and mass of each dead weight supplied with it. Mass of each weight is determined with an uncertainty of 10 ppm and reported with its uncertainty. To determine the effective area of the piston, the Cross method is used. The pressure is supplied to the DWT through a Transfer standard piston gauge and is balanced by the weights of the DWT. However, final adjustment is carried out by adjusting the mass on Transfer piston gauge. Gravitational force applied to the DWT divided by the pressure supplied by the Transfer piston gauge gives effective piston area. The effective area of DWT includes the pressure distortion coefficient.

The DWT is loaded with weights in steps such that loading increases or decreases monotonically. Six sets of observations are taken. Each set consists of monotonically increasing and decreasing loads. From the values of the effective area, geometrical area of the DWT piston and pressure distortion coefficient is determined by least square method.

11.3.3.1 Numerical Example

Let the DWT is loaded with weights shown in column 3 with mass values in column 4 of Table 11.6. Identity of each weight is given in column 1 and its mass value in column 2. The pressure indicated by the Transfer piston gauge is shown in column 5. Effective area is the pressure indicated by the Transfer gauge to the mass value of the corresponding weights and is shown in column 6 of the Table 11.6. Effective area consists of the geometrical area of the piston and pressure distortion coefficient.

Effective area A_{eff} of the DWT piston is

$$A_{\mathrm{eff}} = A_o + A_o l P.$$

Six sets of observations at each combination of weights, both increasing and decreasing loads, are taken. Means of observations in the increasing and decreasing loads for each set are taken; hence, there are 60 data points. Leaving the data point

Table 11.6 Observation sheet for calibration of a DWT (Dead Weight Tester)

Weight No.	Mass of weight in kg	Weights on DWT Carrier	Cumulative mass (kg)	Pressure of transfer gauge (kg/cm^2)	Effective piston area (cm^2)
PWC	0.8067954	PWC	0.8067954	1.0073	0.80091
1	4.033977	PWC + 1	4.8407724	6.01278	0.80508
2	4.034531	PWC + 1–2	8.8753034	11.0172	0.80558
3	4.034841	PWC + 1–3	12.9101444	16.02232	0.80576
4	1.613334	PWC + 1–4	14.5234784	18.02614	0.80569
5	1.613134	PWC + 1–5	16.1366124	20.02483	0.80583
6	1.613303	PWC + 1–6	17.7499164	21.71759	0.80576
7	0.8067011	PWC + 1–7	18.5566175	23.02767	0.80584
8	0.8066785	PWC + 1–8	19.3632964	24.02781	0.80587
9	0.8123599	PWC + 1–9	20.1756563	25.03121	0.80602
9	0.8066785	PWC + 1–9	20.1756563	25.03121	0.80602
8	0.8067011	PWC + 1–8	19.3632964	24.02781	0.80587
7	0.8067011	PWC + 1–7	18.5566175	23.02767	0.80584
6	1.613303	PWC + 1–6	17.7499164	22.02851	0.80577
5	1.613134	PWC + 1–5	16.1366124	20.02483	0.80583
4	1.613334	PWC + 1–4	14.5234784	18.02591	0.80570
3	4.034841	PWC + 1–3	12.9101444	16.02232	0.80576
2	4.034531	PWC + 1–2	8.8753034	11.0172	0.80558
1	4.033977	PWC + 1	4.8407724	6.01278	0.80508
PWC	0.8067954	PWC	0.8067954	1.0073	0.80091

for no weight on the PWC, 54 data points are considered for the least square method. Best estimates of A_o and λ obtained are:

$$A_o = 0.804992\,\text{cm}^2$$

and $\lambda = 0.000039\,\text{cm}^2/\text{kg}$.

Standard deviation by least squares method is 0.00081 cm^2.

Uncertainty components calculated by type B method will be calculated in the same way as for Transfer gauge using the formula

$$A_o(1+\lambda) = \frac{g_{\text{local}} \sum_{i=1} M_i(1-\sigma/\rho) + T \times C + \Gamma_w}{P_{\text{Transfer}}\left[(1+\alpha_p+\alpha_c)(t-t_s)\right]}. \tag{11.57}$$

11.4 Uncertainty in Temperature Measurement and Instruments

Definition: Unit of thermodynamic temperature is kelvin. One kelvin is the fraction 1/276.16 of the thermodynamic temperature of the triple point of water (TPW).

This definition refers to water having isotopic composition defined exactly by the following amount of substance ratios:

0.00015576 mole of ^{2}H per mole of ^{1}H
0.0003799 mole of ^{17}O per mole of ^{16}O
0.0020052 mole of ^{18}O per mole of ^{16}O

Such chemically pure water is known as Vienna Standard Mean Ocean Water (SMOW).

Temperature scale: The temperature scale consists of phase transition temperatures of some well-characterized pure materials.

Realization of TPW: Triple point is realized through a standard triple point water cell, hence may entail the measurement uncertainty.

Sources of uncertainty: The sources of uncertainty may be divided into two groups:

I. Environmental and purity of material such as
   1. Chemical impurities including isotopic composition
   2. Hydrostatic head (surrounding pressure)
   3. Gas pressure
II. Measurement is carried out through standard platinum resistance thermometer and resistance measuring bridge so uncertainty sources are:
   1. Standard resistor
   2. Bridge measurement
   3. Self-heating of standard platinum resistance thermometer (SPRT)
   4. Heat-flux immersion

11.4.1 Uncertainty in Triple Point and Other Fixed Points

Taking into consideration the aforesaid sources of uncertainty, the uncertainty in Triple point of water and phase transition temperatures of Gallium (Ga), Indium (In), Tin (Zn), Aluminium (Al) and Silver as reported by Steffen Rudtsch et al. of PTB Germany [7] is given in Table 11.7.

11.4.2 Temperature Scale and Primary Standards

Having established the triple point of water and other fixed points, intermediate temperatures, in accordance with ITS 90, are realized through the standard platinum resistance thermometers (SPRTs). These SPRTs are calibrated at specific sets of fixed points and specific reference deviation functions are given for interpolation of intervening temperatures. For the measurement of resistance ratio, ac and dc bridges

Table 11.7 Component uncertainties in transition temperatures

Fixed points	H_2O	Ga	In	Sn	Zn	Al	Ag
Uncertainty contribution type B in mK							
Chemical impurities, isotopes	0.031	0.06	0.25	0.31	0.54	0.40	0.65
Hydrostatic head correction	0.004	0.01	0.02	0.02	0.02	0.02	0.08
Error in gas pressure	0.005	0.01	0.1	0.08	0.12	0.3	0.3
Standard resistor	0.05	0/01	0.01	0.01	0.01	0.01	0.01
Bridge measurement	0.015	0.02	0.11	0.12	0.16	0.20	0.25
Uncertainty propagation of TPW	–	0.08	0.09	0.11	0.15	0.20	0.28
Self-heating	0.04	0.05	0.15	0.20	0.20	0.20	0.20
Heat-flux error	0.01	0.01	0.20	0.10	0.10	0.10	0.10
Choice of fixed point	0.01	0.03	0.06	0.06	0.06	0.20	0.20
Combined Type B uncertainty	0.074	0.12	0.40	0.43	0.64	0.65	0.87
Type A component in mK	0.03	0.05	0.20	0.15	0.15	0.30	0.30
Standard combined uncertainty	0.08	0.13	0.45	0.45	0.66	0.71	0.92
Expanded uncertainty $k = 2\,\text{mK}$	0.16	0.26	0.89	0.91	1.31	1.43	1.83

are used. Thus, a national metrology laboratory will have the fixed point cells and SPRTs as primary standards along with a resistance measuring bridge.

11.4.3 Dissemination of Temperature Scale

The dissemination of the temperature scale is carried out by the inter-comparison of fixed point cells and SPRTs of other laboratories against primary standards. The standard uncertainty of the primary standard and uncertainty due to sources mentioned in Table 11.7, where applicable, are calculated, and type A uncertainty due to repeated observations is calculated and are suitably combined to get standard uncertainty. A coverage factor to be multiplied to the standard uncertainty to give the expanded uncertainty at particular level of confidence may also be determined from Students t Tables.

11.4.4 Thermocouples as Temperature Measuring Instruments

Thermocouples (TCs) are inevitably used in industry and many other applications. Thermo-couples, in general, have a lesser accuracy than SPRTs. Thermo-couples are calibrated against the fixed point cells by measuring the voltage between the junctions of the TC. The ice point with an accuracy of better than 5 mK is used to enclose the reference junction. Nano-voltmeter is used for measuring the voltage, which is calibrated against a dc-voltage standard before every use. Typical calibration uncertainty of noble metal type S or Type B thermocouples is given in Table 11.8.

In additions to the fixed points of ITS-90, the melting point of palladium is used for thermo-couples to be used for high temperature.

Table 11.8 Uncertainty of noble metal type S or Type B thermocouples

S No	Fixed points	Temperature in °C	Uncertainty in K
1	Ice point	0.00	0.13
2	Indium (In)	156.60	0.13
3	Tin (Sn)	231.93	0.15
4	Zinc (Zn)	419.53	0.11
5	Aluminium (Al)	660.32	0.15
6	Silver (Ag)	961.78	0.18
7	Gold (Au)	1064.18	0.20
8	Copper (Cu)	1084.62	0.22
9	Palladium (Pd) (air)	1553.40	1.30

11.4.5 Calibration of a Digital Thermometer

Particulars of Instrument under-test and the standard used: The digital thermometer (DT) under-test has a resolution of 0.1°C and accuracy of ±0.6°C. The standard used is a thermocouple (TC), which has a correction of 0.5°C and uncertainty of 2.0°C at a confidence level of 99%. This means that the coverage factor (k) is 2.58. Standard furnace is maintained at a constant temperature of 500°C.

Ten measurements of DT have been taken at the constant temperature of 500°C. *Mathematical model*: Input–output relationship is

$$T = D + C. \tag{11.58}$$

T is the measured temperature through the standard thermocouple (TC), D is the indication of the DT and C is the correction to be applied to the DT under test at 500°C.

Uncertainty components are
Non-repeatability of DT
Type B uncertainty of DT due to digital readout
Type B uncertainty of TC

Ten observations are taken at 500°C, mean and standard deviations are calculated, which give the correction and standard uncertainty by Type A evaluation method. The observation sheet is shown in Table 11.9.

From table 11.7, the correction C is given as

$$C = 500.5 - 501.42 = -0.92\text{°C}. \tag{11.59}$$

11.4.5.1 Uncertainty

$$\text{Standard deviation of the mean} = 0.103/\sqrt{10} = 0.03\text{°C}. \tag{11.60}$$

Table 11.9 Observations and deviations

S No	Observations (T_i)°C	$T_i - \bar{T}$	$(T_i - T)^2$
1	501.5	0.08	0.0064
2	501.4	−0.02	0.0004
3	501.5	0.08	0.0064
4	501.3	−0.12	0.0144
5	501.3	−0.12	0.0144
6	501.4	−0.02	0.0004
7	501.5	0.08	0.0064
8	501.6	0.18	0.0324
9	501.3	−0.12	0.0144
10	501.4	0.02	.0004
Sum	5014.2°C		0.0960 (°C)^2

Mean = 501.42°C
Standard deviation = $(0.096/9)^{1/2\circ}$C = 0.103°C

Standard uncertainty $u_{\mathrm{A}} = \pm 0.03$°C. Degrees of freedom for u_{A} is $10 - 1 = 9$.

Type B uncertainty of DT due to digital readout of 0.1°C: This means that actual value of the observation may lie anywhere in between the range of −0.05°C to +0.05°C with equal probability that is the actual value will follow rectangular distribution. Hence, uncertainty due to its digital read out

$$U_2 = 0.05/\sqrt{3} = \pm 0.029°\mathrm{C}. \tag{11.61}$$

Uncertainty due to standard TC is

$$U_3 = \pm 2.0/2.58 = \pm 0.78°\mathrm{C}. \tag{11.62}$$

Hence, combined standard uncertainty u_{c} is given as

$$u_{\mathrm{c}} = \sqrt{(0.03)^2 + (0.029)^2 + (0.78)^2} = 0.78°\mathrm{C}. \tag{11.63}$$

To find out the coverage factor, first we should find out effective degree of freedom ν_{eff} it is given as

$$\nu_{\mathrm{eff}} = \frac{(0.78)^4}{(0.03)^4/9} = 4112784 \approx \infty.$$

Hence, coverage factor at 95.45% Confidence level from the t distribution is 2. Giving extended uncertainty as

$$\text{Expanded uncertainty} = 2 \times 0.78 = 1.7°\mathrm{C}. \tag{11.64}$$

11.5 Uncertainty in Luminous Flux Measurement

11.5.1 Principle

Luminous flux of a lamp under test (T) is compared with that of the standard lamp (S) of known luminous flux. Each lamp is placed in the centre of the integrating sphere. The substitution method is used, in which the test lamp substitute the standard lamp. The luminous flux of a source is evaluated by comparing the indirect illuminance in the two cases.

11.5.2 Procedure for Calibration

Steps to be followed are:

1. Switch on the measuring equipment and let the auxiliary lamp warm up for 15 min.
2. Mount the standard lamp at the centre of the integrating sphere.
3. After burning in period, the indirect illuminance E_S is measured.
4. Turn off the supply of the standard lamp.
5. The switched on auxiliary lamp moved into the sphere. The indirect illuminance E_{AS} is measured.
6. The standard lamp is taken out of the sphere and the test lamp is mounted into the centre of the sphere with auxiliary lamp still burning. Indirect illuminance E_{AT} is measured.
7. Turn the test lamp on. After burning in period, the indirect illuminance E_T is again measured.

The luminous flux Φ_T of the lamp under test is given as

$$\Phi_T = \Phi_S \frac{E_T}{E_S} \frac{E_{AS}}{E_{AT}}. \tag{11.65}$$

The factor $\frac{E_{AS}}{E_{AT}}$ takes care of the effect of different sizes and types of test and standard lamps. If the lamp under test and the standard are identical in size, shape, electrical parameters and colour temperature, then $\frac{E_{AS}}{E_{AT}}$ and $\frac{E_S}{E_T}$ is 1 giving Φ_T equal to Φ_S.

Here, we see that measurand Φ_T is not a linear function of input quantities E_T, E_S, E_{AS} and E_{AT}; however, the log of the measurement is a linear function of the logarithms of input quantities, hence Taylor's expansion up to second term is exact.

11.5.3 Expression for Uncertainty

From (11.65), the square of the relative uncertainty of under test lamp $\frac{u_c^2(\Phi_T)}{\Phi_T^2}$ is expressed as:

$$\frac{u_c^2(\Phi_T)}{\Phi_T^2} = \frac{u^2(\Phi_S)}{\Phi_S^2} + \frac{u^2(E_S)}{E_S^2} + \frac{u^2(E_T)}{E_T^2} + \frac{u^2(E_{AS})}{E_{AS}^2} + \frac{u^2(E_{AT})}{E_{AT}^2}. \tag{11.66}$$

From the certificate of the standard lamp, one can get Φ_S and $u(\Phi_S)$. E_S, E_T, E_{AS} and E_{AT} are all measured quantities, so the mean values of E_S, E_T, E_{AS} and E_{AT} and their respective uncertainties are calculated by Type A evaluation method.

11.5.4 Example

An example with due corrections is taken from [8]. The mean value of E_{AS} and its standard deviations are calculated as shown in the Table 11.10.

Standard deviation of the mean $= 14.12 \times 10^{-3}/\sqrt{10} = 4.46 \times 10^{-3}$
Hence, standard uncertainty $u_A = \pm 4.46 \times 10^{-3}$
Degrees of freedom for u_A is $10 - 1 = 9$
Giving us

$$\frac{u(\overline{E}_{AS})}{E_{AS}} = \frac{4.46 \times 10^{-3}}{9.276} = \pm 4.8 \times 10^{-4}. \tag{11.67}$$

Similarly, mean and standard deviation of the mean of E_{AT}, E_S and E_T are calculated from each set of 10 observations. Let the calculated values are:

$$\frac{u(\overline{E}_{AT})}{E_{AT}} = \pm 6 \times 10^{-4} \text{ and } \overline{E}_{AT} = 9.20\,\text{lux}, \tag{11.68}$$

$$\frac{u(\overline{E}_S)}{E_S} = \pm 1.2 \times 10^{-2} \quad \overline{E}_S = 81.14\,\text{lux}, \tag{11.69}$$

$$\frac{u(\overline{E}_T)}{E_T} = \pm 1.6 \times 10^{-2} \quad \overline{E}_T = 83.76\,\text{lux}. \tag{11.70}$$

Table 11.10 Observations E_{AS} and deviations

S No.	E_{AS} *lux*	$E_{AS} - \overline{E}_{AS}$	$(E_{AS} - \overline{E}_{AS})^2$
1	9.296	+0.020	400×10^{-6}
2	9.279	+0.003	9×10^{-6}
3	9.254	−0.022	484×10^{-6}
4	9.290	+0.014	196×10^{-6}
5	9.271	−0.005	25×10^{-6}
6	9.272	−0.004	16×10^{-6}
7	9.286	+0.010	100×10^{-6}
8	9.285	+0.009	81×10^{-6}
9	9.254	−0.022	484×10^{-6}
10	9.277	+0.001	1×10^{-6}
Mean $= \overline{E}_{AS} = 9.276$ lux		Standard deviation $= (\sqrt{1796/9})10^{-3} = 14.12 \times 10^{-3}$	

The degree of freedom in each case is 9.

The certificate of the standard lamp states

$$\Phi_S = 1045 \pm 9.12\,\text{lm}.$$

Substituting the value of E_{AT}, E_{AS}, E_S, E_A and Φ_S into (11.65), we get

$$\Phi_T = 1045\frac{9.276}{9.20}\frac{83.76}{81.14} = 1087.7. \tag{11.71}$$

The uncertainty in the assigned value of the standard lamp is given as 9.12 lm. The measured value of flux of the lamp will naturally follow normal distribution; hence, the stated uncertainty will either be equal to the standard deviation or some multiple of it. For example, the coverage factor is 2 if level of confidence at which the flux value is assigned is 95.45%. Let the coverage factor be 2 in assigning the uncertainty value, giving us

$$\text{Type B uncertainty} = 9.12/2 = 4.6\,\text{lm}.$$

Relative standard uncertainty of the standard lamp $\frac{u(\Phi_S)}{\Phi_s} = \frac{4.6}{1045} = \pm 4.4 \times 10^{-3}$.

11.5.5 Combined Relative Uncertainty

Hence, combined relative uncertainty in the calibration of the lamp under test is given by

$$\begin{aligned}\frac{u_c(\Phi_T)}{\Phi_T} &= 10^{-3}\sqrt{(0.48)^2 + (0.6)^2 + (12)^2 + (16)^2 + (4.4)^2} \\ &= 10^{-3}\sqrt{0.1849 + 0.36 + 144 + 256 + 19.36} = 10^{-3}\sqrt{419.909} \\ &= 20.49 \times 10^{-3} = 2.05 \times 10^{-2}.\end{aligned} \tag{11.72}$$

Absolute uncertainty $\pm 2.05 \times 10^{-2} \times 1086\,\text{lm} = \pm 22.3\,\text{lm}$.

Effective degree of freedom ν_{eff} is given as

$$\nu_{eff} = \frac{(2.05 \times 10^{-2})^4}{\left[\frac{(4.4)^4}{\infty} + \frac{(0.48)^4}{9} + \frac{(0.60)^4}{9} + \frac{(12)^4}{9} + \frac{(16)^4}{9}\right] 10^{-12}} \approx 18.$$

The coverage factor k from Students t distribution for 95.45% confidence level for 19 degrees of freedom is 2.10 (Table A.6).

Hence, combined relative uncertainty for 95.42% level of confidence is $2.10 \times 2.05 \times 10^{-2} = 4.3 \times 10^{-2}$ or 46.5 lm.

11.5.6 Expression of Results with Uncertainty

The result of calibration of the lamp under test may be reported as follows:

$$\Phi_T = 1086.8 \pm 22.3\,\text{lm or} \tag{11.73}$$

$$\Phi_T = 1086.8 \pm 16.5\,\text{lm(level of confidence level 95\%)}. \tag{11.74}$$

References

1. N.K. Aggarwal, P.C. Jain, Uncertainties in slip gauge calibration. MAPAN **5**, 27–31 (1990)
2. K.G. Birch, Uncertainties in the measurement of gauge blocks by interferometry, NPL report 29 (National Physical Laboratory, Teddington, 1978)
3. K.P. Birch, R.E. Ward, G. Wilkening, F. Reinboth, Evaluation of the effect of variations in the refractive index of air upon the uncertainty of industrial length measurement", Report **EUR 14103EN** Community Bureau of Reference, Commission of the European Communities, Brussels, 1992
4. P.L.M. Heydemann, B.E. Wech, *Experimental Thermodynamics*, Vol 2 (LeNeindre B and Vodar B) (Butterworth, London, 1975)
5. J.K.N. Sharma, K.K. Jain, A.K. Bandyopadhyay, J. Jager, J. Phys. E Sci. Instrum. **21**,635 (1988)
6. A.K. Bandyopadhyay, A.C. Gupta, Establishment of the national pressure standards. MAPAN **14**, 3–14 (1999)
7. S. Rudtsch, E. Tegeler, J. Fischer, Calibration of contact thermometers above the Triple point of water at PTB. MAPAN **20**, 165–169 (2005)
8. NABL 141, *Guidelines for Estimation and Expression of Uncertainty in Measurement* (Department of Science and Technology, Technology Bhawan, New Delhi, 2000)

Chapter 12
Uncertainty in Calibration of Electrical Instruments

12.1 Uncertainty in Calibration of RF Power Sensor

The examples have been taken from the document NABL-141 by National Accreditation Board for testing and calibration of Laboratories [1].

12.1.1 Principle of Calibration

Basically an RF power sensor is calibrated by substituting with a standard power sensor fed by a well-monitored source of known coefficient of reflection. The calibration is carried out in terms of a factor K, which is the ratio of the incident power at a frequency f to the incident power at a reference frequency. Normally reference frequency is 50 MHz and frequency of calibration in this case is 18 GHz.

12.1.2 Mathematical Modelling

The calibration factor K_x is expressed as

$$K_x = (K_s + D_s) \cdot R_{DC} \cdot R_m \cdot R_{ref}. \tag{12.1}$$

Here

K_s is the calibration factor of the standard sensor with D_s as drift,
R_{DC} is the ratio of DC voltage outputs $= V_x / V_s$,
R_m is the mismatch losses and
R_{ref} is the ratio of power outputs of the reference source.

Here measurand K_x is not a linear function of input quantities so the expansion of K_x by Taylor's theorem to first two terms will only be approximate. However,

S.V. Gupta, *Measurement Uncertainties*, DOI 10.1007/978-3-642-20989-5_12,

logarithm of the measurand K_x is a linear function of logarithms of input quantities, which are independent.

It may be noted that R_{DC}, R_m and R_{ref} are all equal to unity except the uncertainty part.

Partial differential coefficients with respect to above variables are as follows:

$$\begin{aligned}
\frac{\delta K_x}{K_x} &= \frac{\delta K_s}{K_s + D_s},\\
\frac{\delta K_x}{K_x} &= \frac{\delta Ds}{K_s + D_s},\\
\frac{\delta K_x}{K_x} &= \frac{\delta R_{DC}}{R_{DC}},\\
\frac{\delta K_x}{K_x} &= \frac{\delta R_m}{R_m},\\
\frac{\delta K_x}{K_x} &= \frac{\delta R_{ref}}{R_{ref}}.
\end{aligned} \tag{12.2}$$

Applying the law of variances and substituting the square of the corresponding standard uncertainty for variances, we get

$$\frac{u_{K_x}^2}{K_x^2} = \frac{u_{K_s}^2}{(K_s + D_s)^2} + \frac{u_{D_s}^2}{(K_s + D_s)^2} + u_{R_{DC_c}}^2 + u_{R_m}^2 + u_{R_{ref}}^2. \tag{12.3}$$

12.1.3 Type A Evaluation of Uncertainty

For uncertainty through Type A evaluation, the calibration factor K_x is measured 6 times by connecting and disconnecting the under-test sensor and standard sensors turn by turn. The quantity measured is the voltage ratio in each case. Observation data with deviations are given in Table 12.1

Table 12.1 Observations and deviations

S. no.	K_x	$K_{xi} - \overline{K}_x$	$(K_{xi} - \overline{K}_x)^2 \times 10^6$
1	0.957	+0.007	49
2	0.954	+0.004	16
3	0.951	+0.001	1
4	0.946	−0.004	16
5	0.949	−0.001	1
6	0.943	−0.007	49
Mean $\bar{K}_x = 0.950$			

Standard deviation (standard uncertainty) of the mean $= 0.0051/\sqrt{6} = 0.0021$ with 5 degrees of freedom.

$$U_{Kx}(\text{Type A}) = 0.0021. \tag{12.4}$$

12.1.4 Type B Evaluation of Uncertainty

Type B evaluation of uncertainty components is calculated with the help of the following data:

12.1.4.1 Uncertainty in the Calibration Factor of Standard Sensor

In the certificate of the standard sensor, it is stated that uncertainty in calibration factor of the standard sensor at 95% confidence level is ±0.010; the stated value of sensor follows normal distribution with infinite degrees of freedom. Hence uncertainty $u(K_s)$

$$u(K_s) = 0.010/1.96 = 0.0051. \tag{12.5}$$

12.1.4.2 Uncertainty Due to Drift in Calibration Factor of Standard Sensor

Drift in calibration factor is ±0.003 per year and the sensor was calibrated 6 months ago. Hence the correction of $0.003/2 = 0.0015$ is to be added to the stated calibration factor.

12.1.4.3 Uncertainty in the Standard Source

1. From the certificate of the standard source the instability in the ratio output power at 50 MHz is ±0.004 and R_{ref} follows a rectangular distribution with infinite degrees of freedom; it is because the actual R_{ref} is 1 with equal probability. Hence $u(R_{\text{ref}})$ is given as

$$u(R_{\text{ref}}) = 0.004/\sqrt{3} = 0.0023. \tag{12.6}$$

12.1.4.4 Uncertainty Due to Mismatch

2. As the source is not perfectly matched and the phase relation of the reflection coefficients of the source, the under-test and standard sensors are not known,

there is an uncertainty due to mismatch for each sensor at the calibration frequency (18 GHz) and as well as reference frequency (50 MHz). From the well-known formula the uncertainty between source and the sensor is given as $2R_g \cdot R_s$. Here R_g, R_s and R_x are respectively the reflections coefficients of the source, standard sensor and under-test sensor. Their numerical values are given as follows:

R_g at 50 MHz is 0.02 and at 18 GHz is 0.05.

R_s at 50 MHz is 0.02 and at 18 GHz is 0.08.

R_x at 50 MHz is 0.02 and at 18 GHz is 0.09.

Hence uncertainty due to mismatch in

$$\begin{aligned}
&\text{Standard sensor at 50 MHz} = \pm 2 \times 0.02 \times 0.02 = \pm 0.0008,\\
&\text{Standard sensor at 18 GHz} = \pm 2 \times 0.05 \times 0.08 = \pm 0.0080,\\
&\text{Under-test sensor at 50 MHz} = \pm 2 \times 0.02 \times 0.02 = \pm 0.0008,\\
&\text{Under-test sensor at 18 GHz} = \pm 2 \times 0.05 \times 0.09 = \pm 0.0090.
\end{aligned} \tag{12.7}$$

Hence standard uncertainties (standard deviations) of U shape distribution are

$$\begin{aligned}
u(\mathrm{Ms}) &= 0.0008/\sqrt{2} = 0.00056 \text{ at 50 MHz},\\
u(\mathrm{Ms}) &= 0.008/\sqrt{2} = 0.0056 \text{ at 18 GHz},\\
u(\mathrm{Mx}) &= 0.0008/\sqrt{2} = 0.00056 \text{ at 50 MHz},\\
u(\mathrm{Mx}) &= 0.009/\sqrt{2} = 0.0064 \text{ at 18 GHz}.
\end{aligned} \tag{12.8}$$

Degrees of freedom in each case are infinite.

Uncertainty due to mismatch is given as

$$\begin{aligned}
u_{R_\mathrm{m}}^2 &= (0.00056)^2 + (0.0056)^2 + (0.00056)^2 + (0.0064)^2\\
&= 0.000072972
\end{aligned} \tag{12.9}$$

or

$$\mathrm{u_{Rm}} = 0.00854.$$

12.1.4.5 Uncertainty Due to Instrument Non-linearity

3. The uncertainty in nonlinearity of the instrument at 95% confidence level is ±0.001 with infinite degrees of freedom. Hence standard uncertainty due to nonlinearity is

$$u(R_\mathrm{c}) = 0.001/1.96 = 0.0005. \tag{12.10}$$

12.1.5 Combined Standard and Expanded Uncertainty

12.1.5.1 Standard Uncertainty

$$\begin{aligned}(\text{Combined standard uncertainty})^2 &= (0.0021)^2 + (0.0051)^2 + (0.0023)^2 \\ &\quad + (0.0005)^2 + (0.00854)^2 \\ &= 0.0001089072, \\ \text{Combined standard uncertainty} &= 0.0104. \end{aligned} \tag{12.11}$$

12.1.5.2 Expanded Uncertainty

For expanded uncertainty effective degrees of freedom ν_{eff} are to be calculated first given as

Effective degrees of freedom $\nu_{\text{eff}} = \frac{(0.0104)^4}{(0.0021)^4/5+0} \approx 305 \Rightarrow \infty$.

The value of Student's t for infinite degrees of freedom at 95.45% level of confidence is 2, giving

Expanded uncertainty at 95.45% $= 2 \times 0.0104 = 0.0208$.

12.1.6 Statement of Result with Uncertainty

Result of the calibration factor of under-test sensor at 18 GHz is 0.950 ± 0.021 at 95.5% level of confidence. It may be noted that correction due to drift (0.0015) is negligible.

12.2 Uncertainty in Calibration of a Digital Multi-meter

12.2.1 Equipment and Principle of Calibration used

12.2.1.1 Equipment used is

1. Thermal Voltage Converter (TVC) used as transfer standard
2. DC voltage calibrator as standard
3. Highly stable AC voltage supply
4. Nano-voltmeter used as indicator
5. Under-test $6\frac{1}{2}$ digit Digital Multi-Meter (DMM). Example is the calibration of DMM at 0.5 V and at 1 kHz.

12.2.1.2 Outline of Method for Calibration

DMM and TVC are connected in parallel via a coaxial switch and a Tee adapter and the AC voltage from a highly stable AC calibrator is applied to both DMM and TVC, such that DMM indicates exactly 0.500000 V. Let the nano-voltmeter indicate emf across it as V_x. DMM is disconnected. A DC voltage of positive polarity is applied to TVC and supply is adjusted such that indication in nano-voltmeter is again V_x. The output of DC calibrator is noted as V_1. The polarity of DC voltage is reversed and the supply is adjusted such that Nano-voltmeter again reads V_x. Let the DC voltage be V_2 then

$$V_{DC} = \frac{V_1 + V_2}{2} = V_i.$$

The measurement process is repeated at the least 5 times.

12.2.2 *Mathematical Model*

If V_{AC} is the AC voltage estimated for an indicated value of 0.500000 V on DMM, then

$$V_{AC} = (V_{DC} + EV_{DC} + EV_{th})(1 + \delta). \tag{12.12}$$

Here

EV_{DC} is the error in the DC voltage calibrator due to it lack of stability.
EV_{th} is the error due to reversal of polarity in thermal emf.
And δ is the AC/DC transfer correction factor of TVC at the frequency of calibration.
EV_{th} is normally very small say about 1 μV.

Important precautions are that inter-connecting leads are coaxial, are shielded and are small. The reference plane of measurement is brought close to input plane of DMM. The precautions reduce the loading and transmission errors.

12.2.3 *Type A Evaluation of Uncertainty*

The input observations are the mean values of V_1 and V_2 indicated by the DC calibrator and are given in Table 12.2 along with calculations of standard deviation.

$$\text{Standard deviation of the mean} = \text{standard uncertainty (Type A)} = \sqrt{(103/20)}$$
$$= 2.27\,\mu\text{V}. \tag{12.13}$$

Table 12.2 Observations and deviations

S. no.	Observations V_i	$V_i - \overline{V}$	$(V_i - \overline{V})^2 \times 10^{12}$
1	0.4999986	−0.0000034	11.56
2	0.4999982	−0.0000074	54.76
3	0.4999991	0.0000016	2.56
4	0.4999994	0.0000046	21.16
5	0.4999993	0.0000036	12.96
Mean $\overline{V} = 0.4999\,989\,4\,\text{V}$		Standard deviation $= 10^{-6}\sqrt{103/4} = 5\,\mu\text{V}$	

12.2.4 *Type B Evaluation of Uncertainty*

12.2.4.1 DC Calibrator

The certificate of the DC calibrator states that relative uncertainty at 95% level of confidence is $\pm 5.8 \times 10^{-6}$ and distribution is normal.

Relative uncertainty is $\pm 5.8 \times 10^{-6}$ at 95% confidence level; hence coverage factor is 1.96, giving relative standard uncertainty as

$$\text{Relative standard uncertainty} = \pm\, 5.8 \times 10^{-6}/1.96 = \pm 2.96 \times 10^{-6}\text{V},$$

giving

$$\text{Standard uncertainty(absolute)} = 2.96 \times 0.5\,\mu\text{V} = 1.48\,\mu. \tag{12.14}$$

Degrees of freedom are infinity.

Due to Lack of Stability

Lack of stability for a period of 3 months is 5.0×10^{-6} of the output. This means that actual value will follow rectangular distribution, giving uncertainty due to this count

$$5 \times 10^{-6} \times 0.5/\sqrt{3} = 1.44\,\mu\text{V}. \tag{12.15}$$

Degrees of freedom are infinity.

12.2.4.2 TVC

Measurement of Uncertainty

From the certificate of TVC, the AC/DC transfer correction factor δ for TVC is 80×10^{-6} with a relative uncertainty $\pm 1.0 \times 10^{-4}$ at 95% confidence level. It means the coverage factor 1.96, giving

$$\text{Standard uncertainty at } 0.5\,\text{V} = 100 \times 0.5/1.96\,\mu\text{V} = 25.5\,\mu\text{V}. \tag{12.16}$$

Degrees of freedom are infinite.

12.2.5 Combined Standard Uncertainty U_c and Expanded Uncertainty

Combined standard uncertainty U_c is given as

$$\begin{aligned} u_\text{c} &= \sqrt{(2.27)^2 + (1.48)^2 + (1.44)^2 + (25.5)^2} = 25.72\,\mu\text{V}, \\ u_\text{c} &= 25.72\,\mu\text{V}. \end{aligned} \tag{12.17}$$

12.2.5.1 Expanded Uncertainty

For expanded uncertainty, effective degree of freedom ν_eff is to be calculated

Effective degrees of freedom ν_eff

$$\begin{aligned} \nu_\text{eff} &= \frac{(25.68)^4}{(1.48)^4/\infty + 1.44)^4/\infty + (25.5)^4/\infty + (2.27)^4/4} = \frac{4(25.68)^4}{(2.27)^4} \\ &= 65514 \approx \infty. \end{aligned}$$

Coverage factor for expanded uncertainty for 95.45% confidence level with infinitely large degree of freedom is 2

Hence expanded uncertainty u_e at 0.5 V is given as

$$u_\text{e} = 51\,\mu\text{V}. \tag{12.18}$$

12.2.6 Statement of Results

Substituting the values of V_DC, EV_DC, EV_th and δ in (12.12), the value of V_AC is given as

$$V_\text{AC} = [0.499989\,\text{V} + 2.06\,\text{mV} + 1.0\,\text{mV}](1 + 0.000008) = 0.499996\,\text{V}.$$

Observation of the DMM = 0.500000 V, giving the correction C to DMM at 0.5 V as

$$C = -4\,\mu\text{V}. \tag{12.19}$$

The correction is much less than the uncertainty; hence correction to DMM may be taken as negligible at 0.5 V (the point of calibration) with an uncertainty of $\pm 51\ \mu$V.

12.3 Uncertainty in Calibration of a Digital Instrument

The procedure of calibration and ascertaining uncertainty of measurement

12.3.1 Principle of Calibration

Let us consider a $4\frac{1}{2}$ digital voltmeter of 100 V range with last digit representing 10 mV. To calibrate a digital multi-meter (MUC) say at 10 V, a continuous voltage is applied through a voltage calibrator till it reads 10 V. Continue to increase the voltage in steps equal to one tenth of the resolution of MUC till MUC indicates 10.01(one least count more of MUC). Note the indication of the standard. Let it be 10.003 V. Then the error of MUC at 10 V is given by

$$\text{Error}\,(E) = 10.00 + 0.5 \times 0.01 - 10.003 = 0.002\,\text{V}.$$

12.3.2 Type A Evaluation of Uncertainty

Following the above principle of calibration, the measurements are repeated several times (say 5 times) at 10 V. The observations and calculations are shown in Table 12.3.

Standard deviation of the mean = standard uncertainty (Type A) u_a is given as

$$u_a = 5.47 \times 10^{-4}/\sqrt{5} = 2.45 \times 10^{-4}\,\text{V}, \tag{12.20}$$

$$\text{Error in MUC at } 10\,\text{V} = 10.00 + 0.5 \times 0.01 - 10.0034 = +0.0016\,\text{V}. \tag{12.21}$$

Table 12.3 Observations and deviations

S. no.	Indication of standard for last digit jump of MUC *Vi*	Indication of standard $V_i - \overline{V}$	$(V_i - \overline{V})$ 2, 108 V
1	10.003	−0.0004	16
2	10.004	+0.0006	36
3	10.003	−0.0004	16
4	10.003	−0.0004	16
5	10.004	+0.0006	36
Sum = 50.017		Variance $V_{ar\,1} = 30 \times 10^{-8}\,\text{V}^2$	
Mean 10.0034		Standard deviation $5.47 \times 10^{-4}\,\text{V}$ = SD1	

Similarly calibration is carried out at other points of the scale of MUC. Minimum five points such as 10, 30, 50, 70 and 90 V are chosen. Mean of all the variances at the points of calibration divided by 5 will give square of uncertainty calculated by Type A evaluation.

$$\text{uncertainty(Type A)} = \frac{\sqrt{(\text{SD1})^2 + \text{SD2})^2 + (\text{SD3})^2 + (\text{SD4})^2 + (\text{SD5})^2}}{\sqrt{5 \times 5}}. \tag{12.22}$$

Let after calculations of all the five standard deviations, the uncertainty by Type A evaluation u_a be given as

$$u_a = 0.251.$$

The degrees of freedom will be 20.

12.3.3 *Type B Evaluation of Uncertainty*

12.3.3.1 Standard Meter

Let the uncertainty of voltage calibrator in its certificate of calibration be expressed as $(100 + 4.5\,V)\,\mu\text{V}$ at 99% confidence level, V is the output voltage in volts. So uncertainty at 10 V is given as follows:

$(100 + 45)\,\mu\text{V} = 145\,\mu\text{V}$ with 2.58 as coverage factor, giving standard uncertainty at 10 V as

$$u_s = (145/2.58)\,\mu\text{V} = 56.2\,\mu\text{V}. \tag{12.23}$$

Degrees of freedom are infinite.

The uncertainty components at different points of scale are similarly calculated.

12.3.3.2 Due to Resolution of Metre Under Calibration (MUC)

Last digit of MUC is 10 mV. Hence actual value of the voltage may lie anywhere within a semi-range of ± 5 mV with equal probability and hence follows a rectangular distribution, giving us the following:

Uncertainty due resolution [2] in the MUC $5/\sqrt{3}\,\text{mV} = 2.887\,\text{mV}$. However, this component of uncertainty is to be used when this MUC is used for measuring voltage. In its calibration we have taken its observations at its change points.

12.3.4 Combined Uncertainty

$$\text{Combined uncertainty } U_c \text{ at } 10\,\text{V} = \sqrt{(0.0562)^2 + (0.251)^2} = 0.257\,\text{mV}. \tag{12.24}$$

12.3.5 Expanded Uncertainty

For calculation of extended uncertainty, effective degrees of freedom (ν_{eff}) are given as

$$\nu_{\text{eff}} = \frac{(0.257)^4}{0.251/20 + 0.056/\infty} = 21.98 \approx 22. \tag{12.25}$$

Coverage factor for 95.45% of confidence level for 22 degrees of freedom is 2.13. Hence extended uncertainty

$$u_e = 2.13 \times 0.257\,\text{mV} = 0.547\,\text{mV}. \tag{12.26}$$

12.3.6 Statement of Results

The statement of error and uncertainty at 10 V is as follows:

$$\text{The error at } 10\,\text{V is} (1.6 \pm 0.547)\,\text{mV}. \tag{12.27}$$

12.4 Uncertainty Calculation for Correlated Input Quantities

Until now all input variables were independent of each other; now we wish to consider a case of dependent variables. Let us consider the calibration of a 10 kΩ resistor R_{ref} against the ten resistors of 1,000 Ω each [3]. Each of these resistors was calibrated with negligible uncertainty against a standard resistor R_s with standard uncertainty $u(R_s) = 100\,\text{m}\Omega$ as given in the certificate.

For the purpose of calibration, ten resistors are connected in series with wires of negligible resistance to make 10 kΩ.

Mathematical model is as follows:

$$R_{\text{ref}} = \sum_{p=1}^{p=10} R_p + O. \tag{12.28}$$

Here O is the observed difference between the resistor R_{ref} and summation of resistors R_p, and p is from 1 to 10. Here each R_p is expected to be correlated as each has been calibrated against the same standard resistor R_s; hence uncertainty equation is

$$u^2 = \sum_{p=1}^{p=10} u_p^2 \delta R_{\text{ref}}/\delta R_p + 2\sum_{q=1}^{q=10}\sum_{p=q+1}^{p=10} (\delta R_{\text{ref}}/\delta R_p)\delta R_{\text{ref}}/R_q u(R_p)u(R_q)r(R_p, R_q). \tag{12.29}$$

12.4.1 Type A Evaluation of Uncertainty

Uncertainty in O is obtained by Type A evaluation in the usual fashion. Let it be u_A.

12.4.2 Type B Evaluation of Uncertainty

Here we see that

$$\delta R_{\text{ref}}/\delta R_p = 1 \text{ and } u(R_{\text{p}}) = 100\,\text{m}\Omega, \text{ for all values of } p. \tag{12.30}$$

Each resistor R_p has been compared with negligible uncertainty with the standard R_s. So each resistor is fully correlated with each other; hence coefficient of correlation $r(R_p, R_q) = 1$, for all values of p and q from 1 to 10. Equation (12.28) becomes

$$u^2 = \sum_{p=1}^{p=10} u(R_{\text{p}})^2 + 2\sum_{q=1}^{q=10}\sum_{p=q+1}^{p=10} u(R_p)u(R_q) = \left\{\sum_{p=1}^{p=10} u(R_p)\right\}^2, \tag{12.31}$$

giving $u = u_B = 10.\ 100\,\text{m}\Omega = 1\ \Omega$.

Hence combined standard uncertainty u_c is given as

$$u_c^2 = \sqrt{u_A^2 + u_B^2}. \tag{12.32}$$

12.5 Vector Measurands

This example [4] deals with the treatment of multiple measurands or output quantities determined simultaneously in the same measurement (same input quantities). The example is about simultaneous measurements of pure resistance R and the reactance of a circuit fed by AC. Input quantities are the current I, amplitude of the voltage V and change in phase angle ϕ of the alternating current. The measurands are impedance Z, resistance R and the reactance X.

However the impedance $Z = \sqrt{R^2 + X^2}$; therefore, there are only two independent output quantities, namely R and X.

12.5.1 Mathematical Model

$$
\begin{aligned}
R &= \frac{V \cos \varphi}{I} \\
X &= \frac{V \sin \varphi}{I} \quad \text{and} \\
Z &= \frac{V}{I}.
\end{aligned}
\tag{12.33}
$$

It may be noticed that here all input quantities are correlated and output quantities are not linearly related to the input quantities.

Last equation of (12.33) may be written as

$$\mathrm{Log}(Z) = \mathrm{Log}\, V - \mathrm{Log}\, I, \tag{12.34}$$

giving us

$$\frac{1}{Z}\frac{\delta Z}{\delta V} = \frac{1}{V}, \tag{12.35}$$

$$\frac{1}{Z}\frac{\delta Z}{\delta I} = -\frac{1}{I}, \tag{12.36}$$

$$\Delta Z = \frac{Z}{V} - \frac{Z}{I}. \tag{12.37}$$

12.5.2 Combined Uncertainty

Applying the law of variances for correlated input variables and replacing each by the square of it uncertainty, we get

$$u_c^2(Z) = Z^2 \left\{\frac{u(\overline{V})}{V}\right\}^2 + Z^2 \left\{\frac{u(\overline{I})}{I}\right\}^2 - 2Z^2 \left\{\frac{u(\overline{V})}{V}\right\} \left\{\frac{u(\overline{I})}{I}\right\} \cdot r(\overline{V}, \overline{I}). \tag{12.38}$$

Dividing by both sides of (12.38) by Z^2, we get the combined relative uncertainty of Z in terms of relative uncertainties of V and I

Table 12.4 Observations for V, I and ϕ

S. no.	V (V)	I (mA)	ϕ (rad)
1	5.006	20.563	1.0456
2	4.994	20.539	1.0436
3	5.005	20.540	1.0468
4	4.991	20.585	1.0428
5	4.999	20.578	1.0432
Mean	4.999	20.561	1.0444

$$u_{\rm rc}^2(Z) = u_{\rm r}^2(\overline{V}) + u_{\rm r}^2(\overline{I}) - 2u_{\rm r}(\overline{V})u_{\rm r}(\overline{I})r(\overline{V},\overline{I}). \tag{12.39}$$

Let simultaneous measurements of all the input quantities be taken 5 times. Observed values are shown in Table 12.4 and the subsequent calculations and correlation of their estimates are shown in Table 12.5.

$$\begin{aligned} s_{\overline{V}} &= u_V = \sqrt{(174/20)} = 2.95 \times 10^{-3}\,\text{V}, \\ s_{\overline{I}} &= u_I = \sqrt{(1,794/20)} = 9.47 \times 10^{-4}\,\text{mA}, \\ s_{\bar{\varphi}} &= u_{\phi 2.7}\sqrt{(1,184/20)} = 7.69 \times 10^{-4}\,\text{rad}. \end{aligned} \tag{12.40}$$

Also

$$\begin{aligned} \sigma_V &= \sqrt{(174/5)} = 5.9 \times 10^{-3}\,\text{V}, \\ \sigma_I &= \sqrt{(1,794/5)} = 18.9 \times 10^{-4}\,\text{mA}, \\ \sigma_f &= \sqrt{(1,184/5)} = 15.38 \times 10^{-3}\,\text{rad}. \end{aligned} \tag{12.41}$$

12.5.3 Correlation Coefficients

We know that

$$r(x,y) = \frac{\sum\limits_{p=1}^{p=n}(x_p - \bar{x})(y_p - \bar{y})/n}{\sigma_x \sigma_y}.$$

Using the results from Table 12.5 and (12.41), we get

Table 12.5 Calculation sheet

$V-\overline{V}$ $=A$	$I-\overline{I}$ $=B$	$\phi-\overline{\varphi}$ $=C$	$A^2 10^{-6}$	$B^2 10^{-6}$	$C^2 10^{-8}$	AB10^{-6}	AC10^{-7}	BC10^{-7}
7	2	12	49	4	144	14	84	24
−5	−22	−8	25	484	64	110	40	176
6	−21	24	36	441	576	−126	144	−504
−8	24	−16	64	576	256	−192	128	−384
0	17	−12	0	289	144	0	0	−204
174			174×10^{-6}	$1{,}794\times 10^{-6}$	$1{,}184\times 10^{-8}$	-194×10^{-6}	396×10^{-7}	-892×10^{-7}

$$r(\overline{V},\overline{I}) = \frac{-194/5}{2.95.9 \times 918.9} = -0.35, \tag{12.42}$$

$$r(\overline{V},\overline{I}) = \frac{396/5}{5.9 \times 15.38} = 0.87,$$

$$r(\overline{I},\overline{\varphi}) = \frac{-892/5}{18.9 \times 15.38} = -0.62.$$

From the mean values of V and I, we get

$$Z = 4.999/0.020561 = 249.25\,\text{ohm}, \tag{12.43}$$

$$R = \frac{V}{I}\cos\varphi = \frac{4.999}{0.020561}\cos(1.0444 \times 180/3.1416)$$
$$= 249.25 \times 0.5403 = 134.69\,\text{ohm},$$

$$X = \frac{4.999}{0.020561}\sin(57.29^\circ) = 249.25 \times 0.8414 = 209.7\,\text{ohm}.$$

Substituting values from (12.40) and (12.42) in (12.38), combined uncertainty in measurement of Z is given by

$$u_c^2 = (249.25)^2\left\{(2.95 \times 10^{-3}/4.999)^2 + (9.47 \times 10^{-7}/20.561 \times 10^{-3})^2\right.$$
$$\left.-2 \times (2.95 \times 10^{-3}/4.999) \times (9.47 \times 10^{-7}/20.561 \times 10^{-3})(-0.35)\right\}$$
$$= (249.25)^2\{76.8191 + 0.2121 + 1.902\} \times 10^{-8}$$
$$\mathrm{u_c} = 249.25 \times 8.8844 \times 10^{-4} = 0.221\,\text{ohm}$$

or relative combined relative uncertainty $u_{rc} = 8.88 \times 10^{-4}$.

It may be noted that measurands Z, X and R depend upon the same input quantities V, I and ϕ; therefore Z, X and R are also correlated.

References

1. NABL 141, Guidelines for estimation and expression of uncertainty in measurement. Department of Science and Technology, Technology Bhawan, New Delhi 110016 (2000)
2. R.R. Cordero, G. Seckmeyer, F. Labee, Effect of resolution on uncertainty. Metrologia **43**, L33–L38 (2006)
3. BIPM, IEC, IFCC, ISO, IUPAC, IUPAP, andOIML, Guide to the Expression of Uncertainty in Measurement (ISO, Switzerland, 1995)
4. C.M. Wang, H.K. Iyer, Uncertainty analysis for vector measurands using fiducial inference. Metrologia **43**, 486–494 (2006)

Appendix A

Tables

Gaussian Distribution

Table A.1 gives the probability of happening for the given value of the variable z. The probability values for z from 0 to 3.49 (in steps of 0.01) have been tabulated

$$z = \frac{x - \mu}{\sigma}, \; z = 1 \text{ corresponds one standard deviation.}$$

Table A.2 gives the cumulative frequency (area covered) from $-\infty$ to the given value of z. In fact the table gives the cumulative normal distribution against deviation from the mean expressed in terms of standard deviation.

Table A.3 gives the area covered by the variable from 0 to z. In fact Table A.3 can be derived from Table A.2 by subtracting 0.5 from each entry.

Table A.4 give the probability interval for the given value of z. It is the area covered by the variables from $-z$ to $+z$. For given value of z, every entry in Table A.4 is twice the entry in Table A.3.

Table A.5 gives the values of z for the given probability interval.

Student *t* Distribution

Table A.6 is the student t distribution. It gives the values of t for given degrees of freedom and assigned probability interval or the percentage points.

S.V. Gupta, *Measurement Uncertainties*, DOI 10.1007/978-3-642-20989-5,

Table A.1 Probability values of the normal (Gaussian) distribution

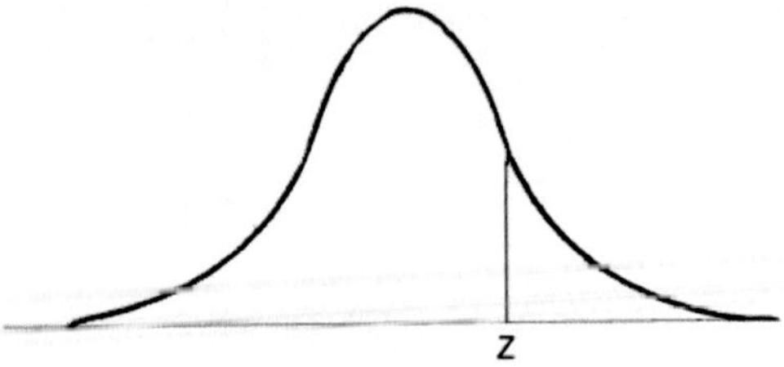

$$Y = \frac{1}{\sqrt{2\pi}} \exp(-z^2/2)$$

z	0.0	0.1	0.2	0.3	0.4	0.5	0.6	0.7	0.8	0.9
0.0	0.3989	0.3989	0.3989	0.3988	0.3986	0.3984	0.3982	0.3980	0.3977	0.3973
0.1	0.3970	0.3965	0.3961	0.3956	0.3951	0.3945	0.3939	0.3932	0.3925	0.3918
0.2	0.3910	0.3902	0.3894	0.3885	0.3876	0.3867	0.3857	0.3847	0.3836	0.3825
0.3	0.3814	0.3802	0.3790	0.3778	0.3765	0.3752	0.3739	0.3725	0.3712	0.3697
0.4	0.3683	0.3668	0.3653	0.3637	0.3621	0.3605	0.3589	0.3572	0.3555	0.3538
0.5	0.3521	0.3503	0.3485	0.3467	0.3448	0.3429	0.3410	0.3391	0.3372	0.3352
0.6	0.3332	0.3312	0.3292	0.3271	0.3251	0.3230	0.3209	0.3187	0.3166	0.3144
0.7	0.3123	0.3101	0.3079	0.3056	0.3034	0.3011	0.2989	0.2966	0.2943	0.2920
0.8	0.2897	0.2874	0.2850	0.2827	0.2803	0.2780	0.2756	0.2732	0.2709	0.2685
0.9	0.2661	0.2637	0.2613	0.2589	0.2565	0.2541	0.2516	0.2492	0.2468	0.2444
1.0	0.2420	0.2396	0.2371	0.2347	0.2323	0.2299	0.2275	0.2251	0.2227	0.2203
1.1	0.2179	0.2155	0.2131	0.2107	0.2083	0.2059	0.2036	0.2012	0.1989	0.1965
1.2	0.1942	0.1919	0.1895	0.1872	0.1849	0.1826	0.1804	0.1781	0.1758	0.1736
1.3	0.1714	0.1691	0.1669	0.1647	0.1626	0.1604	0.1582	0.1561	0.1539	0.1518
1.4	0.1497	0.1476	0.1456	0.1435	0.1415	0.1394	0.1374	0.1354	0.1334	0.1315
1.5	0.1295	0.1276	0.1257	0.1238	0.1219	0.1200	0.1182	0.1163	0.1145	0.1127
1.6	0.1109	0.1092	0.1074	0.1057	0.1040	0.1023	0.1006	0.0989	0.0973	0.0957
1.7	0.0940	0.0925	0.0909	0.0893	0.0878	0.0863	0.0848	0.0833	0.0818	0.0804
1.8	0.0790	0.0775	0.0761	0.0748	0.0734	0.0721	0.0707	0.0694	0.0681	0.0669
1.9	0.0656	0.0644	0.0632	0.0620	0.0608	0.0596	0.0584	0.0573	0.0562	0.0551
2.0	0.0540	0.0529	0.0519	0.0508	0.0498	0.0488	0.0478	0.0468	0.0459	0.0449
2.1	0.0440	0.0431	0.0422	0.0413	0.0404	0.0396	0.0387	0.0379	0.0371	0.0363
2.2	0.0355	0.0347	0.0339	0.0332	0.0325	0.0317	0.0310	0.0303	0.0297	0.0290
2.3	0.0283	0.0277	0.0270	0.0264	0.0258	0.0252	0.0246	0.0241	0.0235	0.0229
2.4	0.0224	0.0219	0.0213	0.0208	0.0203	0.0198	0.0194	0.0189	0.0184	0.0180
2.5	0.0175	0.0171	0.0167	0.0163	0.0158	0.0154	0.0151	0.0147	0.0143	0.0139
2.6	0.0136	0.0132	0.0129	0.0126	0.0122	0.0119	0.0116	0.0113	0.0110	0.0107
2.7	0.0104	0.0101	0.0099	0.0096	0.0093	0.0091	0.0088	0.0086	0.0084	0.0081
2.8	0.0079	0.0077	0.0075	0.0073	0.0071	0.0069	0.0067	0.0065	0.0063	0.0061
2.9	0.0060	0.0058	0.0056	0.0055	0.0053	0.0051	0.0050	0.0048	0.0047	0.0046
3.0	0.0044	0.0043	0.0042	0.0040	0.0039	0.0038	0.0037	0.0036	0.0035	0.0034
3.1	0.0033	0.0032	0.0031	0.0030	0.0029	0.0028	0.0027	0.0026	0.0025	0.0025
3.2	0.0024	0.0023	0.0022	0.0022	0.0021	0.0020	0.0020	0.0019	0.0018	0.0018
3.3	0.0017	0.0017	0.0016	0.0016	0.0015	0.0015	0.0014	0.0014	0.0013	0.0013
3.4	0.0012	0.0012	0.0012	0.0011	0.0011	0.0010	0.0010	0.0010	0.0009	0.0009
3.5	0.0009	0.0008	0.0008	0.0008	0.0008	0.0007	0.0007	0.0007	0.0007	0.0006

Fisher F

Tables A.7 and A.8 respectively gives the value of F (ratio of variances) for given values of degrees of freedom at 5% and 1% points.

χ^2 Table

Table A.9 gives the value of χ^2 for given degrees of freedom for different probability.

Range of Population Mean

Table A.10 gives the confidence limits of the mean. If μ is the population mean and s is the standard deviation then, limits of mean are given as

$$\bar{x} - ks \leq \mu \leq \bar{x} + ks.$$

The values of k for given number of observations n is given for confidence probability of 0.95 (5% points) and 0.99 (1% point).

Range of Standard Deviation

Table A.11 gives the value of $k_{\min}$ and $k_{\max}$ for given number of observations and given confidence probability (covered area) such that

$$k_{\min}s \leq \sigma \leq k_{\max}s.$$

Table A.2 Area under normal curve

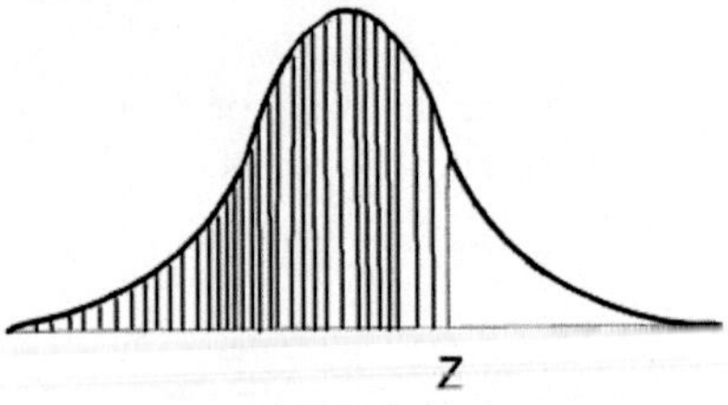

$$\text{Area} = \frac{1}{\sqrt{2\pi}} \int_{-\infty}^{z} \exp(-z^2/2)\mathrm{d}z$$

z	0.0	0.01	0.02	0.03	0.04	0.05	0.06	0.07	0.08	0.09
0.0	0.50000	0.50399	0.50798	0.51196	0.51595	0.51994	0.52392	0.52790	0.53188	0.53585
0.1	0.53982	0.54379	0.54776	0.55171	0.55567	0.55961	0.56356	0.56749	0.57142	0.57534
0.2	0.57926	0.58316	0.58706	0.59095	0.59483	0.59870	0.60256	0.60642	0.61026	0.61409
0.3	0.61791	0.62172	0.62551	0.62930	0.63307	0.63683	0.64057	0.64430	0.64802	0.65173
0.4	0.65542	0.65909	0.66275	0.66640	0.67003	0.67364	0.67724	0.68082	0.68438	0.68793
0.5	0.69146	0.69497	0.69846	0.70194	0.70540	0.70884	0.71226	0.71566	0.71904	0.72240
0.6	0.72574	0.72906	0.73237	0.73565	0.73891	0.74215	0.74537	0.74857	0.75174	0.75490
0.7	0.75803	0.76114	0.76423	0.76730	0.77034	0.77337	0.77637	0.77934	0.78230	0.78523
0.8	0.78814	0.79102	0.79388	0.79672	0.79954	0.80233	0.80510	0.80784	0.81056	0.81326
0.9	0.81593	0.81858	0.82120	0.82380	0.82638	0.82893	0.83146	0.83397	0.83645	0.83890
1.0	0.84133	0.84374	0.84612	0.84848	0.85082	0.85313	0.85541	0.85768	0.85991	0.86213
1.1	0.86432	0.86649	0.86863	0.87075	0.87284	0.87491	0.87696	0.87898	0.88098	0.88296
1.2	0.88491	0.88684	0.88875	0.89063	0.89249	0.89433	0.89615	0.89794	0.89971	0.90145
1.3	0.90318	0.90488	0.90656	0.90822	0.90985	0.91147	0.91306	0.91463	0.91618	0.91771
1.4	0.91922	0.92070	0.92217	0.92362	0.92504	0.92644	0.92783	0.92919	0.93054	0.93186
1.5	0.93316	0.93445	0.93572	0.93696	0.93819	0.93940	0.94059	0.94176	0.94291	0.94405
1.6	0.94517	0.94627	0.94735	0.94842	0.94946	0.95049	0.95151	0.95251	0.95349	0.95445
1.7	0.95540	0.95633	0.95725	0.95815	0.95903	0.95990	0.96076	0.96160	0.96242	0.96323
1.8	0.96403	0.96481	0.96558	0.96633	0.96707	0.96780	0.96852	0.96922	0.96990	0.97058
1.9	0.97124	0.97189	0.97253	0.97315	0.97377	0.97437	0.97496	0.97554	0.97610	0.97666
2.0	0.97720	0.97774	0.97826	0.97878	0.97928	0.97977	0.98025	0.98073	0.98119	0.98164
2.1	0.98209	0.98252	0.98295	0.98337	0.98377	0.98417	0.98456	0.98495	0.98532	0.98569
2.2	0.98605	0.98640	0.98674	0.98708	0.98741	0.98773	0.98804	0.98835	0.98865	0.98894
2.3	0.98923	0.98951	0.98978	0.99005	0.99031	0.99056	0.99081	0.99106	0.99130	0.99153
2.4	0.99175	0.99198	0.99219	0.99240	0.99261	0.99281	0.99301	0.99320	0.99338	0.99356
2.5	0.99374	0.99392	0.99408	0.99425	0.99441	0.99457	0.99472	0.99487	0.99501	0.99516
2.6	0.99529	0.99543	0.99556	0.99568	0.99581	0.99593	0.99605	0.99616	0.99627	0.99638
2.7	0.99649	0.99659	0.99669	0.99679	0.99688	0.99698	0.99707	0.99715	0.99724	0.99732
2.8	0.99740	0.99748	0.99756	0.99763	0.99770	0.99777	0.99784	0.99790	0.99797	0.99803
2.9	0.99809	0.99815	0.99821	0.99826	0.99832	0.99837	0.99842	0.99847	0.99852	0.99856
3.0	0.99861	0.99865	0.99869	0.99874	0.99878	0.99882	0.99885	0.99889	0.99892	0.99896
3.1	0.99899	0.99902	0.99905	0.99908	0.99911	0.99914	0.99917	0.99920	0.99922	0.99925
3.2	0.99927	0.99929	0.99932	0.99934	0.99937	0.99939	0.99940	0.99942	0.99944	0.99946
3.3	0.99948	0.99949	0.99951	0.99953	0.99955	0.99956	0.99957	0.99959	0.99960	0.99961
3.4	0.99962	0.99963	0.99964	0.99966	0.99967	0.99968	0.99969	0.99970	0.99972	0.99973
3.5	0.99974	0.99974	0.99975	0.99975	0.99976	0.99977	0.99977	0.99978	0.99978	0.99979
3.6	0.99980	0.99980	0.99981	0.99981	0.99982	0.99983	0.99983	0.99984	0.99984	0.99985

Table A.3 Area under the normal Gaussian distribution

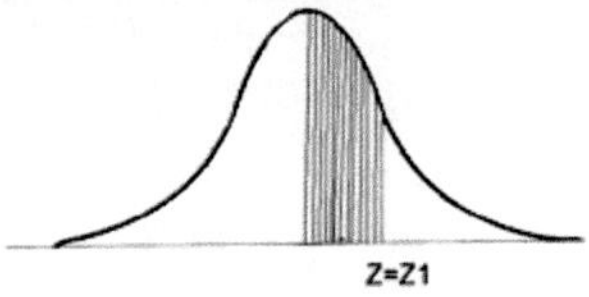

$$\text{Area} = \frac{1}{\sqrt{2\pi}} \int_0^z \exp(-z^2/2)\mathrm{d}z$$

z	0.0	0.1	0.2	0.3	0.4	0.5	0.6	0.7	0.8	0.9
0.0	0.0000	0.0040	0.0080	0.0120	0.0160	0.0199	0.0239	0.0279	0.0319	0.0359
0.1	0.0398	0.0438	0.0478	0.0517	0.0557	0.0596	0.0636	0.0675	0.0714	0.0753
0.2	0.0793	0.0832	0.0871	0.0910	0.0948	0.0987	0.1026	0.1064	0.1103	0.1141
0.3	0.1179	0.1217	0.1255	0.1293	0.1331	0.1368	0.1406	0.1443	0.1480	0.1517
0.4	0.1554	0.1591	0.1628	0.1664	0.1700	0.1736	0.1772	0.1808	0.1844	0.1879
0.5	0.1915	0.1950	0.1985	0.2019	0.2054	0.2088	0.2123	0.2157	0.2190	0.2224
0.6	0.2258	0.2291	0.2324	0.2357	0.2389	0.2422	0.2454	0.2486	0.2518	0.2549
0.7	0.2580	0.2612	0.2642	0.2673	0.2704	0.2734	0.2764	0.2794	0.2823	0.2852
0.8	0.2882	0.2910	0.2939	0.2967	0.2996	0.3023	0.3051	0.3079	0.3106	0.3133
0.9	0.3159	0.3186	0.3212	0.3238	0.3264	0.3290	0.3315	0.3340	0.3365	0.3389
1.0	0.3414	0.3438	0.3461	0.3485	0.3508	0.3531	0.3554	0.3577	0.3599	0.3621
1.1	0.3643	0.3665	0.3686	0.3708	0.3729	0.3749	0.3770	0.3790	0.3810	0.3830
1.2	0.3849	0.3869	0.3888	0.3907	0.3925	0.3944	0.3962	0.3980	0.3997	0.4015
1.3	0.4032	0.4049	0.4066	0.4082	0.4099	0.4115	0.4131	0.4147	0.4162	0.4177
1.4	0.4192	0.4207	0.4222	0.4236	0.4251	0.4265	0.4279	0.4292	0.4306	0.4319
1.5	0.4332	0.4345	0.4357	0.4370	0.4382	0.4394	0.4406	0.4418	0.4429	0.4441
1.6	0.4452	0.4463	0.4474	0.4484	0.4495	0.4505	0.4515	0.4525	0.4535	0.4545
1.7	0.4554	0.4564	0.4573	0.4582	0.4591	0.4599	0.4608	0.4616	0.4625	0.4633
1.8	0.4641	0.4648	0.4656	0.4664	0.4671	0.4678	0.4686	0.4693	0.4699	0.4706
1.9	0.4713	0.4719	0.4726	0.4732	0.4738	0.4744	0.4750	0.4756	0.4761	0.4767
2.0	0.4772	0.4778	0.4783	0.4788	0.4793	0.4798	0.4803	0.4808	0.4812	0.4817
2.1	0.4821	0.4826	0.4830	0.4834	0.4838	0.4842	0.4846	0.4850	0.4854	0.4857
2.2	0.4861	0.4864	0.4868	0.4871	0.4874	0.4878	0.4881	0.4884	0.4887	0.4890
2.3	0.4893	0.4895	0.4898	0.4901	0.4904	0.4906	0.4909	0.4911	0.4913	0.4916
2.4	0.4918	0.4920	0.4922	0.4924	0.4926	0.4929	0.4930	0.4932	0.4934	0.4936
2.5	0.4938	0.4940	0.4941	0.4943	0.4945	0.4946	0.4948	0.4949	0.4951	0.4952
2.6	0.4953	0.4955	0.4956	0.4957	0.4959	0.4960	0.4961	0.4962	0.4963	0.4964
2.7	0.4965	0.4966	0.4967	0.4968	0.4969	0.4970	0.4971	0.4972	0.4973	0.4974
2.8	0.4974	0.4975	0.4976	0.4977	0.4977	0.4978	0.4979	0.4979	0.4980	0.4981
2.9	0.4981	0.4982	0.4983	0.4983	0.4984	0.4984	0.4985	0.4985	0.4986	0.4986
3.0	0.4987	0.4987	0.4987	0.4988	0.4988	0.4989	0.4989	0.4989	0.4990	0.4990
3.1	0.4990	0.4991	0.4991	0.4991	0.4992	0.4992	0.4992	0.4992	0.4993	0.4993
3.2	0.4993	0.4993	0.4994	0.4994	0.4994	0.4994	0.4994	0.4995	0.4995	0.4995
3.3	0.4995	0.4995	0.4996	0.4996	0.4996	0.4996	0.4996	0.4996	0.4996	0.4997
3.4	0.4997	0.4997	0.4997	0.4997	0.4997	0.4997	0.4997	0.4997	0.4998	0.4998
3.5	0.4998	0.4998	0.4998	0.4998	0.4998	0.4998	0.4998	0.4998	0.4998	0.4998
3.6	0.4998	0.4999	0.4999	0.4999	0.4999	0.4999	0.4999	0.4999	0.4999	0.4999
3.7	0.4999	0.4999	0.4999	0.4999	0.4999	0.4999	0.4999	0.4999	0.4999	0.4999
3.8	0.4999	0.4999	0.4999	0.4999	0.4999	0.4999	0.4999	0.5000	0.5000	0.5000

Table A.4 The area covered of Gaussian curve between the ordinates $\pm z$

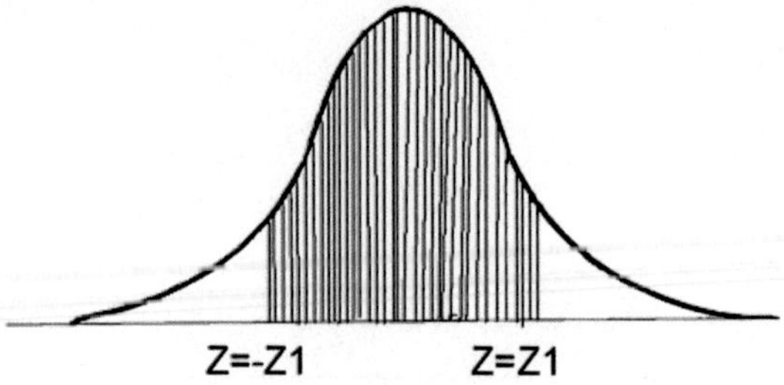

$$\text{Area} = P = \frac{1}{\sqrt{2\pi}} \int_{-z_1}^{z_1} \exp(-z^2/2)\mathrm{d}z$$

z_1	P	$P_r - P_{r-1}$	z_1	P	$P_r - P_r - 1$	z_1	P	$P_r - P_{r-1}$
0.00	0.00000		0.35	0.27366	0.00752	0.7	0.51607	0.00626
0.01	0.00798	0.00798	0.36	0.28115	0.00749	0.8	0.57619	0.06012
0.02	0.01596	0.00798	0.37	0.28861	0.00747	0.9	0.63 I 88	0.05559
0.03	0.01393	0.00797	0.38	0.29605	0.00743	1.0	0.68269	0.05081
0.04	0.03191	0.00798	0.39	0.30346	0.00741	1.1	0.71867	OM598
0.05	003988	0.00797	0.40	0.31084	0.00738	t.2	0.76986	0.041
0.06	0.04784	0.00796	0.41	0.31819	0.00735	t.3	0.80640	0.064
0.07	0.05581	0.00797	0.42	0.32551	0.00732	1.4	0.83849	0.03209
0.08	0.06376	0.00795	0.43	0.33280	0.00729	1.5	0.86639	0.02790
0.09	0.07171	0.00795	0.44	0.34006	0.00716	1.6	0.89040	0.024 a I
0.10	0.07966	0.00795	0.45	0.34729	0.00723	1.7	0.91087	0.02047
0.11	0.08759	0.00793	0.46	0.35448	0.00719	1.8	0.928 ! 4	0.01727
0.12	0.09552	0.00793	0.47	0.36 I 64	0.00716	1.9	0.94257	0.01443
0. 13	0.10343	0.00791	0.48	0.36877	0.007 I 3	2.0	0.95450	0.01193
0.14	0.1134	0.00791	0.49	0.37587	0.007 I a	2.1	0.96427	0.00977
0.15	0.11924	0.00790	0.50	0.38292	0.00705	2.2	0.97219	0.00792
0.16	0.12712	0.00788	0.51	0.38995	0.00703	2.3	0.97855	0.00636
0.17	0.13499	0.00787	0.52	0.39694	0.00699	2.4	0.98360	0.00505
0.18	0.14285	0.00786	0.53	0A0389	0.00695	2.5	0.98758	0.00398
0.19	0.15069	0.00784	0.54	0.41080	0.00691	2.6	0.99068	0.00310
0.20	0.15852	0.00783	0.55	0.41768	0.00688	2.7	0.99307	0.00239
0.21	0.16633	0.00781	0.56	0.42452	0.00684	2.8	0.99489	0.00182
0.22	0.17413	0.00780	0.57	0.43132	0.00680	2.9	0.99627	0.00138
0.23	0.18191	0.00778	0.58	0.43809	0.00677	3.0	0.99730	0.00103
0.24	0.18967	0.00776	0.59	0.44481	0.00672	3.1	0.99806	0.00076
0.25	0.19741	0.00774	0.60	0.45149	0.00668	3.2	0.99863	0.00057
0.26	0.20514	0.00773	0.61	0.45814	0.00665	3.3	0.99903	0.00040
0.27	0.21284	0.00770	0.62	0.46474	0.00660	3.4	0.99933	0.00030
0.28	0.22052	0.00768	0.63	0.47131	0.00657	3.5	0.99953	0.00020
0.29	0.22818	0.00766	0.64	0.47783	0.00652	3.6	0.99968	0.00015
0.30	0.23582	0.00764	0.65	0.48431	0.00648	3.7	0.99978	0.00010
0.31	0.24344	0.00762	0.66	0.49075	0.00644	3.8	0.99986	0.00008
0.32	0.25103	0.00759	0.67	0.49714	0.00639	3.891	0.99990	
0.33	0.25860	0.00757	0.68	0.50350	0.00636			
0.34	0.26614	0.00754	0.69	0.50981	0.00631			

Table A.5 Values of z for the given the probability (area covered)

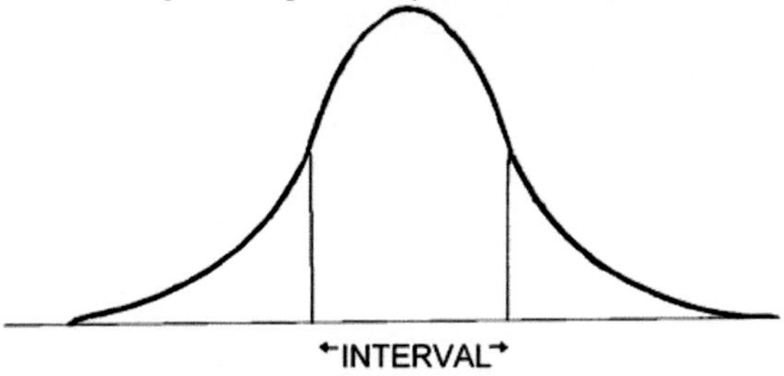

$$P = \frac{1}{\sqrt{2\pi}} \int_{-z}^{z} \exp(-z^2/2)\mathrm{d}z$$

P	z	Diff	P	z	Diff	P	z	Diff
0.00	0.0000		0.35	0.4538	0.0139	0.70	1.036	0.021
0.01	0.0125	0.0125	0.36	0.4677	0.0140	0.71	1.058	0.022
0.02	0.025	0.0126	0.37	0.4817	0.0141	0.72	1.080	0.022
0.03	0.0376	0.0125	0.38	0.4959	0.0142	0.73	1.103	0.023
0.04	0.0502	0.0126	0.39	0.5101	0.0142	0.74	1.126	0.023
0.05	0.0627	0.0125	0.40	0.5244	0.0143	0.75	1.150	0.024
0.06	0.0753	0.0126	0.41	0.5388	0.0144	0.76	1.1	0.025
0.07	0.0878	0.0125	0.42	0.5534	0.0146	0.77	1.200	0.025
0.08	0.1004	0.0126	0.43	0.5681	0.0147	0.78	1.227	0.027
0.09	0.1130	0.0126	0.44	0.5828	0.0147	0.79	1.254	0.027
0.10	0.1257	0.0127	0.45	0.5978	0.0150	0.80	1.282	0.028
0.11	0.1383	0.0126	0.46	0.6128	0.0150	0.81	1.311	0.029
0.12	0.1510	0.0127	0.47	0.6280	0.0152	0.82	1.341	0.030
0.13	0.1637	0.0127	0.48	0.6433	0.0153	0.83	1.372	0.031
0.14	0.1764	0.0127	0.49	0.6588	0.0155	0.84	1.405	0.033
0.15	0.1891	0.0127	0.50	0.6745	0.0157	0.85	1.440	0.D35
0.16	0.2019	0.0128	0.51	0.6903	0.0158	0.86	1.476	0.036
0.17	0.2147	0.0128	0.52	0.7063	0.0160	0.87	1.514	0.038
0.18	0.2275	0.0128	0.53	0.7225	0.0162	0.88	1.555	0.041
0.19	0.2404	0.0129	0.54	0.7388	0.0163	0.89	1.598	0.043
0.20	0.2533	0.0129	0.55	0.7554	0.0166	0.90	1.645	0.047
0.21	0.2663	0.0130	0.56	0.7722	0.0168	0.91	1.695	0.050
0.22	0.2793	0.0130	0.57	0.7892	0.0170	0.92	1.751	0.056
0.23	0.2924	0.0131	0.58	0.8064	0.0172	0.93	1.812	0.061
0.24	0.3055	0.0131	0.59	0.8239	0.0175	0.94	1.881	0.069
0.25	0.3186	0.0133	0.60	0.8416	0.0177	0.95	1.960	0.079
0.26	0.3319	0.0134	0.61	0.8596	0.0180	0.96	2.054	0.094
0.27	0.3451	0.0134	0.62	0.8779	0.0183	0.97	2.170	0.116
0.28	0.3585	0.0134	0.63	0.8965	0.0186	0.98	2.326	0.156
0.29	0.3719	0.0134	0.64	0.9154	0.0189	0.99	2.576	0.250
0.30	0.3853	0.0136	0.65	0.9346	0.0192	0.995	2.807	0.231
0.31	0.3989	0.0136	0.66	0.9542	0.0196	0.999	3.291	0.484
0.32	0.4125	0.0136	0.67	0.9741	0.0199			
0.33	0.4261	0.0138	0.68	0.9945	0.0204			
0.34	0.4399	00139	0.69	1.015	0.0205			

Table A.6 Student t distribution μ and s known

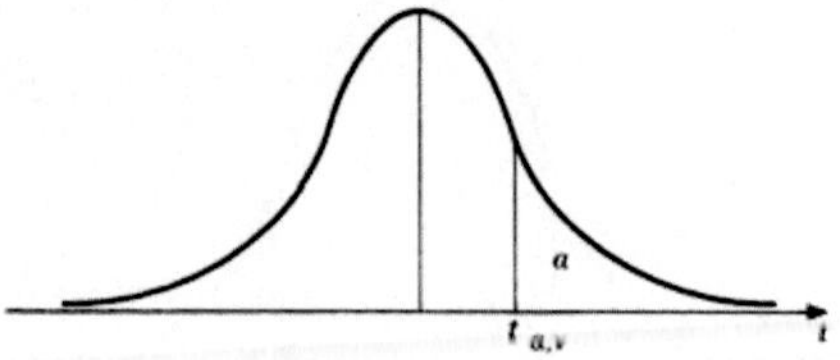

$$P_{tp} = \int_{-t_\nu}^{t_\nu} f(t_\nu)$$

Degree of freedom $\nu = n-1$	Area covered					
	0.90	0.95	0.98	0.99	0.995	0.999
	Percentage points					
	0.05	0.025	0.01	0.005	0.0025	0.0005
	t_ν	t_ν	t_ν	t_ν	t_ν	t_ν
1	6.3138	12.706	31.821	63.657	127.32	636.619
2	2.9200	4.3027	6.965	9.9248	14.089	31.598
3	2.3534	3.1825	4.541	5.8409	7.4533	12.924
4	2.1318	2.7764	3.747	4.6041	5.5976	8.610
5	2.0150	2.5706	3.365	4.0321	4.7733	6.869
6	1.9432	2.4469	3.143	3.7074	4.3168	5.959
7	1.8946	2.3646	2.998	3.4995	4.0293	5.408
8	1.8595	2.3060	2.896	3.3554	3.8325	5.041
9	1.8331	2.2622	2.821	3.2498	3.6897	4.781
10	1.8125	2.2281	2.764	3.1693	3.5814	4.587
11	1.7959	2.2010	2.718	1.1058	3.4966	4.437
12	1.7823	2.1788	2.681	3.0545	3.4284	4.318
13	1.7709	2.1604	2.650	3.0123	13725	4.221
14	1.7613	2.1448	2.624	2.9768	3.3257	4.140
15	1.7530	2.1315	2.602	2.9467	3.2860	4.073
16	1.7459	2.1199	2.583	2.9208	3.2520	4.015
17	1.7396	2.1098	2.567	2.8982	3.2225	3.965
18	1.7341	2.1009	2.552	2.8784	3.1966	3.922
19	1.7291	2.0930	2.539	2.8609	3.1737	3.883
20	1.7247	2.0860	2.528	2.8453	3.1534	3.850
25	1.7081	2.0595	2.485	2.7874	10782	1725
30	1.6973	2.0423	2.457	2.7500	3.0298	3.646
35	1.6996	2.0301	2.438	2.7239	2.9962	3.5915
40	1.6839	2.0211	2.423	2.7045	2.9713	3.5511
45	1.6794	2.0141	2.412	2.6896	2.9522	3.5207
50	1.6759	2.0086	2.403	2.6778	2.9370	3.4965
60	1.6707	2.0003	2.390	2.6603	2.9146	3.4606
70	1.6669	1.9945	2.381	2.6480	2.8988	3.4355
80	1.6641	1.9901	2.374	2.6388	2.8871	3.4[69
90	1.6620	1.9867	2.368	2.6316	2.8779	3.4022
100	1.6602	1.9840	2.364	2.6260	2.8707	13909
150	1.6551	1.9759	2.351	2.6090	2.8492	3.3567
200	1.6525	1.9719	2.345	2.6006	2.8386	3.3400

The values of t_ν for given area covered or the area beyond it

Table A.7 F test: Upper limits for *F* for Probability0.05

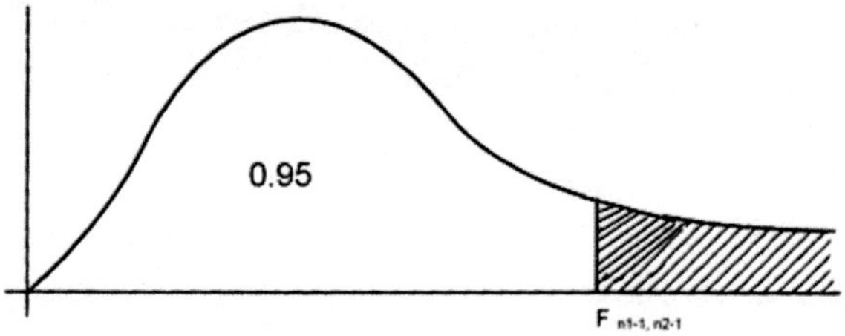

v	*I*	2	3	4	5	6	7	8	9	10	15	20	24	30	40	50	60	80	100
1	161.44	200	216	225	230	234	237	239	241	242	246	248	249	250	251	252	252	252	253
2	18.51	19.0	19.2	19.2	19.3	19.3	19.4	19.4	19.4	19.4	19.4	19.4	19.5	19.5	19.5	19.5	19.5	19.5	19.5
3	10.13	9.55	9.28	9.12	901	8.94	8.89	8.85	'1.81	8.79	8.70	8.66	8.64	8.62	8.59	8.58	8.57	8.56	8.55
4	7.71	694	6.59	6.39	6.26	6.16	6.09	604	6.00	596	5.86	5.80	5.77	5.75	5.72	5.70	5.69	5.67	5.66
5	6.61	5.79	5.41	5.19	5.05	4.95	4.88	4.82	4.77	4.74	4.62	4.56	4.53	4.50	4.46	4.44	4.43	4.41	4.41
6	5.99	5.14	4.76	4.53	4.39	4.28	4.21	4.15	4.10	4.06	3.94	3.87	3.84	3.81	3.77	3.75	3.74	3.72	3.71
7	5.59	4.74	4.35	4.12	3.97	3.87	3.79	3.73	3.68	3.64	3.51	3.44	3.41	3.38	3.34	3.32	3.30	3.29	3.27
8	5.32	4.46	4.07	3.84	3.69	3.58	3.50	3.44	3.39	3.35	3.22	3.15	3.12	3.08	3.04	3.02	3.01	2.99	2.97
9	5.12	4.26	3.86	3.63	348	3.37	3.29	3.23	3.18	3.14	3.01	294	2.90	2.86	2.83	2.80	2.79	2.77	2.76
10	4.96	4.10	3.71	3.48	3.33	3.22	3.14	3.07	3.02	298	2.85	2.77	2.74	2.70	2.66	2.64	2.62	2.60	2.59
15	4.54	3.68	3.29	3.06	2.90	2.79	2.71	2.64	2.59	2.54	2.42	2.33	2.31	2.25	2.20	2.18	2.16	2.14	2.12
20	4.35	3.49	3.10	2.87	2.71	2.60	2.51	2.45	2.39	2.35	2.20	2.12	2.08	2.04	1.99	1.97	1.95	1.92	1.91
25	4.24	3.39	2.99	2.76	2.60	2.49	2.40	2.34	2.28	2.24	2.11	2.02	1.96	1.92	1.87-	1.84	1.82	1.80	1.78
30	4.17	3.32	2.92	2.69	2.53	2.42	2.33	2.27	2.21	2.16	2.01	1.93	1.89	1.85	1.79	1.76	1.74	1.71	1.70
40	4.08	3.23	2.84	2.61	2.45	2.34	2.25	2.18	2.12	2.08	1.92	1.84	1.81	1.74	1.69	1.66	1.64	1.62	1.59
50	4.03	3.18	2.79	2.56	2.40	2.29	2.20	2.13	2.07	2.03	1.87	1.78	1.74	1.69	1.63	1.60	1.58	1.54	1.52
60	4.00	3.15	2.76	2.53	2.37	2.25	2.17	2.10	2.04	1.99	1.84	1.75	1.70	1.65	1.59	1.56	1.53	1.50	1.48
80	3.96	3.11	2.72	2.49	2.33	2.21	2.13	2.06	2.00	1.95	1.79	1.70	1.65	1.60	1.54	1.51	1.48	1.45	1.43
100	3.94	3.09	2.70	2.46	2.31	2.19	2.10	2.03	1.97	1.93	1.77	1.68	1.63	1.57	1.52	1.48	1.45	1.41	1.39

Table A.8 *F* test: Upper limits for *F* for Probability0.01

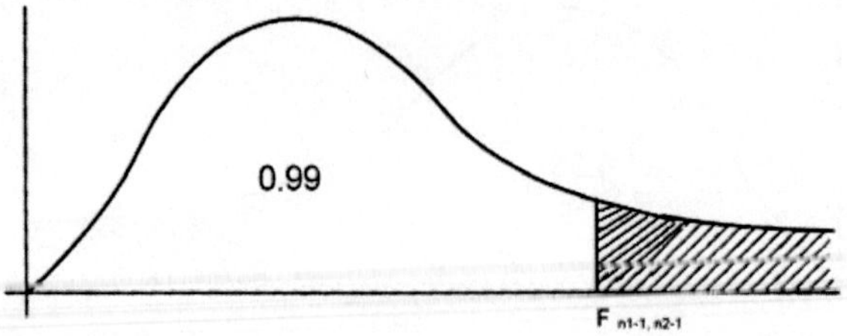

ν	1	2	3	4	5	6	7	8	9	10	15	20	24	30	40	50	60	80	100
2	98.50	99.0	99.2	99.2	99.3	99.4	99.4	99.4	99.4	99.4	99.4	99.4	99.5	99.5	99.5	99.5	99.5	99.5	99.5
3	34.12	30.8	29.5	28.7	28.2	27.9	27.7	27.5	27.3	27.2	26.9	26.7	26.6	26.5	26.4	26.4	26.3	26.3	26.2
4	21.20	18.0	16.7	16.0	15.5	152	15.0	14.8	14.7	14.5	14.2	14.0	13.9	13.9	13.7	13.7	13.7	13.6	13.6
5	16.26	13.3	12.1	11.4	11.0	10.7	10.5	10.3	10.2	10.1	9.72	9.55	9.47	9.38	9.29	9.24	9.20	9.16	9.13
6	13.75	10.9	9.78	915	875	8.47	8.26	8.10	7.98	7.87	7.56	7.40	7.31	7.23	7.14	7.09	7.06	7.01	6.99
7	12.25	9.55	8.45	7.85	7.46	7.19	6.99	684	6.72	6.62	6.31	6.16	6.07	5.99	5.91	5.86	5.82	5.78	5.75
8	11.26	8.65	7.59	7.01	6.63	6.37	6.18	6.03	5.91	581	5.52	5.36	5.28	5.20	5.12	5.07	5.03	4.99	4.96
9	10.56	8.02	6.99	6.42	6.06	5.80	5.61	5.47	5.35	5.26	4.96	4.81	4.73	4.65	4.57	4.52	4.48	4.44	4.42
10	10.04	7.56	6.55	5.99	5.64	5.39	5.20	5.06	4.94	4.85	4.56	4.41	4.33	4.25	4.17	4.12	4.08	4.04	4.01
15	8.68	6.36	5.42	4.89	4.56	4.32	4.14	4.00	3.89	3.80	3.52	3.37	3.29	3.21	3.13	3.08	3.05	3.00	2.98
20	8.10	5.85	4.94	4.43	4.10	3.87	3.70	3.56	3.46	3.37	3.09	2.94	2.86	2.78	2.69	2.64	2.61	2.56	2.54
25	7.77	5.57	4.68	4.18	3.86	3.63	3.46	3.32	3.22	3.13	2.85	2.70	2.62	2.54	2.45	2.40	2.36	2.32	2.29
30	7.56	5.39	.4.51	4.02	3.70	3.47	3.30	3.17	3.07	2.98	2.70	2.55	2.47	2.39	2.30	2.25	2.21	2.16	2.13
40	7.31	5.18	4.31	3.83	3.51	3.29	312	2.99	2.89	2.80	2.52	2.37	2.29	2.20	2.11	2.06	2.02	1.97	1.94
50	7.17	5.06	4.20	3.72	3.41	3.19	3.02	2.89	2.79	2.70	2.42	2.27	2.18	2.10	2.01	1.95	1.91	1.86	1.82
60	7.08	4.98	4.13	3.65	3.34	3.12	2.95	282	2.72	2.63	2.35	2.20	2.12	*i03*	1.94	1.88	1.84	1.78	1.75
80	6.96	4.88	4.04	3.56	3.26	3.04	2.87	2.74	2.64	2.55	2.27	2.12	2.03	1.94	1.85	1.79	1.75	1.69	1.66
100	6.90	4.82	3.9B	3.51	3.21	2.99	282	269	2.59	2.50	2.22	2.07	1.9B	1.89	1.80	1.73	1.69	1.63	1.60

Table A.9 Significant values $\chi^2(\alpha)$ of Chi-square distribution (Right tail areas for given probability α)

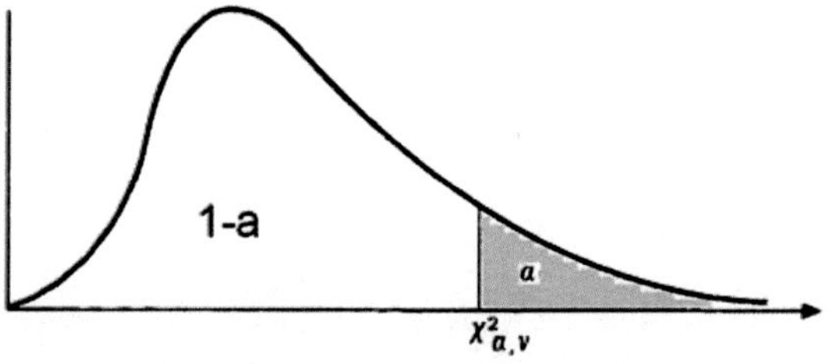

$$P = P_r(\chi^2 > \varphi\chi^2(\alpha) = \alpha$$

Degree of freedom	Probability (Level of significance) Right tail probability areas						
ν	0.99	0.95	0.50	0.10	0.05	0.02	0.01
1	−000157	0.00393	0.455	2.706	3.841	5.214	6.635
2	0.0201	0.103	1.386	4.605	5.991	7.824	9.210
3	0.115	0.352	2.366	6.251	7.815	9.837	11.341
4	0.297	0.711	3.357	7.779	9.488	11.668	13.277
5	0.554	1.145	4.351	9.236	11.070	13.388	15.086
6	0.872	1.635	5.348	10.645	12.592	15.033	16.812
7	1.239	2.167	6.346	12.017	14.067	16.622	18.475
8	1.646	2.733	7.344	13.362	15.507	18.1680	20.09.0
9	2.088	3.325	8.343	14.684	16.919	c19.679	21.666
10	2.558	3.940	9.340	15.987	18.307	21.161	23.209
11	3.053	4.575	10.341	17.275	19.675	22.618c	24.725
12	3.571	5.226	11.340	18.549	21.026	24.054	26.217
13	4.107	5.892	12.340	19.812	22.362	25.172	27.688
14	4.66.0	6.571	13.339	21.064	23.685	26.873	29.141
15	4.229	7.261	14.339	22.307	24.996	28.259	30.578
16	5.812	7.962	15.338	23.542	26.296	29–633	32.000
17	6.408	8.672	16.338	24.769	27.587	30.995	33.409
18	7.015	9.390	17.338	25.989	28.869	32346	34.805
19	7.633	10.117	18.338	27.204	30.144	33.687	36.191
20	8.260	10.851	19.337	28.412	31.410	35.020	37.566
21	8.897	11.591	20.337	29.615	32.671	36.348	38.932
22	9.542	12.338	21.337	3.0.813	33.924	37.659	40.289
23	10.196	13.091	22.337	32.007	35.172	38.968	41.638
24	10.856	13.848	23.337	32.196	36.415	40.270	42.980
25	11.524	14.6 f 1	24.33.1	34.382	37.652	41.566	44.314
26	12.198	15.379	25.336	35363	38.885	41.856	45.642
27	12.879	16.151	26.336.	36.741	40.113	44.140	46.963
28	13.565	16.928	27.336	37.916	41.337	45.419	48.278
29	14.256	17.708	28.336	39.087	42.557	46.693	49.588
30	14.953	18.493	29.336	40.256	43.773	47.962	50.892

Table A.10 Confidence limits for the Mean μ

Confidence level 95%						Confidence level 99%					
n	k	n	k	n	k	n	k	n	k	n	k
1	-	30	0.3734	95	0.2037	1		30	0.5033	95	0.2698
2	8.9845	31	0.3668	100	0.1964	2	4.5012	31	0.4939	100	0.2627
3	2.4842	32	0.3605	110	0.1890	3	5.7301	32	0.4851	110	0.2500
4	1.5913	33	0.3546	120	0.1808	4	2.9205	33	0.4767	120	0.2390
5	1.2461	34	0.3489	130	0.1735	5	2.0590	34	0.4688	130	0.2293
6	1.0494	35	0.3435	140	0.1671	6	1.6461	35	0.4612	140	0.2207
7	0.9248	36	0.3384	150	0.1674	7	1.4013	36	0.4540	\11S0	0.2131
8	0.8360	37	0.3334	160	0.1561	8	1.2373	37	0.4471	160	0.2061
9	0.7687	38	0.3287	170	0.1514	9	1.1185	38	0.4405	170	0.1998
10	0.7154	39	0.3242	180	0.1471	10	1.0277	39	0.4342	180	0.1941
11	0.6718	40	0.3198	190	0.1431	11	0.9556	40	0.4282	190	0.1888
12	0.6354	41	0.3156	200	0.1394	12	0.8966	41	0.4224	200	0.1839
13	0.6043	42	0.3116	250	0.1240	13	0.8472	42	0.4168	250	0.1629
14	0.5774	43	0.3078	300	0.1132	14	0.8051	43	0.4115	300	0.1487
15	0.5538	44	0.3040	350	0.1048	15	0.7686	44	0.4063	350	0.1377
16	0.5329	45	0.3004	400	0.0980	16	0.7367	45	0.4013	400	0.1288.
17	0.5142	46	0.2970	450	0.0924	17	0.7084	46	0.3966	450	0.1214
18	0.4973	47	0.2936	500	0.0877	18	0.6831	47	0.3919	500	0.1152
19	0.4820	48	0.2904	550	0.0836	19	0.6604	48	0.3875	550	0.1098
20	0.4680	49	0.2872	600	0.0800	20	0.6397	49	0.3832	600	0.1052
21	0.4552	50	0.2842	650	0.0769	21	0.6209	50	0.3790	650	0.1010
22	0.4434	55	0.2703	700	0.0741	22	0.6037	55	0.3600	700	0.0974
23	0.4324	60	0.2583	750	0.0716	23	0.5878	60	0.3436	750	0.0941
24	0.4223	65	0.2478	800	0.0693	24	0.5730	65	0.3293	800	0.0911
25	0.4128	70	0.2385	850	0.0672	25	0.5594	70	0.3166	850	0.0884
26	0.4039	75	0.2301	900	0.0653	26	0.5467	75	0.3053	900	0.0859
27	0.3956	80	0.2225	950	0.0636	27	0.5348	80	0.2951	950	0.0836
28	0.3878	85	0.2157	–	–	28	0.5236	85	0.2859	–	–
29	0.3804	90	0.2095	1,000	0.620	29	0.5131	90	0.2775	1,000	0.0815

The tabulated values give values of the confidence factor k defining the confidence limits $\bar{x} \pm ks$ for the mean μ. Values of k are given for two confidence levels namely 95% and 99% The value of $k = t_\nu/\sqrt{n}$, n being the size of the sample

Table A.11 Limits of standards deviation of population in terms of Standard deviation of the sample for given probability

$$k_{\min} s \leq \sigma \leq k_{\max} s$$

P	0.90		0.95		0.98		0.99	
n	k_{min}	k_{max}	k_{min}	k_{max}	k_{min}	k_{max}	k_{min}	k_{max}
2	0.5102	15.947	0.4463	31.910	0.3882	79.789	0.3562	159.58
3	0.5777	4.416	0.5207	6.285	0.4660	9.974	0.4344	14.124
4	0.6196	2.920	0.5665	3.729	0.5142	5.111	0.4834	6.468
5	0.6493	2.372	0.5991	2.874	0.5489	3.669	0.5188	4.396
6	0.6720	2.089	0.6242	2.453	0.5757	3.003	0.5464	3.485
7	0.6903	1.915	0.6444	2.202	0.5974	2.623	0.5688	2.980
8	0.7054	1.797	0.6612	2.035	0.6155	2.377	0.5875	2.660
9	0.7183	1.711	0.6755	1.916	0.6310	2.204	0.6036	2.439
10	0.7293	1.645	0.6878	0.286	0.6445	2.076	0.6177	2.278
11	0.7391	1.593	0.6987	1.755	0.6564	1.977	0.6301	2.154
12	0.7477	1.551	0.7084	1.698	0.6670	1.898	0.6412	2.056
13	0.7555	1.515	0.7171	1.651	0.6765	1.833	0.6512	1.976
14	0.7625	1.485	0.7250	1.611	0.6852	1.799	0.6603	1.910
15	0.7688	1.460	0.7321	1.577	0.6931	1.733	0.6686	1.854
16	0.7747	1.437	0.7387	1.548	0.7004	1.694	0.6762	1.806
17	0.7800	1.418	0.7448	1.522	0.7071	1.659	0.6833	1.764
18	0.7850	1.400	0.7504	1.499	0.7133	1.629	0.6899	1.727
1.9	0.7896	1.384	0.7556	1.479	0.7191	1.602	0.6960	1.695
20	0.7939	1.370	0.7604	1.461	0.7246	1.578	0.7018	1.666
25	0.8118	1.316	0.7808	1.391	0.7473	1.487	0.7258	1.558
30	0.8255	1.280	0.7964	1.344	0.7647	1.426	0.7444	1.487
35	0.8364	1.253	0.8089	1.310	0.7788	1.382	0.7594	1.435
40	0.8454	1.232	0.8192	1.284	0.7904	1.349	0.7718	1.397
45	0.8529	1.215	0.8279	1.263	0.8002	1.323	0.7823	1.366
50	0.8594	1.202	0.8353	1.246	0.8087	1.301	0.7914	1.341
55	0.8651	1.190	0.8419	1.232	0.8161	1.283	0.7994	1.320
60	0.8701	1.180	0.8476	1.220	0.8227	1.268	0.8065	1.303
65	0.8746	1.172	0.8528	1.209	0.8286	1.255	0.8128	1.287
70	0.8786	1.165	0.8574	1.200	0.8339	1.243	0.8185	1.274
75	0.8822	1.158	0.8616	1.192	0.8387	1.233	0.8237	1.263
80	0.8855	1.152	0.8655	1.184	0.8431	1.224	0.8284	1.252
85	0.8885	1.147	0.8690	1.178	0.8471	1.216	0.8328	1.243
90	0.8913	1.142	0.8722	1.172	0.8508	1.209	0.8368	1.235
95	0.8939	1.138	0.8752	1.167	0.8543	1.202	0.405	1.227
100	0.8963	1.134	0.8780	1.162	0.8575	1.196	0.8440	1.220
∞	1.0000		1000		1000		1000	

Bibliography

Research papers

1. I. Lira, D. Grientschnig, Bayesian assessment of uncertainty in metrology: a tutorial. Metrologia **47**(3), R1–R14 (2010)
2. C. Elster, B. Toman, Analysis of key comparisons: estimating laboratories' biases by a fixed effects model using Bayesian model averaging. Metrologia **47**(3), 113–119 (2010)
3. R.N. Kacker, J.F. Lawrence, Rectangular distribution whose end points are not exactly known: curvilinear trapezoidal distribution. Metrologia **47**(3), 120–126 (2010)
4. F. Pennecchi, L. Oberto, Uncertainty evaluation for the estimate of a complex-valued quantity modulus. Metrologia **47**(3), 157–166 (2010)
5. L. Lira, The probability distribution of a quantity with given mean and variance. Metrologia **46**(6), L27–L28 (2009)
6. J. Lovell-Smith, The propagation of uncertainty for humidity calculations. Metrologia **46**(6), 607–615 (2009)
7. D. Calonico, F. Levi, L. Lorini, G. Mana, Bayesian estimate of the zero-density frequency of a Cs fountain. Metrologia **46**(6), 629–636 (2009)
8. R.B. Frenkel, Fiducial inference applied to uncertainty estimation when identical readings are obtained under low instrument resolution. Metrologia **46**(6), 661–667 (2009)
9. N. Pousset, B. Rougié, A. Razet, Uncertainty evaluation for measurement of LED colour by Monte Carlo simulations. Metrologia **46**(6), 704–718 (2009)
10. M. Müller, C. Rink, On the convergence of the Monte Carlo block design. Metrologia **46**(5), 404–408 (2009)
11. R. Willink, A formulation of the law of propagation of uncertainty to facilitate the treatment of shared influences. Metrologia **46**(3), 145–153 (2009)
12. R. Willink, Representing Monte Carlo output distributions for transferability in uncertainty analysis: modelling with quantile functions. Metrologia **46**(3), 154–166 (2009)
13. Y. Fujita, Y. Kurano, K. Fujii, Evaluation of uncertainty in viscosity measurements by capillary master viscometers. Metrologia **46**(3), 237–248 (2009)
14. C. Elster, B. Toman, Bayesian uncertainty analysis under prior ignorance of the measurand versus analysis using the Supplement 1 to the *Guide*: a comparison. Metrologia **46**(3), 261–266 (2009)
15. I. Lira, Comment on Rectangular distribution whose width is not exactly known: isocurvilinear trapezoidal distribution. Metrologia **46**(3), L20 (2009)
16. M.G. Cox, G.B. Rossi, P.M. Harris, A. Forbes, A probabilistic approach to the analysis of measurement processes. Metrologia **45**(5), 493–502 (2008)

S.V. Gupta, *Measurement Uncertainties*, DOI 10.1007/978-3-642-20989-5,

17. J. Blázquez, A. García-Berrocal, C. Montalvo, M. Balbás, The coverage factor in a Flatten-Gaussian distribution. Metrologia **45**(5), 503–506 (2008)
18. R.N. Kacker, A. Forbes, R. Kessel, K.D. Sommer, Bayesian posterior predictive p-value of statistical consistency in inter-laboratory evaluations. Metrologia **45**(5), 512–523 (2008)
19. I. Lira, The generalized maximum entropy trapezoidal probability density function. Metrologia **45**(4), L17–L20 (2008)
20. I. Lira, On the long-run success rate of coverage intervals. Metrologia **45**(4), L21–L23 (2008)
21. C. Elster, A. Link, Uncertainty evaluation for dynamic measurements modelled by a linear time-invariant system. Metrologia **45**(4), 464–473 (2008)
22. R. Willink, Estimation and uncertainty in fitting straight lines to data: different techniques. Metrologia **45**(3), 290–298 (2008)
23. A.C. Baratto, Measurand: a cornerstone concept in metrology. Metrologia **45**(3), 299–307 (2008)
24. B.D. Hall, Evaluating methods of calculating measurement uncertainty. Metrologia **45**(2), L5–L8 (2008)
25. G.A. Kyriazis, Comparison of GUM Supplement 1 and Bayesian analysis using a simple linear calibration model. Metrologia **45**(2), L9–L11 (2008)
26. C.H. Sim, M.H. Lim, Evaluating expanded uncertainty in measurement with a fitted distribution. Metrologia **45**(2), 178–184 (2008)
27. J. Lovell-Smith, An expression for the uncertainty in the water vapour pressure enhancement factor for moist air. Metrologia **44**(6), L49–L52 (2007)
28. R. Kacker, Comments on 'Bayesian evaluation of comparison data'. Metrologia **44**(6), L57–L61 (2007)
29. B.D. Hall, Some considerations related to the evaluation of measurement uncertainty for complex-valued quantities in radio frequency measurements. Metrologia **44**(6), L62–L67 (2007)
30. A. Possolo, B. Toman, Assessment of measurement uncertainty via observation equations. Metrologia **44**(6), 464–475 (2007)
31. J. Hannig, H.K. Iyer, C.M. Wang, Fiducial approach to uncertainty assessment accounting for error due to instrument resolution. Metrologia **44**(6), 476–483 (2007)
32. M.W. Keller, N.M. Zimmerman, A.L. Eichenberger, Uncertainty budget for the NIST electron counting capacitance standard, ECCS-1. Metrologia **44**(6), 505–512 (2007)
33. R. Kacker, K.D. Sommer, R. Kessel, Evolution of modern approaches to express uncertainty in measurement. Metrologia **44**(6), 513–529 (2007)
34. F. Pavese, The definition of the measurand in key comparisons: lessons learnt with thermal standards. Metrologia **44**(5), 327–339 (2007)
35. R. Willink, A generalization of the Welch–Satterthwaite formula for use with correlated uncertainty components. Metrologia **44**(5), 340–349 (2007)
36. L. Martins, M.N. Frota, I. Lira, Uncertainty associated with the energy content in flow measurement of natural gas including real-time correction for fluid properties. Metrologia **44**(5), 350–355 (2007)
37. I. Lira, C. Elster, W. Wöger, Probabilistic and least-squares inference of the parameters of a straight-line model. Metrologia **44**(5), 379–384 (2007)
38. F.O. Bochud, C.J. Bailat, J.P. Laedermann, Bayesian statistics in radionuclide metrology: measurement of a decaying source. Metrologia **44**(4), S95–S101 (2007)
39. R. Willink, Uncertainty and data-fitting procedures. Metrologia **44**(3), L33–L35 (2007)
40. R. Willink, Uncertainty of functionals of calibration curves. Metrologia **44**(3), 182–186 (2007)
41. M.G. Cox, The evaluation of key comparison data: determining the largest consistent subset. Metrologia **44**(3), 187–200 (2007)
42. G. Mana, F. Pennecchi, Uncertainty propagation in non-linear measurement equations. Metrologia **44**(3), 246–251 (2007)
43. B. Toman, Statistical interpretation of key comparison degrees of equivalence based on distributions of belief. Metrologia **44**(2), L14–L17 (2007)

44. C. Elster, Calculation of uncertainty in the presence of prior knowledge. Metrologia **44**(2), 111–116 (2007)
45. R.N. Kacker, J.F. Lawrence, Trapezoidal and triangular distributions for Type B evaluation of standard uncertainty. Metrologia **44**(2), 117–127 (2007)
46. W. Hässelbarth, W. Bremser, Measurement uncertainty for multiple measurands: characterization and comparison of uncertainty matrices. Metrologia **44**(2), 128–145 (2007)
47. A.F. Obaton, J. Lebenberg, N. Fischer, S. Guimier, J. Dubard, Two procedures for the estimation of the uncertainty of spectral irradiance measurement for UV source calibration. Metrologia **44**(2), 152–160 (2007)
48. R.R. Cordero, G. Seckmeyer, Effect of the resolution on the uncertainty evaluation. Metrologia **43**(6), L33–L38 (2006)
49. R. Willink, On using the Monte Carlo method to calculate uncertainty intervals. Metrologia **43**(6), L39–L42 (2006)
50. L.A. Chen, H.N. Hung, Extending the discussion on coverage intervals and statistical coverage intervals. Metrologia **43**(6), L43–L44 (2006)
51. B.D. Hall, Computing uncertainty with uncertain numbers. Metrologia, **43**(6), L56–L61 (2006)
52. C.M. Wang, H.K. Iyer, Uncertainty analysis for vector measurands using fiducial inference. Metrologia **43**(6), 486–494 (2006)
53. R. Willink, Uncertainty analysis by moments for asymmetric variables. Metrologia **43**(6), 522–530 (2006)
54. S. Nadarajah, Exact calculation of the coverage interval for the convolution of two Student's *t* distributions. Metrologia **43**(5), L21–L22 (2006)
55. A. Balsamo, G. Mana, F. Pennecchi, The expression of uncertainty in non-linear parameter estimation. Metrologia **43**(5), 396–402 (2006)
56. W. Bich, M.G. Cox, P.M. Harris, Evolution of the "guide to the expression of uncertainty in measurement". Metrologia **43**(4), S161–S166 (2006)
57. R. Kacker, B. Toman, D. Huang, Comparison of ISO-GUM, draft GUM supplement 1 and Bayesian statistics using simple linear calibration. Metrologia **43**(4), S167–S177 (2006)
58. M.G. Cox, B.R.L. Siebert, The use of a Monte Carlo method for evaluating uncertainty and expanded uncertainty. Metrologia **43**(4), S178–S188 (2006)
59. R. Kessel, R. Kacker, M. Berglund, Coefficient of contribution to the combined standard uncertainty. Metrologia **43**(4), S189–S195 (2006)
60. W. Bich, L. Callegaro, F. Pennecchi, Non-linear models and best estimates in the GUM. Metrologia, **43**(4), S196–S199 (2006)
61. K.D. Sommer, B.R.L. Siebert, Systematic approach to the modelling of measurements for uncertainty evaluation. Metrologia **43**(4), S200–S210 (2006)
62. R. Willink, Principles of probability and statistics for metrology. Metrologia **43**(4), S211–S219 (2006)
63. I. Lira, Bayesian evaluation of comparison data. Metrologia **43**(4), S231–S234 (2006)
64. G. Ratel, Median and weighted median as estimators for the key comparison reference value (KCRV). Metrologia **43**(4), S244–S248 (2006)
65. I. Lira, W. Wöger, Comparison between the conventional and Bayesian approaches to evaluate measurement data. Metrologia **43**(4), S249–S259 (2006)
66. M.L. Rastello, A. Premoli, Least squares problems with element-wise weighting. Metrologia **43**(4), S260–S267 (2006)
67. M.G. Cox, C. Eiø, G. Mana, F. Pennecchi, The generalized weighted mean of correlated quantities. Metrologia **43**(4), S268–S275 (2006)
68. N.F. Zhang, Calculation of the uncertainty of the mean of autocorrelated measurements. Metrologia **43**(4), S276–S281 (2006)
69. N.F. Zhang, Calculation of the uncertainty of the mean of autocorrelated measurements. Metrologia **43**(4), S276–S281 (2006)
70. A.B. Forbes, Uncertainty evaluation associated with fitting geometric surfaces to coordinate data. Metrologia **43**(4), S282–S290 (2006)

71. M.J.T. Milton, P.M. Harris, I.M. Smith, A.S. Brown, B.A. Goody, Implementation of a generalized least-squares method for determining calibration curves from data with general uncertainty structures. Metrologia **43**(4), S291–S298 (2006)
72. N.F. Zhang, The uncertainty associated with the weighted mean of measurement data. Metrologia **43**(3), 195–204 (2006)
73. A. Hornikova, N.F. Zhang, M.J. Welch, S. Tai, An application of combining results from multiple methods – statistical evaluation of uncertainty for NIST SRM 1508a. Metrologia **43**(3), 205–212 (2006)
74. B.D. Hall, Monte Carlo uncertainty calculations with small-sample estimates of complex quantities. Metrologia **43**(3), 220–226 (2006)
75. R.N. Kacker, Bayesian alternative to the ISO-GUM's use of the Welch-Satterthwaite formula. Metrologia **43**(1), 1–11 (2006)
76. R. Willink, Forming a comparison reference value from different distributions of belief. Metrologia **43**(1), 12–20 (2006)
77. B. Toman, Linear statistical models in the presence of systematic effects requiring a Type B evaluation of uncertainty. Metrologia **43**(1), 27–33 (2006)
78. S.V. Crowder, R.D. Moyer, A two-stage Monte Carlo approach to the expression of uncertainty with non-linear measurement equation and small sample size. Metrologia **43**(1), 34–41 (2006)
79. P. Fotowicz, An analytical method for calculating a coverage interval. Metrologia **43**(1), 42–45 (2006)
80. M.J. Korczynski, M.G. Cox, P.M. Harris, in *Convolution and Uncertainty Evaluation*", Advanced Mathematical Tools in Metrology VII (World Scientific, Singapore, 2006), pp. 188–195
81. M.G. Cox, P.M. Harris, *Software Specifications for Uncertainty Evaluation*. Tech. Rep. DEM-ES-010 (National Physical Laboratory, Teddington, 2006)
82. B.A. Wichmann, I.D. Hill, Generating good pseudo-random numbers. Comput. Stat. Data Anal. **51**, 1614–1622 (2006)
83. OIML, *Conventional Value of the Result of Weighing in Air*. Tech. Rep. OIML D 28 (OIML, Paris, 2004)
84. R. Willink, Coverage intervals and statistical coverage intervals. Metrologia **41**, L5–L6 (2004)
85. R. Kacker, A. Jones, On use of Bayesian statistics to make the Guide to the expression of uncertainty in measurement consistent. Metrologia, **40**, 235–248 (2003)
86. M.G. Cox, The evaluation of key comparison data. Metrologia **39**, 589–595 (2002)
87. B.D. Hall, R. Willink, Does Welch-Satterthwaite make a good uncertainty estimate? Metrologia **38**, 9–15 (2001)
88. L.J. Gleser, Assessing uncertainty in measurement. Stat. Sci. **13**, 277–290 (1998)
89. K. Weise, W. Woger, A Bayesian theory of measurement uncertainty. Meas. Sci. Technol. **3**, 1–11 (1992)
90. See Ref [89].
91. W. Woger, Probability assignment to systematic deviations by the Principle of Maximum Entropy. IEEE Trans. Instrum. Meas. **IM-36**, 655–658 (1987)
92. W.A. Fuller, *Measurement Error Models* (Wiley, New York, 1987)
93. D.W. Allan, IEEE Trans. Instrum. Meas. **IM-36**, 646–654 (1987)
94. CIPM, BIPM Proc.-Verb. Com. Int. Poids et Mesures **54**, 14, 35 (1986) (in French); P. Giacomo, Metrologia **24**, 49–50 (1987) (in English)
95. B.A. Wichmann, I.D. Hill, Correction. Algorithm AS183. An efficient and portable pseudo-random number generator. Appl. Stat. **33** 123 (1984)
96. J.W. Müller, in *Precision Measurement and Fundamental Constants* II, ed. by B.N. Taylor, W.D. Phillips. Natl. Bur. Stand. (U.S.) Spec. Publ. 617 (US GPO, Washington, DC, 1984), pp. 375–381
97. A. Chan, G. Golub, R. LeVeque, Algorithms for computing the sample variance: analysis and recommendations. Am. Stat. **37**, 242–247 (1983)

98. B.A. Wichmann, I.D. Hill, Algorithm AS183. An efficient and portable pseudo-random number generator. Appl. Stat. **31**, 188–190 (1982)
99. R. Kaarls, BIPM Proc.-Verb. Com. Int. Poids et Mesures **49**, A1–A12 (1981) (in French); P. Giacomo, Metrologia **17**, 73–74 (1981) (in English)
100. CIPM, BIPM Proc.-Verb. Com. Int. Poids et Mesures **49**, 8–9, 26 (1981) (in French); P. Giacomo, Metrologia **18**, 43–44 (1982) (in English)
101. CIPM, Rapport BIPM-80/3, Report on the BIPM enquiry on error statements. BIPM Proc. Verb. Com. Int. Poids et Mesures **48**, C1–C30 (1980)
102. J.W. Müller, Nucl. Instrum. Methods **163**, 241–251 (1979)
103. A. Kinderman, J. Monahan, J. Ramage, Computer methods for sampling from Student's t-distribution. Math. Comput. **31**, 1009–1018 (1977)
104. D.C. Dowson, A. Wragg, Maximum entropy distributions having prescribed first and second order moments. IEEE Trans. **IT-19**, 689–693 (1973)
105. R. Price, A useful theorem for nonlinear devices having Gaussian inputs. IEEE Trans. Inform. Theory **IT-4** (1958)
106. A. Papoulis, On an extension of Price's theorem. IEEE Trans. Inform. Theory **IT-11** (1965)
107. R.S. Scowen, Quickersort, Algorithm 271. Comm. ACM **8** 669 (1965)
108. R.W. Beatty, Insertion loss concepts. Proc. IEEE **52**, 663–671 (1964)
109. G.E.P. Box, M. Muller, A note on the generation of random normal variates. Ann. Math. Stat. **29**, 610–611 (1958)
110. C.E. Shannon, A mathematical theory of information. Bell Syst. Tech. J. **27**, 623–656 (1948)
111. F.E. Satterthwaite, Psychometrika **6**, 309–316 (1941); Biometrics Bull.**2**(6), 110–114 (1946)
112. B.L. Welch, J. R. Stat. Soc. Suppl. **3**, 29–48 (1936); Biometrika **29**, 350–362 (1938); Biometrika **34**, 28–35 (1947)
113. H. Fairfield-Smith, J. Counc. Sci. Ind. Res. (Australia) **9**(3), 211 (1936)
114. M.G. Cox, The numerical evaluation of B-splines. J. Inst. Math. Appl. **10**, 134–149 (1972)
Note: Papers at S No. 111, 112 and 113 are about the effective degrees of freedom.

Documents

Uncertainty

International documents

115. BIPM, IEC, IFCC, ILAC, ISO, IUPAC, IUPAP, and OIML, JCGM, Evaluation of measurement data-Guide to the Expression of uncertainty in measurement. (Latest versions of 1995 GUM) **100** (2008)
116. BIPM, IEC, IFCC, ILAC, ISO, IUPAC, IUPAP, and OIML, Evaluation of measurement data "Applications of the least-squares method". JCGM **107**, under preparation
117. BIPM, IEC, IFCC, ILAC, ISO, IUPAC, IUPAP, and OIML, Evaluation of measurement data - Concepts and basic principles. JCGM **105**, in preparation
118. BIPM, IEC, IFCC, ILAC, ISO, IUPAC, IUPAP, and OIML, Evaluation of measurement data – Supplement 2 to the Guide to the expression of uncertainty in measurement" – Models with any number of output quantities. JCGM **102**, in preparation
119. BIPM, IEC, IFCC, ILAC, ISO, IUPAC, IUPAP, and OIML, Evaluation of measurement data – Supplement 3 to the Guide to the expression of uncertainty in measurement – Modeling. JCGM **103**, in preparation
120. BIPM, IEC, IFCC, ILAC, ISO, IUPAC, IUPAP, and OIML, Evaluation of measurement data – The role of measurement uncertainty in conformity assessment. JCGM **106**, in preparation

121. IEC, IEC Guide 115 *Application of Uncertainty of Measurement to Conformity Assessment Activities in the Electro-Technical Sector* (IEC Geneva, 2007)
122. ISO, ISO 10576-1. *Statistical Methods – Guidelines for the Evaluation of Conformity with Specified Requirements*. Part 1: General Principles (ISO, Geneva, 2003)
123. ISO/IEC, ISO/IEC 17025, *General Requirements for the Competence of Testing and Calibration Laboratories* (ISO, Geneva, 2005)
124. ISO, ISO 3534-1 *Statistics – Vocabulary and Symbols*. Part 1: Probability and General Statistical Terms (ISO, Geneva)
125. ISO, ISO 5725 *Precision of Test Methods – Determination of Repeatability and Reproducibility for a Standard Test Method by Inter-Laboratory Tests* (ISO, Geneva, 1986)
The standard ISO 5725:1986 has been replaced by ISO 5725 consisting of the following parts, under the general title "Accuracy (trueness and precision) of measurement methods and results".
Part 1: General principles and definitions
Part 2: Basic method for the determination of repeatability and reproducibility of a standard measurement method
Part 3: Intermediate measures of the precision of a standard measurement method
Part 4: Basic methods for the determination of the trueness of a standard measurement method
Part 5: Alternative methods for the determination of the precision of a standard measurement method
Part 6: Use in practice of accuracy values
126. ISO, *International Vocabulary of Basic and General Terms in Metrology*, 2nd edn. (ISO, Geneva, 1993)
127. ISO 3534–1:1993, *Statistics – Vocabulary and Symbols* – Part 1: Probability and General Statistical Terms (ISO, Geneva, 1993)
128. ISO Guide 35:1989, *Certification of Reference Materials – General and Statistical Principles*, 2nd edn. (ISO, Geneva, 1989)
129. EURACHEM/CITAC, *Quantifying Uncertainty in Analytical Measurement*, 2nd edn. (EURACHEM, 2000)
130. EA, *Expression of the Uncertainty of Measurement in Calibration*. Tech. Rep. EA-4/02, European Co-operation for Accreditation (EA, 1999)

National Documents

131. S. Bell, *Measurement Good Practice Guide No 11 – A Beginner's Guide to Uncertainty of Measurement* (National Physical Laboratory, Teddington, 1999)
132. M.G. Cox, P.M. Harris, *SSfM Best Practice Guide No 6 – Uncertainty Evaluation*. Tech. Rep. DEM-ES-011 (National Physical Laboratory, Teddington, 2006)
133. B.N. Taylor, C.E. Kuyatt, *Guidelines for Evaluating and Expressing the Uncertainty of NIST Measurement Results*. Tech. Rep. TN1297 (National Institute of Standards and Technology, Gaithersburg, 1994)
134. M.G. Cox, P.M. Harris, *Software Specifications for Uncertainty Evaluation*. Tech. Rep. DEM-ES-010 (National Physical Laboratory, Teddington, 2006)
135. NABL, *Guidelines for Estimation and Expression of Uncertainty in Measurement*. NABL document No 141 (NABL, New Delhi, 2000)

Computational Documents

136. B.A. Wichmann, I.D. Hill, Algorithm AS183 1982 "An efficient and portable pseudo-random number generator". Appl. Stat. **31**, 188–190 (1982)

137. B.A. Wichmann, I.D. Hill, Correction. algorithm AS183 "An efficient and portable pseudo-random number generator". Appl. Stat. **33**, 123 (1984)
138. B.A. Wichmann, I.D. Hill, Generating good pseudo-random numbers. Comput. Stat. Data Anal. **51**, 1614–1622g(2006)
139. P. L'Ecuyer, R. Simard, TestU01: a software library in ANSI C for empirical testing of random number generators. http://www.iro.umontreal.ca/~simardr/testu01/tu01.html
140. M. Matsumoto, T. Mersenne, T. Nishimura, A 623-dimensionally equidistributed uniform pseudo-random number generator. ACM Trans. Model. Comput. Simul. **8**, 3–30 (1998)
141. B.D. McCullough, B. Wilson, On the accuracy of statistical procedures in Microsoft Excel 2003. Comput. Stat. Data Anal. **49**, 1244–1252 (2004)
142. Netlib. The Netlib repository of freely available software, documents, and databases of interest to the numerical, scientific computing, and other communities contains facilities for sampling from probability distributions, http://www.netlib.org
143. NIST. The NIST Digital Library of Mathematical Functions contains facilities for sampling from probability distributions, http://dlmf.nist.gov
144. J. Leydold, Automatic sampling with the ratio-of-uniforms method. ACM Trans. Math. Software **26**, 78–98 (2000)

Books

Bayesian Statistics

145. J. Bernardo, A. Smith, *Bayesian Theory* (Wiley, New York, 2000)
146. A. Gelman, J.B. Carlin, H.S. Stern, D.B. Rubin, *Bayesian Data Analysis* (Chapman and Hall, London, 2004)
147. S.J. Press, *Bayesian Statistics: Principles, Models, and Applications* (Wiley New York, 1989)
148. K. Weise, W. Woger, A Bayesian theory of measurement uncertainty. Meas. Sci. Technol. **3**, 1–11 (1992)
149. I.H. Lira, W. Woger, Bayesian evaluation of the standard uncertainty and coverage probability in a simple measurement model. Meas. Sci. Technol. **12**, 1172–1179 (2001)
150. R. Kacker, A. Jones, On use of Bayesian statistics to make the guide to the expression of uncertainty in measurement consistent. Metrologia **40**, 235–248 (2003)

Monte Carlo Method

151. C.P. Robert, G. Casella, *Monte Carlo Statistical Methods* (Springer, New York, 1999)
152. M.G. Cox, B.R.L. Siebert, The use of a Monte Carlo method for evaluating uncertainty and expanded uncertainty. Metrologia **43**, S178–S188 (2006)

Entropy

153. E.T. Jaynes, *Where Do We Stand on Maximum Entropy*? Papers on Probability, Statistics, and Statistical Physics (Kluwer Academic, Dorderecht, 1989), pp. 210–314
154. D.C. Dowson, A. Wragg, Maximum entropy distributions having prescribed first and second order moments. IEEE Trans. **IT-19**, 689–693 (1973)
155. W. Woger, Probability assignment to systematic deviations by the Principle of Maximum Entropy. IEEE Trans. Instrum. Meas. **IM-36**, 655–658 (1987)

Statistics

156. G.C. Casella, R.L. Berger, *Statistical Inference*, 2nd edn. (Duxbury, Pacific Grove, 2001)
157. W. Feller, *An Introduction to Probability Theory and Its Applications*, vol. 1 (Wiley, Chichester, 1968)
158. W. Feller, *An Introduction to Probability Theory and Its Applications*, vol. 2 (Wiley, Chichester, 1971)
159. J.R. Rice, *Mathematical Statistics and Data Analysis*, 2nd edn. (Duxbury, Belmont, 1995)
160. P.M. Berthouex, L.C. Brown, *Statistics for Environmental Engineers* (CRC, Boca Raton, 1994)
161. S.D. Conte, C. de Boor, *Elementary Numerical Analysis: An Algorithmic Approach* (McGraw-Hill, New York, 1972)
162. H.A. David, *Order Statistics* (Wiley, New York, 1981)
163. T.J. Dekker, *Finding a Zero by Means of Successive Linear Interpolation in Constructive Aspects of the Fundamental Theorem of Algebra* (Wiley Interscience, New York, 1969)
164. L. Devroye, *Non-Uniform Random Number Generation* (Springer, New York, 1986)
165. M. Evans, N. Hastings, B. Peacock, *Statistical Distributions* (Wiley, New York, 2000)
166. R.B. Frenkel, *Statistical Background to the ISO 'Guide to the Expression of Uncertainty in Measurement*. Publication number TIP P1242 (National Measurement Laboratory, CSIRO, Canberra, 2002)
167. N.J. Higham, *Accuracy and Stability of Numerical Algorithms* (SIAM Philadelphia, 1996)
168. C.B. Moler, *Numerical Computing with MATLAB* (SIAM, Philadelphia, 2004)
169. J.R. Rice, Mathematical Statistics and Data Analysis, 2nd edn. (Duxbury, Belmont, 1995)
170. M.J. Salter, N.M. Ridler, M.G. Cox, *Distribution of Correlation Coefficient for Samples Taken from a Bivariate Normal Distribution*. Tech. Rep. CETM 22 (National Physical Laboratory, Teddington, 2000)
171. I.J. Schoenberg, Cardinal interpolation and spline functions. J. Approx. Theory **2**, 167–206 (1969)
172. G. Strang, K. Borre, *Linear Algebra, Geodesy and GPS* (Wiley, Wellesley-Cambridge, New York, 1997)
173. T.B. Barker, *Quality by Experimental Design* (Marcel Dekker, New York, 1985)
174. D.B. Hibbert, *Quality Assurance for the Analytical Chemistry Laboratory* (Oxford University Press, Oxford, 2007)
175. D.M. Kerns, R.W. Beatty, *Basic Theory of Waveguide Junctions and Introductory Microwave Network Analysis* (Pergamon, London, 1967)
176. G.E.P. Box, W.G. Hunter, J.S. Hunter, *Statistics for Experimenters* (Wiley, New York, 1978)
177. H. Jeffreys, *Theory of Probability*, 3rd edn. (Oxford University Press, Oxford, 1983)

Uncertainty

178. C.F. Dietrich, *Uncertainty, Calibration and Probability* (Adam Hilger, Bristol, 1991)
179. N.M. Ridler, M.J. Salter, Propagating S-parameter uncertainties to other measurement quantities. The 58th ARFTG (Automatic RF Techniques Group) Conference Digest (2001)
180. I. Lira, *Evaluating the Uncertainty of Measurement – Fundamentals and Practical Guidance* (Institute of Physics, Bristol, 2002)

Index

S.V. Gupta, *Measurement Uncertainties*, DOI 10.1007/978-3-642-20989-5,